CIRCULATING NUCLEIC ACIDS IN PLASMA AND SERUM IV

ANNALS OF THE NEW YORK ACADEMY OF SCIENCES

Volume 1075

CIRCULATING NUCLEIC ACIDS IN PLASMA AND SERUM IV

Edited by R. Swaminathan, Asif Butt, and Peter Gahan

Published by Blackwell Publishing on behalf of the New York Academy of Sciences

Boston, Massachusetts

2006

Library of Congress Cataloging-in-Publication Data

International Symposium on Circulating Nucleic Acids in Plasma/Serum (4th : 2005 : London, England)
Circulating nucleic acids in plasma and serum IV / edited by Ramasamyiyer Swaminathan, Asif Butt, Peter Gahan.
p. ; cm. – (Annals of the New York Academy of Sciences, ISSN 0077-8923 ; v. 1074)
Includes bibliographical references.
ISBN-13: 978-1-57331-627-9 (alk. paper)
ISBN-10: 1-57331-627-X (alk. paper)
1. Nucleic acids–Congresses. 2. Blood–Analysis–Congresses. 3. Tumor markers–Congresses. I. Swaminathan, R., Professor. II. Butt, Asif. III. Gahan, Peter B. IV. New York Academy of Sciences. V. Title. VI. Series.
[DNLM: 1. Nucleic Acids–blood–Congresses. 2. Tumor Markers, Biological–blood–Congresses. W1 AN626YL v.1074 2006 / QU 58 I606 2006]

QP620.I53 2005
616.07'561–dc22

2006018862

The *Annals of the New York Academy of Sciences* (ISSN: 0077-8923 [print]; ISSN: 1749-6632 [online]) is published 28 times a year on behalf of the New York Academy of Sciences by Blackwell Publishing, with offices located at 350 Main Street, Malden, Massachusetts 02148 USA, PO Box 1354, Garsington Road, Oxford OX4 2DQ UK, and PO Box 378 Carlton South, 3053 Victoria Australia.

Information for subscribers: Subscription prices for 2006 are: Premium Institutional: \$3850.00 (US) and £2139.00 (Europe and Rest of World).
Customers in the UK should add VAT at 5%. Customers in the EU should also add VAT at 5% or provide a VAT registration number or evidence of entitlement to exemption. Customers in Canada should add 7% GST or provide evidence of entitlement to exemption. The Premium Institutional price also includes online access to full-text articles from 1997 to present, where available. For other pricing options or more information about online access to Blackwell Publishing journals, including access information and terms and conditions, please visit www.blackwellpublishing.com/nyas.

Membership information: Members may order copies of the *Annals* volumes directly from the Academy by visiting www.nyas.org/annals, emailing membership@nyas.org, faxing 212-888-2894, or calling 800-843-6927 (US only), or +1 212 838 0230, ext. 345 (International). For more information on becoming a member of the New York Academy of Sciences, please visit www.nyas.org/membership.

Journal Customer Services: For ordering information, claims, and any inquiry concerning your institutional subscription, please contact your nearest office:
UK: Email: customerservices@blackwellpublishing.com; Tel: +44 (0) 1865 778315; Fax +44 (0) 1865 471775
US: Email: customerservices@blackwellpublishing.com; Tel: +1 781 388 8599 or 1 800 835 6770 (Toll free in the USA); Fax: +1 781 388 8232
Asia: Email: customerservices@blackwellpublishing.com; Tel: +65 6511 8000; Fax: +61 3 8359 1120
Members: Claims and inquiries on member orders should be directed to the Academy at email: membership@nyas.org or Tel: +1 212 838 0230 (International) or 800-843-6927 (US only).

Mailing: The *Annals of the New York Academy of Sciences* are mailed Standard Rate. **Postmaster:** Send all address changes to *Annals of the New York Academy of Sciences*, Blackwell Publishing, Inc., Journals Subscription Department, 350 Main Street, Malden, MA 01248-5020. Mailing to rest of world by DHL Smart and Global Mail.

Annals are available to subscribers online at the New York Academy of Sciences and also at Blackwell Synergy. Visit www.annalsnyas.org or www.blackwell-synergy.com to search the articles and register for table of contents e-mail alerts. Access to full text and PDF downloads of *Annals* articles are available to nonmembers and subscribers on a pay-per-view basis at www.annalsnyas.org.

The paper used in this publication meets the minimum requirements of the National Standard for Information Sciences Permanence of Paper for Printed Library Materials, ANSI Z39.48_1984.

ISSN: 0077-8923 (print); 1749-6632 (online)
ISBN-10: 1-57331-627-X (paper); ISBN-13: 978-1-57331-627-9(paper)

A catalogue record for this title is available from the British Library.

Digitization of the Annals of the New York Academy of Sciences:

An agreement has recently been reached between Blackwell Publishing and the New York Academy of Sciences to digitize the entire run of the *Annals of the New York Academy of Sciences* back to volume one, issue one.

The back files, which have been defined as all of those issues published before 1997, will be sold to libraries as part of Blackwell Publishing's Legacy Sales Program and hosted on the Blackwell Synergy website.

Copyright of all material will remain with the rights holder. Contributors: Please contact Blackwell Publishing if you do not wish an article or picture from the *Annals of the New York Academy of Sciences* to be included in this digitization project.

ANNALS OF THE NEW YORK ACADEMY OF SCIENCES

Volume 1075
September 2006

CIRCULATING NUCLEIC ACIDS IN PLASMA AND SERUM IV

Editors
R. SWAMINATHAN, ASIF BUTT, AND PETER GAHAN

This volume is the result of a meeting entitled **Fourth International Conference on Circulating Nucleic Acids in Plasma/Serum (CNAPS-IV)**, held on September 4-6, 2005 in the Franklin-Wilkins Building of Kings College in London, UK.

CONTENTS

Part II. Fetal Nucleic Acids

Part III. Nucleic Acids and Cancer

Part IV. Nucleic Acids in Other Diseases

Part V. Methodology

Financial assistance was received from:

- PreAnalytiX, UK
- QIAGEN Ltd., UK

Welcoming Remarks

It is a great pleasure to be able to welcome you all to King's College, within the University of London, for the fourth international meeting on Circulating Nucleic Acids in Plasma and Serum, CNAPS.

I understand that the first meeting in this series was in France in 1999, the second in Hong Kong in 2001, and the last one in 2003 in Santa Monica. Now, 2 years later, CNAPS IV is taking place in our Franklin–Wilkins Building at the Waterloo Campus.

I hope it would not be too presumptuous to suggest that King's College and this specific building provide a particularly appropriate venue for you, given the historical scientific activities associated with this location, and its relevance to your group's central areas of interest. Work that was particularly linked with the MRC Biophysics Unit here from the 1950s onward included:

- The defining X-ray crystallographic studies by Rosalind Franklin and Maurice Wilkins, which were seminal in bringing about the Watson and Crick model for the structure of DNA;
- Work by Pelc on stripping film techniques for tissue autoradiography and the concept of the cell cycle;
- Jean Hansen's celebrated investigations that lead to the establishment of the Hansen–Huxley model for actin–myosin interactions in muscle contraction; and
- Joe Chayen and Peter Gahan's work on cytoplasmic DNA.

Given this context, what more appropriate locale could there be for your deliberations from now until Tuesday? Those previous staff members of King's College, in honor of whom this building has been named, would have been fascinated by your subject matter.

If you will allow me a personal reminiscence, I was here, as the academic head of the new building, at the beginning of the present millennium, when we officially opened and named it. The ceremony took place in this room and was led by Princess Anne, who is the Chancellor of the University of London. The real scientific stars of the occasion, however, were James Watson and Maurice Wilkins, who were here in person, and Frances Crick who sent us a video message from California, rather as absentee Oscar winners do who cannot be present at the award ceremony. If my children ever produce grandchildren for me, I will bore them in years to come by telling them how their aged grandfather

Address for correspondence: Prof. Philip Whitfield, King's College, London, James Clerk Maxwell Building, 57 Waterloo Road, London SE1 8WA, UK.
e-mail: phil.whitfield@kcl.ac.uk

Ann. N.Y. Acad. Sci. 1075: xiii–xiv (2006). © 2006 New York Academy of Sciences.
doi: 10.1196/annals.1368.048

once had lunch in Waterloo sitting between two of the Nobel Laureates who discovered the structure of DNA.

Enough of reminiscence—Professor Swaminathan has provided me with the details of your program and, in addition, like a good scientist, I have scanned PubMed, Web of Knowledge, and Google Scholar to look at the recent work produced and cited in your field.

Having done so, I am amazed at the way in which the concept of circulating nucleic acids has permeated so many biological and clinical subfields. It is clear that it has fundamental and applied significance in relation to:

- oncology,
- prenatal diagnosis,
- organ transplantation,
- posttrauma monitoring, and
- infectious agent detection, among many others.

Incidentally, I am myself a parasitologist and much of my work is on the bloodstream-inhabiting fluke, *Schistosoma*. I would be intrigued to know whether schistosome nucleic acids are released by the worms and are identifiable in the blood of the 250 million people who are infected with this parasite worldwide. If this export were occurring it could open up a fascinating new approach to diagnosis.

Your subject area has moved a long way in the 50 and more years in which it has had some coherence as a discipline, but, as in all vigorously developing areas of science, you continue to confront clusters of questions that remain to be answered, including:

- the real *in vivo* origins of your nucleic acids;
- their modes of release from the cells in which they are synthesized;
- the natural fate of the nucleic acids; and
- their normal *in vivo* functionality.

I am sure that you will successfully engage with these and other questions during the conference and I hope very much that you enjoy your time in London, King's College and the Franklin–Wilkins Building.

—PHILIP WHITFIELD
Professor and Vice Principal
King's College London
University of London
London, UK

Circulating Nucleic Acids in Plasma and Serum

Recent Developments

R. SWAMINATHAN AND ASIF N. BUTT

Department of Chemical Pathology, St Thomas' Hospital, Lambeth Palace Road, London, SE1 7EH, UK

ABSTRACT: Nucleic acids (DNA and RNA) have been detected in plasma, serum, urine, and other body fluids from healthy subjects as well as in patients. The ability to detect and quantitate specific DNA and RNA sequences has opened up the possibility of diagnosis and monitoring of diseases. With the recent developments in the field of circulating nucleic acids the application in the diagnostic field has increased. The recent discovery of epigenetic changes in placental/fetal DNA and the detection of fetal/placental-specific RNAs have made it possible to use this technology in all pregnancies irrespective of the gender of the fetus. With the application of mass spectrometry and other techniques to this field, it is now possible to detect very small amounts of specific DNA in the presence of excess of other nonspecific nucleic acids (e.g., detection of mutations in fetal DNA in the presence of excess of maternal DNA). Circulating nucleic acids have now been shown to be useful in other conditions, such as diabetes mellitus, trauma, stroke, and myocardial infarction. In oncology, detection and monitoring of tumors is now possible by the detection of tumor-derived nucleic acids. In spite of these advances questions regarding the origin and biologic significance of circulating nucleic acids remain to be answered. Furthermore preanalytical and analytical aspects of this field remain to be standardized.

KEYWORDS: circulating DNA; RNA; oncology; prenatal diagnosis; transplantation

INTRODUCTION

The fourth international conference on circulating nucleic acids in plasma/serum was recently held in London, England at King's College. The conference was held at the Franklin–Wilkins Building where, in the MRC Bio-

Address for correspondence: R. Swaminathan, Department of Chemical Pathology, 5th Floor, North Wing, St Thomas' Hospital, Lambeth Palace Road, London, SE1 7EH, UK. Voice: +440-20-7188-1285; fax: +44-0-20-7928-4226.
e-mail: r.swaminathan@kcl.ac.uk

Ann. N.Y. Acad. Sci. 1075: 1–9 (2006). © 2006 New York Academy of Sciences.
doi: 10.1196/annals.1368.001

physics unit, Rosalind Franklin and Maurice Wilkins described the crystallographic structure of DNA. The aim of the fourth conference was to bring together scientists and clinicians interested in this field to present and to discuss the latest advances in circulating nucleic acids. Areas of discussion included the biology of nucleic acids in circulation, circulating nucleic acids in fetal medicine, oncology, diabetes, and other diseases, as well as analytical and preanalytical aspects. The conference was attended by over 200 delegates and there were 20 symposium presentations, 31 oral presentations, and 79 poster presentations.

In 1947 Mendel and Metais discovered DNA in plasma and serum.[1] This was before Watson and Crick described the helical structure of the DNA. Mandel and Metais, using a simple perchloric acid precipitation method, showed the presence of nucleic acids in healthy subjects as well as in ill patients. Interestingly they reported high amounts of DNA in serum from pregnant women. This discovery of nucleic acids in plasma and serum was forgotten until the 1960s, when it was shown that patients with systemic lupus erythematosus (SLE) had higher levels of circulating DNA.[2] Further work in 1970 showed that increased levels of DNA in circulation could be detected in other diseases, such as rheumatoid arthritis, glomerular nephritis, pancreatitis, inflammatory bowel disease, and hepatitis.[3] This was followed by the discovery of higher concentration of circulating DNA in patients with cancer compared to patients without cancer.[4] Furthermore, it was shown that concentration of circulating DNA was higher in patients with metastases compared to patients with localized disease and that the circulating DNA concentration decreased by up to 90% following radiotherapy.[4] The cellular origin of the extracellular DNA was not determined until the 1980s, when Stroun and Anker showed that circulating free DNA was derived from tumor cells. They showed that certain characteristics of tumor DNA were also present in circulating DNA.[5] Further work in the 1990s showed that oncogene mutations, loss of heterozygosity (LOH), and microsatellite shifts can be found in the plasma and serum matching those occurring in primary tumors.[6] The next major breakthrough in the field of circulating nucleic acids was the discovery of fetal DNA in maternal plasma by Lo *et al.*[7] This opened up a new field of investigation of noninvasive prenatal diagnosis. Measurement of circulating fetal DNA in maternal plasma has been used as a noninvasive method for fetal gender determination, assessing chromosome ploidy and monitoring various pregnancy-related complications. A summary of the highlights of the fourth international conference on circulating nucleic acids in plasma/serum is presented below.

Biology of Circulating Nucleic Acids

In a series of elegant experiments Stroun and co-workers showed that after three generations of grafting between two varieties of eggplant, hereditary characteristics could be transmitted between plants.[8] They proposed that DNA circulates in the free form, which can enter somatic and reproductive cells. Further experiments showed that the DNA released from living cells was accompanied by DNA polymerase and that there seemed to be preferential release of newly synthesized DNA.[9] These early studies showed the presence of circulating nucleic acids in the free form and that they have the potential to initiate synthesis of RNA.[10] One of the functions of circulating DNA may be to act as a messenger between cells and tissues within the organism.

Anker *et al.* suggested that circulating DNA may be involved in the immune response. They showed that when nude mice were injected with DNA released from T lymphocytes, which had previously been exposed to inactivated polio virus, they produced neutralizing antibodies.[11] Furthermore, mice injected with DNA released by human T cells were shown to produce antibodies carrying human allotypes as shown by neutralization by antiallotype serum. These studies suggest circulating DNA may have a role in the immune response. However, at this time it is difficult to define the exact role of circulating DNA in the immune response.

In addition to DNA it is known that RNA can be detected in the circulation. It is now well documented that RNA can be detected in plasma, serum, and other body fluids, as well as from cell-free supernatants of *in vitro* cultivated cells. The RNA appears to be complexed with other molecules, making it resistant to digestion by RNase.[12] The mechanism by which RNA is released into the circulation is still not clear. Possible mechanisms suggested are an active process and a nonspecific release from apoptotic or necrotic cells. Like DNA, RNA in the circulation can also be taken up by other cells. Over the past several years messenger RNA detected in the circulation includes fetal genes in the circulation of pregnant women, housekeeping genes in healthy persons as well as in patients with disease, and genes overexpressed in a variety of tumors.[13] In spite of such a large literature on circulating RNA, the biological significance of circulating RNA is not yet known. Measurement of circulating messenger RNA has been used in the field of oncology as possible tumor markers and in the field of fetal medicine to detect fetal abnormalities and pregnancy complications. However, the full potential of measurement of circulating RNA in circulation is yet to be realized and this will depend on new methods and further diagnostic work.

Nucleic Acids in Fetal Medicine

Since the discovery of fetal nucleic acids in maternal plasma there have been a large number of publications on the possible applications and the significance of fetal nucleic acids. However, there are very few studies on the factors that influence cell-free DNA levels in normal pregnancies. Bianchi and co-workers have examined in detail the effect of gestational age, race, parity, smoking history, and type of conception (natural or assisted). They showed that gestational age is a significant factor affecting fetal DNA levels. Fetal DNA levels increase by about 21% per week in the first trimester. Smoking and maternal age have no influence on fetal DNA concentrations; however, there was a significant inverse correlation between maternal weight and fetal DNA levels.[14] There was a difference in the fetal DNA levels according to the type of conception.[15] Using three-dimensional ultrasound examination the relationship between placental volume and cell-free DNA levels were examined and no association was found between these variables.[16]

Most of the studies on fetal DNA are based on the detection of Y chromosome signal in the presence of a male fetus. The progress in the application of circulating nucleic acids in fetal medicine is hampered by the lack of a universal fetal DNA marker. Lo and co-workers have recently shown the potential value of fetal epigenetic markers. In an important study, Poon *et al.*[17] showed that a region of DNA, which is methylated, if inherited from the father, and unmethylated, if inherited from the mother, could be used to differentiate fetal origin of DNA in circulation. These workers then went on to show that the maspin gene is hypomethylated in the placenta, but hypermethylated in maternal blood cells.[18] Hypomethylated maspin sequences were shown to be present in the plasma of pregnant women and this was rapidly cleared after delivery. The concentration of hypomethylated maspin in the circulation was elevated nearly six times in the plasma of women suffering from pre-eclampsia, showing that hypomethylated maspin could be a very useful fetal marker. This has opened up a new field of noninvasive prenatal diagnosis that can be applied to all pregnancies.

Another approach to detect fetal cell-free DNA in maternal plasma is to examine the size.[19] Li *et al.* have shown that fetal cell-free DNA is smaller than comparable maternal cell-free DNA molecules.[19] They extracted DNA from maternal plasma and subjected it to gel electrophoresis. Following gel electrophoresis they carefully sliced discreet fragments of different sizes and examined the fragments for fetal DNA using the SRY locus. This showed that the majority of fetal DNA had a molecular size of less than 300 bp whereas most of the maternal DNA was greater than 300 bp. This finding suggests that fetal cell-free DNA could be enriched on the basis of its size. The applicability of this to clinical diagnosis was shown by the ability to detect achrondroplasia and β-thalassemia prenatally.[20] By isolating the cell-free DNA from maternal plasma according to size and using an allele-specific real-time PCR-based approach, they demonstrated the feasibility of detecting β-thalassemia in maternal plasma.

Fetal DNA has been successfully used in the fetal blood group genotyping. Daniels *et al.* have shown that by using fetal DNA in maternal plasma Rh status of the fetus can be successfully determined. Using real-time quantitative PCR they were able to detect RHD exons 4, 5, and 10.[21] By incorporating SRY and in RHD-negative females eight biallelic polymorphisms, they provide an internal positive control to overcome inadequacies in PCR amplification. This technique is greater than 99% accurate in detecting Rh status of the fetus according to experience from the National Blood Group Reference Laboratory in Bristol and Sanquin Laboratory in Amsterdam. Prospects for the detection of other blood types, such as RHc, Rhe, K, FY look very promising.[22]

Nucleic acids in maternal plasma are also useful in the detection of maternal complications, such as pre-eclampsia and intrauterine growth retardation (IUGR). In a longitudinal study of 130 pregnant women using total DNA (β-globin) and fetal DNA (SRY), Butt and co-workers were able to show that higher levels of β-globin and SRY genes were detected in those pregnant mothers who went on to develop IUGR or pre-eclampsia, showing that circulating nucleic acids may be a useful way of predicting abnormal outcome of pregnancy. In addition to fetal DNA, fetal RNA could be detected in maternal plasma and this has opened up another approach for noninvasive prenatal diagnosis. Measurement of circulating fetal RNA has the advantage that it can be applied to all pregnancies irrespective of the fetal gender or polymorphisms between mother and fetus. Recent studies have shown that fetus-specific circulating RNA has greatly increased the number of markers that could be used for prenatal diagnosis and monitoring.[23] Using microarray technology, Tsui *et al.* identified new placental-derived plasma RNA markers. It was shown that the concentration of mRNA for corticotrophin-releasing hormone (CRH) is increased 10-fold in pre-eclamptic women, which was related to the severity of pre-eclampsia.[24]

An important aspect that has emerged from these studies on the use of fetal DNA in maternal plasma for prenatal diagnosis is the lack of assay standardization, wide differences in sample collection and preparations, as well as the lack of a proper external quality assurance scheme. Recently, the special noninvasive advances in fetal and neonatal evaluation network (SAFE) was established with funds from the European Union with the aim of implementing routine cost-effective, noninvasive prenatal diagnostic methods.[25]

Nucleic Acids in Oncology

Development of malignancy is associated with many genetic and epigenetic alterations. These include allelic loss, microsatellite instability, point mutations, amplification, translocation, and gene promoter region hypermethylation. These genetic and epigenetic alterations in the tumor have also been detected in circulating DNA. The detection of the circulating nucleic acids has provided a potential tool for the early diagnosis, monitoring, and prognosis of cancer patients. Studies showing the potential application of circulating nucleic acid have been done for many tumors, including gastrointestinal, breast, lung, head and neck, urological, gynecological, and skin. Hoon's group has shown that in melanoma patients, the outcome of the disease could be predicted by the measurement of loss of heterogeneity in many different genes.[26] This is especially true in patients in late stages of melanoma. Similarly, the detection of methylation status of circulating DNA was also shown to be a predictor of outcome in melanoma subjects.[27] Sidransky's group have shown that aberrant DNA methylation can be detected in the serum and plasma of lung cancer patients. Using tumor and matched serum and plasma DNA, these investigators examined aberrant methylation for several gene promoters and showed that in the majority of lung cancer patients hypermethylation can be detected.[28] This approach could therefore be used for early detection and surveillance of lung cancer. Recent studies from Woll's group suggest that LOH and promoter methylation can be detected not only in lung cancer patients, but also with increasing frequency in smokers and those with dysplastic changes.[29] This suggests that circulating DNA may be able to identify individuals who are at risk of developing lung cancer.

Nucleosomes, which are complexes of DNA with several protein components, can also be detected in the circulation. Holdenrieder *et al.* have identified circulating nucleosomes in patients with cancer and in patients with autoimmune disease. They showed that measurement of nucleosomes together with other tumor markers can be useful in predicting the outcome of chemotherapy and radiotherapy.[30]

Nucleic Acids and Other Diseases

Measurement of circulating DNA and organ-specific RNA has also been found to be useful in situations other than oncology and fetal medicine. These include trauma, stroke, diabetes mellitus, and autoimmune disease. Diabetic retinopathy, which is a serious and major complication of diabetes, is difficult to detect. Recent studies have shown that messenger RNA for rhodopsin in the circulation was significantly higher in patients with diabetic retinopathy and preliminary results suggest that it may be a useful test for the prediction of long-term diabetic retinopathy complication.[31] In patients presenting to the emergency department, plasma nucleic acid concentration was found to be elevated within an hour of the injury and the concentration of DNA is related to the severity of the injury.[32] Furthermore, measurement of DNA at the patient's time of admission was able to predict the outcome in terms of organ failure, acute lung injury, acute respiratory syndrome as well as death. Similarly, in patients presenting with stroke, measurement of β-globin DNA concentration was found to be higher and this was a good predictor of death as a result of the stroke.[33] In patients with chest pain (acute coronary syndrome) plasma DNA concentration was elevated and the degree of elevation was related to the extent of injury; it was highest in those patients who died within 2 years.

The measurement of circulating nucleic acids is also useful in the detection of transplant rejection. Despite HLA matching and the introduction of new immunosuppressive drugs, acute allograft rejection is one of the major complications of transplantation. Accurate and early diagnosis of transplant rejection remains a stumbling block to the success of organ transplantation schemes. Measurement of donor-specific DNA or measurement of mRNA for granulysin and FOXP3 in urine has been found to be a useful predictor of kidney transplant rejection.[34,35]

Methodology Aspects

Analysis of circulating nucleic acids in plasma and serum is conventionally done by real-time PCR because of the low copy number of the target DNA and RNA. The presence of a vast excess of background DNA and RNA makes this a difficult task. The application of MALDI-TOF mass spectrometry to PCR has enabled the detection of nucleic acids with great sensitivity. Ding *et al.* have developed a single allele primer extension reaction (SABER) method to achieve one base specificity and with the combination of the SABER method and MALDI-TOF mass spectrometry, they were able to show noninvasive prenatal diagnosis of β-thalassemia, where there is a single base point mutation.[36,37] Cirigliano has approached the problem of low sensitivity by the use of quantitative fluorescent PCR(QF-PCR). QF-PCR is based on the application of highly polymorphic chromosome-specific short tandem repeats using fluorescent primers, and the products are analyzed using an automated DNA sequencer.[38] These workers have introduced this technique for the prenatal diagnosis of several disorders. On the basis of experience several thousand samples with, they show that this method is accurate and could reduce the need for amniocentesis or chorionic villus sampling. The main advantages of QF-PCR are its accuracy, speed, automation, and low cost.

CONCLUSIONS

Since the discovery of nucleic acid in plasma and serum, the field has advanced and has produced many exciting new developments. Plasma DNA is already in routine use for fetal blood group typing in many countries. With the rapid development in molecular biology techniques, the application of circulating nucleic acids in diagnostics, especially in cancer and prenatal diagnosis, is expected to be revolutionized. Some of these new developments and new insights into the biology of circulating nucleic acids were presented at the CNAPS IV conference, and are presented in this volume. In the next few years further developments will determine the clinical applications of circulating nucleic acids in diagnosis and monitoring of diseases.

REFERENCES

1. MANDEL, P. & P. MÉTAIS. 1948. Les acides nucléiques du plasma sanguin chez l'homme. C. R. Acad. Sci. Paris **142:** 241–243.

2. TAN, E.M. *et al.* 1966. Deoxyribonucleic acid (DNA) and antibodies to DNA in the serum of patients with systemic lupus erythematosus. J. Clin. Invest. **45:** 1732–1740.
3. KOFFLER, D. *et al.* 1973. The occurrence of single-stranded DNA in the serum of patients with systemic lupus erythematosus and other diseases. J. Clin. Invest. **52:** 198–204.
4. LEON, S.A. *et al.* 1977. Free DNA in the serum of cancer patients and the effect of therapy. Cancer Res. **37:** 646–650.
5. STROUN, M. *et al.* 1989. Neoplastic characteristics of the DNA found in the plasma of cancer patients. Oncology **46:** 318–322.
6. NAWROZ, H. *et al.* 1996. Microsatellite alterations in serum DNA of head and neck cancer patients. Nat. Med. **2:** 1035–1037.
7. LO, Y.M.D. *et al.* 1989. Prenatal sex determination by DNA amplification from maternal peripheral blood. Lancet **2:** 1363–1365.
8. STROUN, M. *et al.* 1963. Modifications transmitted to the offspring, provoked by heterograft in *Solanum melongena*. Arch. Sci. Genève **16:** 1–21.
9. STROUN, M. & P. ANKER. 1971. Bacterial nucleic acid synthesis in plants following bacterial contact. Mol. Gen. Genet. **113:** 92–98.
10. GAHAN, P.B. *et al.* 2003. Evidence that direct DNA uptake through cut shoots leads to genetic transformation of *Solanum aviculare* Forst. Cell Biochem. Func. **21:** 11–17.
11. ANKER, P.D. *et al.* 1984. Nude mice injected with DNA released by antigen stimulated human T lymphocytes produce specific antibodies expressing human characteristics. Cell Biochem. Func. **2:** 33–37.
12. ROSI, A. *et al.* 1988. RNA-lipid complexes released from the plasma membrane of human colon carcinoma cells. Cancer Lett. **39:** 153–160.
13. LLEDO, S.M. *et al.* 2004. Real time quantification in plasma of human telomerase reverse transcriptase (hTERT) mRNA in patients with colorectal cancer. Colorectal Dis. **6:** 236–242.
14. WATAGANARA, T. *et al.* 2004. Inverse correlation between maternal weight and second trimester circulating cell-free fetal DNA levels. Obstet. Gynecol. **104:** 545–550.
15. PAN, P.D. *et al.* 2005. Cell-free fetal DNA levels in pregnancies conceived by IVF. Hum. Reprod. **20:** 3152–3156.
16. WATAGANARA, T. *et al.* 2005. Placental volume, as measured by 3-dimensional sonography and levels of maternal plasma cell-free fetal DNA. Am. J. Obstet. Gynecol. **193:** 496–500.
17. POON, L.L.M. *et al.* 2002. Differential DNA methylation between fetus and mother as a strategy for detecting fetal DNA in maternal plasma. Clin. Chem. **48:** 35–41.
18. CHIM, S.S.C. *et al.* 2005. Detection of the placental epigenetic signature of the maspin gene in maternal plasma. Proc. Natl. Acad. Sci. USA **102:** 14753–14758.
19. LI, Y. *et al.* 2004. Size separation of circulatory DNA in maternal plasma permits ready detection of fetal DNA polymorphisms Clin. Chemistry **50:** 1002–1011.
20. LI, Y. *et al.* 2005. Detection of paternally inherited fetal point mutations for beta-thalassemia using size-fractionated cell-free DNA in maternal plasma. JAMA **293:** 843–849.
21. FINNING, K. *et al.* 2004. A clinical service in the UK to predict fetal Rh (Rhesus) D blood group using free fetal DNA in maternal plasma. Ann. N.Y. Acad. Sci. **1022:** 119–123.

22. FINNING, K. *et al.* Prediction of fetal K, c and E blood groups from free fetal DNA in maternal plasma. Transfus. Med. 2006. In press.
23. TSUI, N.B.Y. *et al.* 2004. Systematic micro-array based identification of placental mRNA in maternal plasma: towards non-invasive prenatal gene expression profiling. J. Med. Genet. **41:** 461–467.
24. WONG, B.C.K. *et al.* 2005. Circulating placental RNA in maternal plasma is associated with a preponderance of 5′ mRNA fragments: implications for non-invasive prenatal diagnosis and monitoring. Clin. Chem. **51:** 1786–1795.
25. HULTEN, M. 2005. The SAFE network of excellence. Clin. Chem. **51:** 7.
26. TABACK, B. & D.S. HOON. 2004. Circulating nucleic acids in plasma and serum: past, present and future. Curr. Opin. Mol. Ther. **6:** 273–278.
27. MORI, T. *et al.* 2005. Predictive utility of circulating methylated DNA in serum of melanoma patients. J. Clin. Oncol. **23:** 9351–9358.
28. PAN, H. *et al.* 2005. Loss of heterozygosity patterns provide fingerprints for genetic heterogeneity in multistep cancer progression of tobacco smoke-induced non-small cell lung cancer. Cancer Res. **65:** 1664–1669.
29. KHAN, S. *et al.* 2004. Genetic abnormalities in plasma DNA of patients with lung cancer and other respiratory diseases. Int. J. Cancer **110:** 891–895.
30. HOLDENRIEDER, S. *et al.* 2005. Circulating nucleosomes and cytokeratin 19-fragments in patients with colorectal cancer during chemotherapy. Anticancer Res. **25:** 1795–1801.
31. HAMAOUI, K. *et al.* 2004. Concentration of circulating rhodopsin mRNA in diabetic retinopathy. Clin. Chem. **50:** 2152–2155.
32. LAM, N.Y. *et al.* 2003. Time course of early and late changes in plasma DNA in trauma patients. Clin. Chem. **49:** 1286–1291.
33. LAM, N.Y. *et al.* 2005. Plasma DNA as a prognostic marker for stroke patients with negative neuroimaging within the first 24h of symptom onset. Resuscitation Nov 30 [Epub ahead of print].
34. KOTSCH, K. *et al.* 2004. Enhanced granulysin mRNA expression in urinary sediment in early and delayed acute renal allograft rejection. Transplantation **77:** 1866–1875.
35. MUTHUKUMAR, T. *et al.* 2005. Messenger RNA for FOXP3 in the urine of renal-allograft recipients. N. Engl. J. Med. 353: 2342–2351.
36. DING, C. & C.R. CANTOR. 2003. A high-throughput gene expression analysis technique using competitive PCR and matrix-assisted laser desorption ionization time-of-flight MS. Proc. Natl. Acad. Sci. USA **100:** 3059–3064.
37. DING, C. *et al.* 2004. MS analysis of single-nucleotide differences in circulating nucleic acids: Application to non-invasive prenatal diagnosis. Proc. Natl. Acad. Sci. USA **101:** 10762–10767.
38. CIRIGLIANO, V. *et al.* 2004. Rapid prenatal diagnosis of common chromosome aneuploidies by QF-PCR. Assessment on 18,000 consecutive clinical samples. Mol. Hum. Reprod. **10:** 839–846.

Prehistory of the Notion of Circulating Nucleic Acids in Plasma/Serum (CNAPS)

Birth of a Hypothesis

MAURICE STROUN AND PHILIPPE ANKER

OncoXL, 1228 Plan les Ouates/Geneva, Switzerland

ABSTRACT: In the late 50s and early 60s of the last century, a theoretical fight was taking place between Western and Russian scientists about the theory explaining the mechanism of evolution. According to neo-Darwinism, evolution was the result of hazard and necessity, that is, mutations arriving by chance favoring the survival of the fittest. For the Russian geneticists, acquired characteristics were the basis of evolution, that is, the environment modified the characteristics of the gene. One of the main experiments on which the Russian geneticists based their theory was the transmission of hereditary characteristics by a special technique of grafting between two varieties of plants—a mentor plant and a pupil plant. The pupil variety being entirely dependent on the development of the mentor plant its hereditary characteristics were modified accordingly. In the Western world these experiments were regarded with doubt. We were among the few who tried to repeat this kind of experiment. After three generations of grafting between two varieties of eggplant, we succeeded in obtaining hereditary modifications of the pupil plants, which acquired some of the characteristics of the mentor variety. The linkage between some hereditary characteristics of the mentor plant were broken, the segregation of the offspring was abnormal, dominant characteristics appearing in the offspring of a recessive plant. Rather than adopting the views of the Russian scientists about acquired characteristics, we suggested that DNA was circulating between the mentor and pupil plants and assumed that some nucleic acid molecules bearing genetic information could enter the somatic and reproductive cells of the pupil plant at a propitious moment and remain active.

KEYWORDS: **evolution; DNA; neo-Darwinism**

Address for correspondence: Maurice Stroun, OncoXL, 14 Chemin des Aulx, 1228 Plan les Ouates/Geneva, Switzerland. Voice: +41-22-7369786; fax: 41-22-7368421.
e-mail: mauricestroun@bluewin.ch

Ann. N.Y. Acad. Sci. 1075: 10–20 (2006). © 2006 New York Academy of Sciences.
doi: 10.1196/annals.1368.002

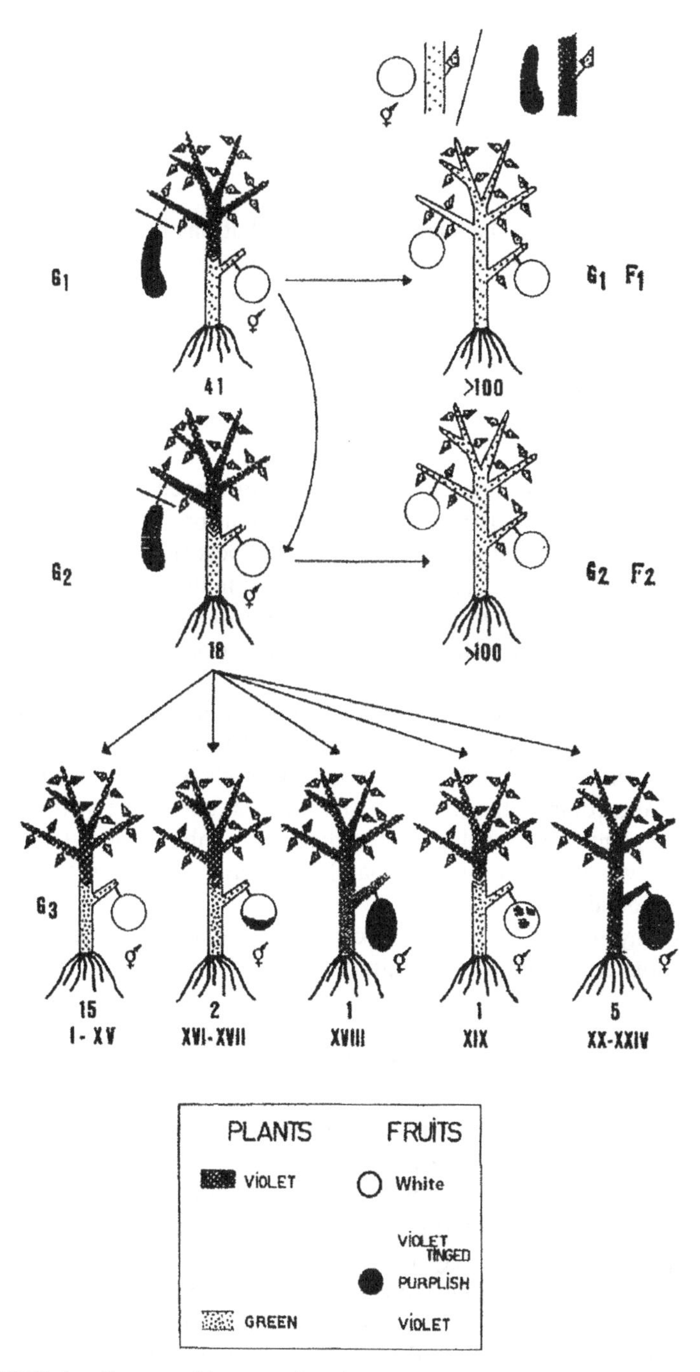

FIGURE 1. Heterograft between *S. melongena* variety Long violet (mentor plant) and the variety White round (pupil plant).

INTRODUCTION

In the 1950s there was an important debate between the Western geneticists and the Russian geneticists about the mechanism of evolution. It was the time of the Cold War and the discussion was carried on as an ideological fight.

The theory of the mechanism of evolution that has been accepted and is still considered as the explanation of the phenomenon is neo-Darwinism. According to this theory, two factors govern evolution: mutations "the singular unpredictable accident of the structure of the DNA" as defined by Nobel Prize winner Jacques Monod, and natural selection, which takes advantage of the whims of chance (the mutations) according to the necessities of the environment. It can be summed up in two words: chance and necessity.

In addition to this so called neo-Darwinist approach, Lamarck explained evolution by acquired characteristics under the influence of the environment. In the early 1950s the Russian geneticists brought back this theory that the environment modified the genome, selecting the best fit.

We studied biology because we were interested in the problem of evolution. So, as soon as one of us finished his Ph.D., he went to Russia to meet the main geneticists, who explained the mechanism of evolution through acquired characteristics.

The fundamental experiment on which the Russian scientists based their theory was through plant grafting. Publications[1] had reported results in which some characteristics of grafted plants had been shown to be altered in the offspring. According to the Russians, the fact that one of the symbionts was totally dependent for its development on the other symbionts and was modified accordingly, was the demonstration that the reproductive cells were dependent on their surroundings.

This is why we started a study[2,3] on the problem of the transmission of altered characteristics by means of grafts according to the Russian technique.

MATERIALS AND METHODS

We made intervariety grafts in *Solanum melongena* and interspecific grafts between a variety of *S. melongena* and a strain of *S. nigrum*.

For the intervariety heterografts, we used as mentor plant a variety that is characterized by a long violet fruit, brown-edged stamens, and a violet stem. As the pupil plant we used a variety, which is white and round, with yellow-edged stamens and a green stem. Long violet fruit as well as brown-edged stamens are dominant; there is a linkage between violet fruits and brown-edged stamens.

The following operating procedures were used: the mentor was the epibiota and the pupil was the hypobiota (FIG. 1). We plucked off all the leaves during the development of the pupil plant, the one we wanted to influence. On the

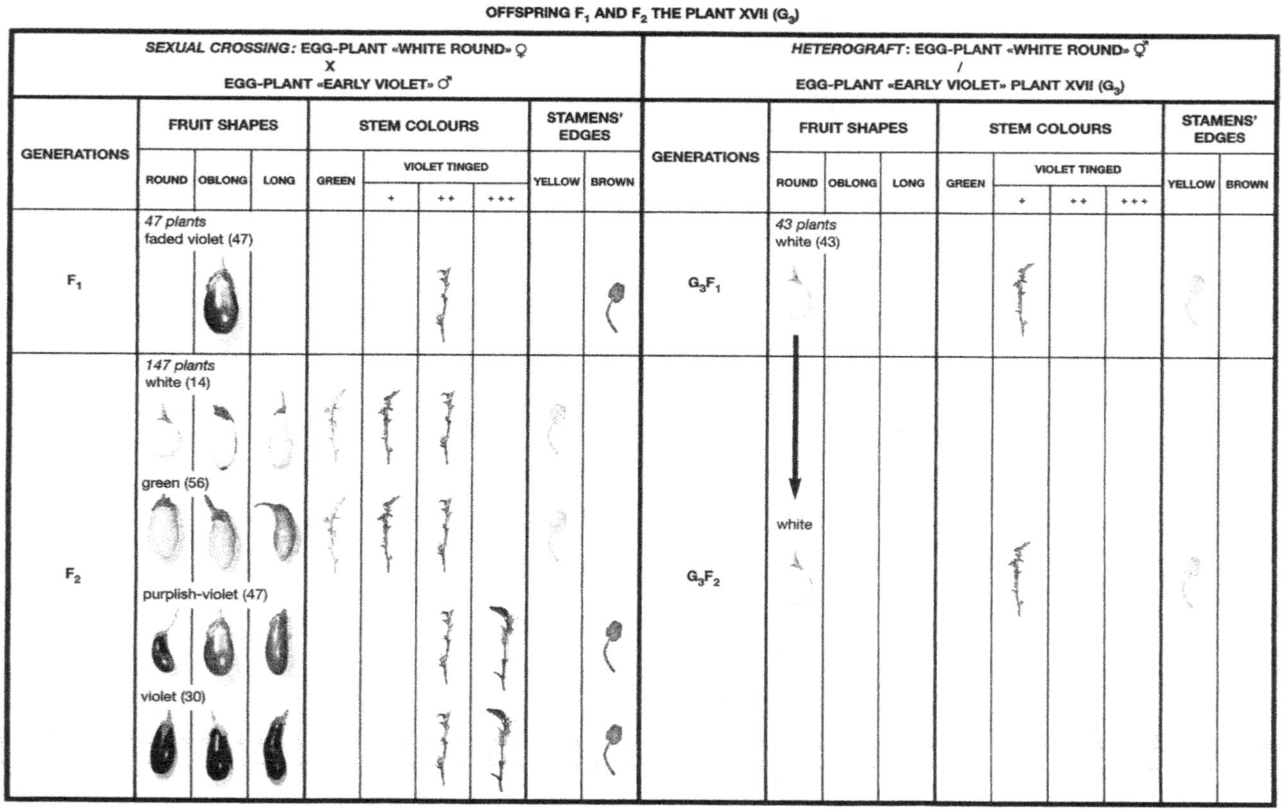

OFFSPRING F_1 AND F_2 THE PLANT XVII (G_3)

SEXUAL CROSSING: EGG-PLANT «WHITE ROUND» ♀
X
EGG-PLANT «EARLY VIOLET» ♂

GENERATIONS
FRUIT SHAPES: ROUND, OBLONG, LONG
STEM COLOURS: GREEN, VIOLET TINGED (+, ++, +++)
STAMENS' EDGES: YELLOW, BROWN

F_1
47 plants
faded violet (47)

F_2
147 plants
white (14)
green (56)
purplish-violet (47)
violet (30)

HETEROGRAFT: EGG-PLANT «WHITE ROUND» ⚥
/
EGG-PLANT «EARLY VIOLET» PLANT XVII (G_3)

GENERATIONS
FRUIT SHAPES: ROUND, OBLONG, LONG
STEM COLOURS: GREEN, VIOLET TINGED (+, ++, +++)
STAMENS' EDGES: YELLOW, BROWN

G_3F_1
43 plants
white (43)

G_3F_2
white

FIGURE 2. Offspring F1 and F2 of the plant XVII (FIG. 1) of the third generation of heterograft.

other hand the mentor kept its foliage, but its flower buds were cut off as soon as they appeared.

The flowers of the pupil plant were self-pollinated. At each generation, each fruit used to obtain the next generation was divided into two parts. The seeds of one part were used to give pupil-plantlets that were then grafted to mentor plants and the other half gave rise to plants that developed without being grafted. We called G the graft generation and GF the generations coming from symbionts, but which developed without being grafted.

We used homografts of the white round variety as control plants.

RESULTS

Among the 24 heterografts between long violet and white round of the third generation of grafts (G3) that bore fruit, 9 showed various modifications (FIG. 1): two had a violet tinge on part of a fruit; one had violet spots. One was purplish oblong and five were violet long.

If we examined the G3F1 of a fruit with a violet tinge on part of the fruit (FIG. 1, G3 XVII) we could see (FIG. 2) that only the stem color had been modified. In G3F1 and in G3F2 the same modification without segregation could be observed.

FIGURE 3. Fruits of the pupil plant XIX of the third generation of heterograft.

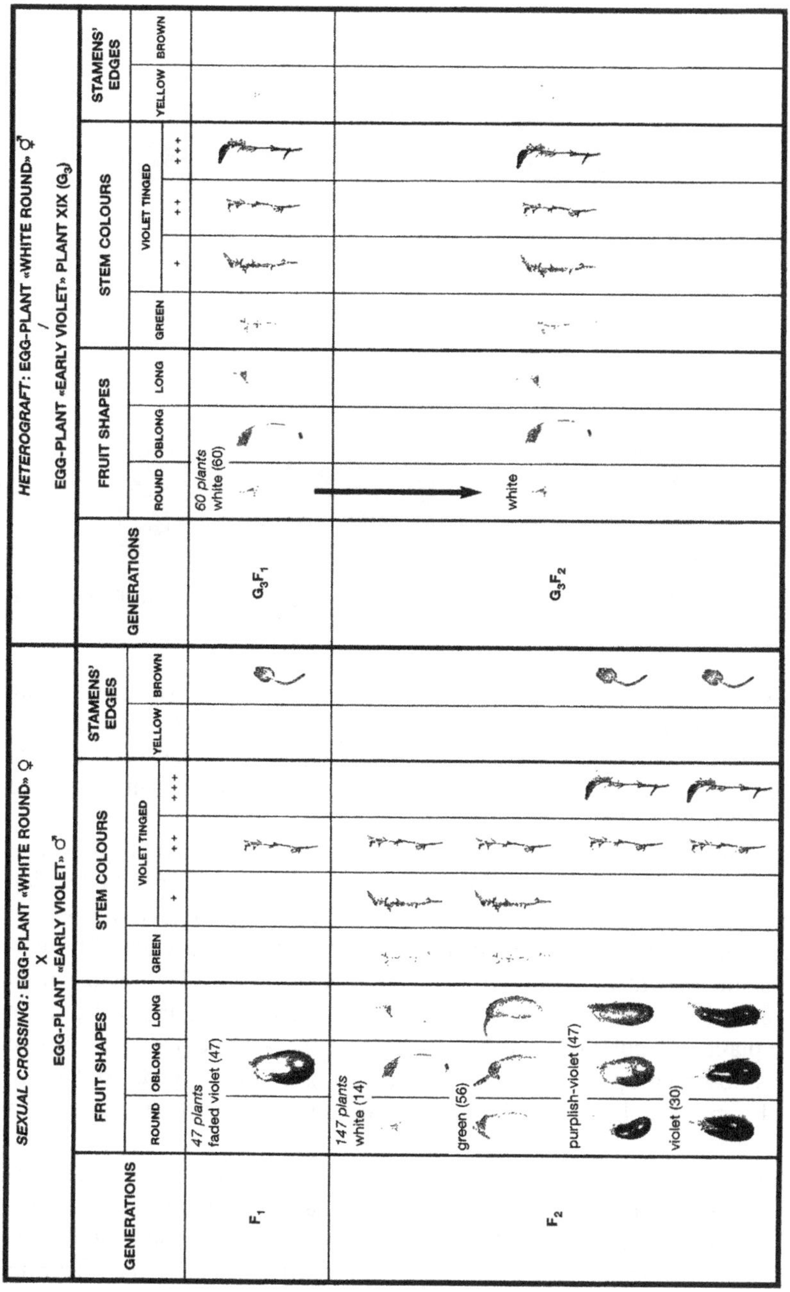

OFFSPRING F_1 AND F_2 OF THE PLANT XIX (G_3)

SEXUAL CROSSING: EGG-PLANT «WHITE ROUND» ♀
X
EGG-PLANT «EARLY VIOLET» ♂

GENERATIONS
FRUIT SHAPES
ROUND
OBLONG
LONG
STEM COLOURS
GREEN
VIOLET TINGED
+
++
+++
STAMENS' EDGES
YELLOW
BROWN

F_1
47 plants
faded violet (47)

F_2
147 plants
white (14)
green (56)
purplish-violet (47)
violet (30)

HETEROGRAFT: EGG-PLANT «WHITE ROUND» ⚥
/
EGG-PLANT «EARLY VIOLET» PLANT XIX (G_3)

GENERATIONS
FRUIT SHAPES
ROUND
OBLONG
LONG
STEM COLOURS
GREEN
VIOLET TINGED
+
++
+++
STAMENS' EDGES
YELLOW
BROWN

G_3F_1
60 plants
white (60)

G_3F_2
white

FIGURE 4. Offspring F1 and F2 of the plant XIX (FIG. 1) of the third generation of heterograft.

FIGURE 5. Fruit of the pupil plant XVIII (FIG. 1) of the third generation of heterograft.

The G3F1 of another plant (XIX) with violet spots (FIG. 3) gave a segregation of both fruit shape and stem color (FIG. 4). Out of the offspring G3F2 of a round fruit, which is a recessive characteristic, a segregation with regard to the shape was observed. From a recessive round shape, dominant long fruits appeared in the offspring.

The offspring coming from a G3 fruit, which was oblong and purple velvet (FIG. 5), presented in G3F1 a segregation (FIG. 6) quite similar to that which can be seen in an F2 resulting from a sexual crossing with, however, two important differences: (i) a linkage rupture shown by the absence of brown-edged stamens, which are always linked in sexual hybrid plants with violet fruits, and (ii) the presence of black-violet fruits and stems absent from the mentor plant. The color of the offspring of these black-violet fruits was stable and was observed up to G3F2. The anthocyanins of these plants studied by paper chromatography were different from the anthocyanins of the velvet long fruit.

The homografts control plants, white round/white round of the third generation of grafts (G3) as well as G3F1 and G3F2 remained similar to the standard.

The interspecific heterograft between *S. melongena* white round and *S. nigrum* black night Shade (violet fruit and violet stem) appeared in the offspring of the third graft generation in G3F1, although the G3 plants themselves were not altered. The offspring of two of the G3 plants showed (FIG. 7) an alteration

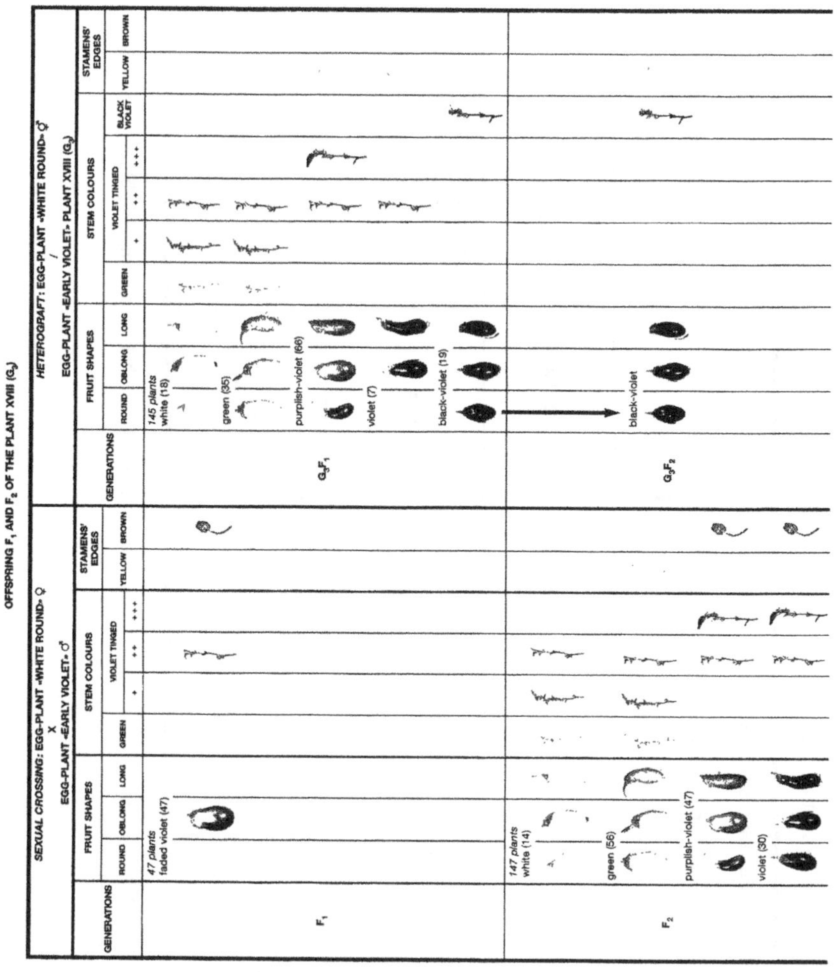

OFFSPRING F_1 AND F_2 OF THE PLANT XVIII (G_3)

SEXUAL CROSSING: EGG-PLANT «WHITE ROUND» ♀ X EGG-PLANT «EARLY VIOLET» ♂

GENERATIONS; FRUIT SHAPES: ROUND, OBLONG, LONG; STEM COLOURS: GREEN, VIOLET TINGED +, ++, +++; STAMENS' EDGES: YELLOW, BROWN

F_1: 47 plants; faded violet (47)

F_2: 147 plants; white (14); green (56); purplish-violet (47); violet (30)

HETEROGRAFT: EGG-PLANT «WHITE ROUND» ♂ / EGG-PLANT «EARLY VIOLET» PLANT XVIII (G_3)

GENERATIONS; FRUIT SHAPES: ROUND, OBLONG, LONG; STEM COLOURS: GREEN, VIOLET TINGED +, ++, +++, BLACK VIOLET; STAMENS' EDGES: YELLOW, BROWN

G_3F_1: 145 plants; white (18); green (35); purplish-violet (66); violet (7); black-violet (19)

G_3F_2: black-violet

FIGURE 6. Offspring F1 and F2 of the plant XVIII (FIG. 1) of the third generation of heterograft.

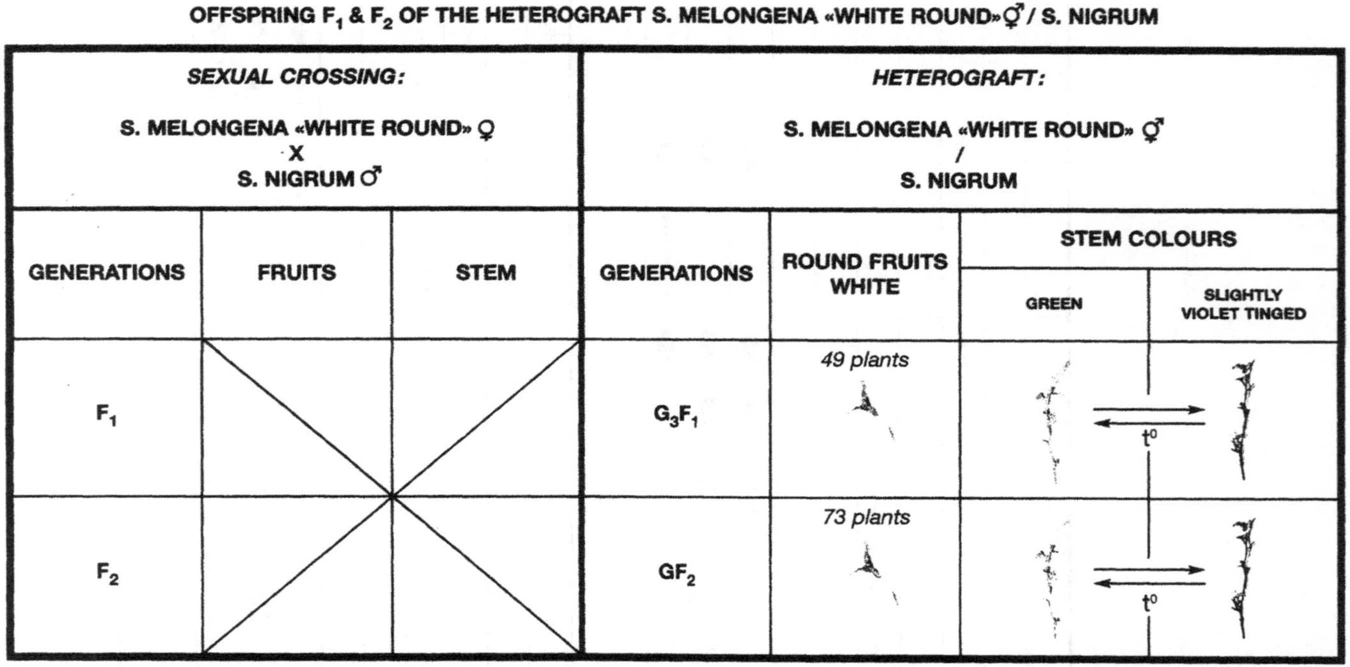

OFFSPRING F_1 & F_2 OF THE HETEROGRAFT S. MELONGENA «WHITE ROUND»⚥ / S. NIGRUM

SEXUAL CROSSING: S. MELONGENA «WHITE ROUND» ♀ X S. NIGRUM ♂			*HETEROGRAFT:* S. MELONGENA «WHITE ROUND» ⚥ / S. NIGRUM				
					STEM COLOURS		
GENERATIONS	FRUITS	STEM	GENERATIONS	ROUND FRUITS WHITE	GREEN		SLIGHTLY VIOLET TINGED
F_1			G_3F_1	*49 plants*		⇄ t°	
F_2			GF_2	*73 plants*		⇄ t°	

FIGURE 7. Offspring G3F1 and G3F2 of the interspecific heterograft between *S. nigrum* black night shade (mentor plant) and *S. melongena* white round eggplant (pupil plant).

of the color of the stems that was slightly violet tinged. The pigmentation, however, was very thermo-labile and disappeared when the temperature rose.

DISCUSSION

It is impossible to attribute these results to a sexual crossing. The different segregations in G3F1 and G3F2 are completely different from an F1 or F2 sexual crossing. Moreover, a sexual crossing would be impossible between plants of our interspecific heterograft, white round *S. melongena* versus black *S. nigrum*

It is difficult to explain this phenomenon by Lamarck's theory of acquired characteristics. Indeed, the amount of modifications observed could not be the result of the survival of the pupil plant thanks to the photosynthesis assumed by the leaves of the mentor plant. Such a rapid transformation of so many characteristics would imply a very weak stability for the species! It does not mean that in some rare cases acquired characteristics also play a role.[4] Of course such a rare phenomenon does not shake neo-Darwinism fundamentally.

The appearance of a new characteristic (black violet) could be due to some gene shuffling. However, it would be strange that so many modifications in the direction of the mentor plant would be the result of shuffling genes.

Finally, the idea of a transformation or a transduction such as can be found in microorganisms should be considered. Thus we assumed[3] that DNA bearing genetic information can circulate and can enter the somatic and reproductive cells at a propitious moment and remain active.

It took us 11 years to discover the active release of nucleic acids by living cells.[5,6] We also observed that it is accompanied by a DNA polymerase.[7,8] There seems to be a preferential release because it is composed of newly synthesized DNA and contains more Alu repeat sequences than the cellular DNA.[9] There is also a spontaneous release of RNA by living cells.[10]

In the meantime we discovered the phenomenon we called *transcession* that has been largely ignored. The released DNA from bacteria comes with its DNA-dependent RNA polymerase.[11] When the bacterial DNA enters plant cells,[12] animal cells,[13] or human cells[14] there is a transient synthesis of bacterial RNA. This phenomenon plays a role in the crown gall, the tumor of a plant produced by *Agrobacterium tumefaciens.*[15]

After many years, the phenomenon of active release of nucleic acids by living cells aroused great interest in different medical fields, which is evidenced in this volume.

ACKNOWLEDGMENTS

For financial help we thank Mrs. Kate Marx in the name of her late husband David Marx.

REFERENCES

1. GLOUCHTCHENKO, I.E. 1948. Vegetative hybridization in plants [in Russian]. Academy Nauk SSR, Moscow.
2. STROUN, M. 1962. Modifications des caractères à la suite de greffes intervariétales chez le *Solanum melongena*. C. R. Acad. Sci. Paris **255:** 361–363.
3. STROUN, M. *et al.* 1963. Modifications transmitted to the offspring, provoked by heterograft in *Solanum melongena*. Arch. Sci. Genève **16:** 1–21.
4. AMZALLAG, G.N. 2000. Maternal transmission of adaptive modifications in salt-treated *Sorghum bicolor:* a first stage in ecotypic differentiation? New Physiol. **146:** 483–492.
5. ANKER, P. *et al.* 1975. Spontaneous release of DNA by human blood lymphocytes shown in an in vitro system. Cancer Res. **30:** 2375–2382.
6. STROUN, M. *et al.* 1977. Spontaneous release of newly synthesized DNA from frog auricles. Arch. Sci. Genève **30:** 229–241.
7. ANKER, P. *et al.* 1975. Spontaneous extracellular synthesis of DNA released by human blood lymphocytes. Cancer **36:** 2832–2839.
8. ANKER, P. & M. STROUN. 1977. Spontaneous extracellular synthesis of DNA released by frog auricles. Arch. Sci. Genève **35:** 263–278.
9. STROUN, M. *et al.* 2001. Alu repeat sequences are present in increased proportions compared to a unique gene plasma/serum. Ann. N. Y. Acad. Sci. **945:** 258–264.
10. STROUN, M. *et al.* 1978. Presence of RNA in the nucleoprotein complex spontaneously released by human lymphocytes and frog auricles in culture. Cancer Res. **38:** 3346–3554.
11. STROUN, M. 1971. On the nature of the polymerase responsible for the transcription of the released bacterial DNA in plant cells. Biochem. Biophys. Res. Com. **44:** 571–578.
12. STROUN, M. & P. ANKER. 1971. Bacterial nucleic acid synthesis in plants following bacterial contact. Mol. Gen. Genet. **113:** 92–98.
13. ANKER, P. & M. STROUN. 1972. Bacterial ribonucleic acid in the frog brain after a bacterial peritoneal infection. Science **78:** 621–623.
14. ANKER, P. *et al.* 2004. Transcession of DNA from bacteria to human cells in culture: a possible role in oncogenesis. Ann. N. Y. Acad. Sci. **1022:** 195–201.
15. STROUN, M. *et al.* 1971. *Agrobacterium tumefaciens* ribonucleic acid synthesis in tomato cells and crown gall induction. J. Bacteriol. **106:** 634–639.

Circulating DNA

Intracellular and Intraorgan Messenger?

P.B. GAHAN

Anatomy and Human Sciences, King's College London, London Bridge, London SE1 1UL, UK

ABSTRACT: The circulation of both foreign and endogenous DNA within plants and its ability to be expressed in the host plants and FI generation is described. These data, together with those from animal systems are used to support the concept that a DNA fraction can act as a messenger between cells and tissues.

KEYWORDS: messenger; DNA; plant; animal; metabolic DNA; transcession; metastases

INTRODUCTION

The spontaneous release of DNA from living cells has been demonstrated frequently during the past 50 years. The initial studies of Stroun *et al.*[1] on eggplants showed that information passed from the tutor (host plant) to the pupil (graft), leading to changes in the shape and color of the fruit carried on the pupil. Chayen and Gahan[2–4] demonstrated that DNA occurred in the cytoplasm of cells of *Vicia faba* and *Allium cepa* and that a part of this cytoplasmic DNA could move to the nucleus.[5] This led to the concept that a fraction of the cytoplasmic DNA could be acting as a messenger[6]—a concept subsequently developed by Pelc,[7] Bell,[8] Adams,[9] and Gahan.[10]

Transcession—Early Studies on DNA Movement in Eukaryotes

During the 1960s, unpublished studies by Stroun and Anker, and Gahan and Sheikh indicated that DNA could move in plants and influence biological events. Stroun and Anker germinated seeds of *Lycopersicon esculentum* in the presence of DNA isolated from three different bacteria (TABLE 1). Although the F1 control plants produced only one plant with abnormal fruit from 33 plants,

Address for correspondence: P.B. Gahan, Anatomy and Human Sciences, King's College London, London Bridge, London SE1 1UL. Voice: 44-(0) 208-959-8511; fax: 44-(0) 208-959-8511.
e-mail: pgahan@aol.com

Ann. N.Y. Acad. Sci. 1075: 21–33 (2006).
doi: 10.1196/annals.1368.003

TABLE 1. Variation in the shape and structure of fruits of *Lycopersicon esculentum* (tomato) F1 plants after germination of seeds in the presence of bacterial DNA

Treatment	Control	*Micrococcus lysodecticus*	*Agrobacterium tumefaciens*	*Pseudomonas fluorescens*
Number of unmodified plants	33 round bilocular	0	0	0
Number of modified plants	1 irregular multilocular	29 irregular multilocular	32 multilocular irregular	37 bilocular lemon-shaped

all the plants germinated in the presence of bacterial DNA produced aberrant fruits in the F1 generation (TABLE 1; FIG. 1). In addition, the F1 plants derived from those germinated in the presence of *Agrobacterium tumefaciens* DNA showed a further flowering modification, having either four or seven sepals as compared with the control plants' five sepals (FIG. 1). Thus, it was clear that exogenous DNA could enter embryos and have an effect upon the plants so derived, as seen in both flowering and fruiting. The germination of seeds of *Medicago sativa* in the presence of DNA demonstrated a general technique for the transformation of plants.[11] Thus, overall, these studies confirm the concept that the bacterial DNA could have entered the plant cells and be integrated into the genome from where it was expressed to yield the modified tomato fruits.

In a parallel series of studies, Gahan and Sheikh found that *L. esculentum* cv. Moneymaker plants bearing crown gall tumors induced by inoculating the main stems with *A. tumefaciens* B6 (FIG. 2) could produce aberrant fruits. If the plants were inoculated in the main stem, aberrant fruits were borne by the branches arising from below the tumor (FIG. 2), that is, from branches formed prior to the induction of the tumor and which, therefore, comprised normal diploid tissues. Normally, since the bacteria cannot enter a healthy plant cell, the tumor is induced by the bacterium through a transfer of its Ti plasmid into the host cell, where it integrates into the host genome and is expressed to induce the tumor.

However, the events observed were at a distance from the tumor and although bacteria may circulate via the plant vascular system, they cannot enter a healthy plant cell, which means that informative molecules must have been released from the bacteria to enter the plant cells and result in the formation of aberrant fruit. Although it is possible that the Ti plasmids may have been involved in the induction of the aberrant fruit, it is also likely that the bacteria released another form of DNA to circulate in the plant since Stroun and Anker showed aberrant fruit induction by nonplasmid bacterial DNA (described above). Stroun and Anker[12] and Stroun *et al.*[13,14] also demonstrated that although bacteria could not enter the cortical tissues from the vascular tissues of eggplants, RNA of either *A. tumefaciens* or *Pseudomonas fluorescens* or *Escherichia coli* could be identified in the cortical tissues after separation of the cortex from the vascular tissues. This confirmed that bacterial DNA

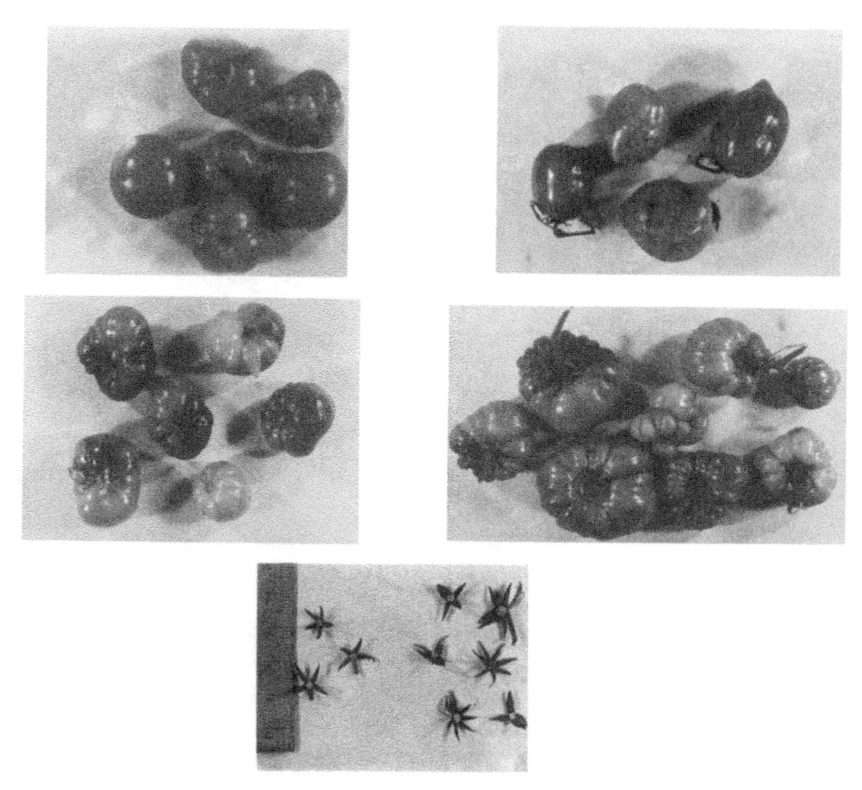

FIGURE 1. *L. esculentum* fruits from F1 of control plants (*top left*); and plant germinated with P. fluorescens DNA (*top right*); with *M. lysodecticus* DNA (*middle left*); and with *A. tumefaciens* DNA (*middle right*). Developing fruit in F1 plants showed changed sepal number (*bottom*) [control, *left*, and *A. tumefaciens*, *right*].

had passed to the cortex, thus resulting in the production of bacterial RNA in the plant host cells. The observed production of aberrant tomato fruits was shown not to be due to transient expression of the bacterial DNA in the tumor-bearing plants on germinating seeds from such fruits. They developed into either (*a*) normal plants, (*b*) plants with modified leaves, or (*c*) plants that produced one leaf and ceased growth (FIG. 2) . The fully developed plants with modified leaves produced aberrant fruits, though fewer in number compared to the normal plants on account of the sterility of a number of flowers (FIG. 2). The modified leaflets were rounded and hairless (FIG. 3), as opposed to the seedlings, which developed as normal plants with pointed, hairy leaflets (FIG. 3). These characteristics appear to be gene, linked (D. Lewis, private communication). Thus, it appeared that there was information flow from the *A. tumefaciens* used to induce the tumor from the crown gall to the point at which the aberrant fruits were produced, and that such information flow was likely on account of DNA movement.

FIGURE 2. Whole *L. esculentum* plant bearing tumor and aberrant fruit (*top*); range of aberrant fruit (*bottom left*). Different forms of F1 seedlings derived from seeds of aberrant fruits in Figure 1 (*bottom right*; see text for details).

This was confirmed by further studies on the *L. esculentum* after feeding cut tips of plants with *A.tumefaciens* B6 when no crown gall tumors developed. After exclusion of the fed zone to avoid bacteria, RNA was extracted from the plants that had been subsequently fed for 7 h with ^{3}H-uridine. Hybridization studies showed that newly synthesized bacterial ^{3}H-RNA was present in the plant tissues, thus demonstrating that DNA transcription had occurred in the plant cells.[15,16] The fact is that a calculated equivalent number of bacteria in culture to that possibly present in the plants yielded a much lower amount of ^{3}H-RNA than that found in the plants. This supports the concept of DNA release from the bacteria and its subsequent transcription. The question of the ability of prokaryote DNA to circulate in higher plants was tested by feeding cut shoots of *L. esculentum* var. Tuckwood with bacterial DNA for 48 h. One to 5 days later shoots were placed in a solution of ^{3}H-thymidine for 24 h, after

FIGURE 3. Leaf of F1 control type plant (*top left*); F1 modified leaf (*top right*). Modified F1 plant showing modified leaf, aberrant fruit, and the sterile flowers (*bottom*).

which the DNA was extracted, purified, and analyzed by cesium chloride gradient centrifugation. Plants that had been fed DNA from *M. lysodekticus*, *A. tumefaciens,* and *E. coli* all had radioactive peaks, which were intermediary between peaks for either the plant or the bacterial DNAs. This material was DNase-sensitive and heat-denatured. After dialysis and sonication, further cesium chloride gradient separation allowed the demonstration of two radioactive peaks, one corresponding to each of the plant and bacterial DNAs. However, if the experiment was performed with the DNA from *Clostridium perfringens*, the radioactivity was found only in the bacterial and the plant DNAs.[17] The implication was that the DNA from either *M. lysodekticus* or *A. tumefaciens* or *E. coli* had entered the plant and the nuclei where it had replicated. Part of the DNA had become attached to the host DNA, from which it was separated by sonication.

In further experiments, cut shoots were fed with ^{3}H-DNA from either *E. coli* or *Bacillus subtilis* for 6 h followed by chasing with water for 48 h. The DNA was extracted, purified, and analyzed on cesium chloride gradient. Two radioactive peaks were identified that corresponded with the bacterial and the plant DNAs. More than 30% of the DNA sedimented at the bacterial site thus confirming that even after 48 h in the plant the bacterial DNA had remained intact.[18] The labeling present in the plant DNA was considered to be due to the breakdown and re-utilization of the bacterial DNA by the host cells. However, when autographic studies were performed after feeding cut shoots of *L. esculentum* var. Tuckwood with *E. coli* ^{3}H-DNA for 6 h and chased with water, labeled radioactivity was shown to be present in the nuclei of both dividing and nondividing cells, the labeling being DNase-sensitive and RNase-resistant.[19] Since the level of labeling was heavy in nuclei from both the dividing and nondividing cells, it would appear that the labeling was not due solely to the incorporation of breakdown products from the bacterial ^{3}H-DNA, but to the DNA itself. This is confirmed by studies on the uptake of DNA by *Vicia faba*, when entry into roots showed a seven fold increase in endogenous acid DNase activity.[20] However, it was not possible to detect DNase activity in the shoots, thereby confirming that DNA fed to cut shoots is not readily broken down. Labeled DNA was also found in mitochondria and chloroplasts. A much lower level of labeling was found if the plants were fed with ^{3}H-thymidine of a similar specific activity to that of the ^{3}H-DNA. Thus, after feeding for 6 h, 20% of the parenchymal cells were labeled with ^{3}H-DNA, but only 5% were labeled with ^{3}H-thymidine. Similarly, 18% of collenchymal cells were labeled with ^{3}H-DNA at 6 h but none was labeled with ^{3}H-thymidine. Phloem (including the actively dividing cambium) showed 40% of cells labeled with ^{3}H-DNA at 6 h, but only 18% with ^{3}H-thymidine. These results would imply that the ^{3}H-DNA was being taken up directly by the nuclei of many cells, especially in the case of collenchyma, where no mitotic figures can be found even after treating the plants for 3 days with colchicine.[21] Hence there would be no need for the synthesis of DNA for cell division in these cells as seen by the lack of incorporation of ^{3}H-thymidine.

In order to confirm that bacterial DNA could enter plant cells integrate, into the plant cell genome, and be expressed, an experiment was performed using bacterial DNA carrying marker genes.[22] Sterile cut shoots of *Solanum aviculare* were fed for 6 h with *E. coli* DNA carrying three marker genes, namely:

1. GUS gene (B-glucuronidase)
2. BAR gene (phosphinotricin, an inhibitor of glutamate synthase to yield resistance to bialophos)
3. NPT II gene (neomycin phosphotransferase II, which blocks the antibiotic interaction with ribosomes, thus leading to resistance to kanamycin).

The cut shoots were chased with water for 24 h prior to rooting and growing under sterile conditions. All three genes were found to be expressed in all tissues of the established plants. In separate experiments, fully transformed roots regenerated from callus derived from tissues of plants transformed with the GUS gene, while the F1 generation from plants transformed with all three genes was also found to carry all three genes. The stable insertion of these genes into the host genome was confirmed by Southern blotting.[22]

Thus, it is clear that prokaryote DNA can be released from bacteria and enter plant cells, where it can enter the nuclei, integrate into the host genome, and be expressed. Similar information on animal cells shows the expression of bacterial DNA in frogs.[23] Frogs were injected intraperitoneally with either *E. coli* or *B. subtilis,* or *A. tumefaciens.* In order to establish whether the bacteria released DNA, which could be expressed in the frog system, the frogs were injected intraperitoneally with antibiotics to destroy the bacteria prior to injecting ^{3}H-uridine intraperitoneally (TABLE 2).

The frogs' brains were subsequently removed and screened for sterility prior to the extraction of total RNA.[23] Hybridization of the RNA against bacterial DNA held on filters revealed the presence of bacterial RNA in the brains of the frogs. This implied that the DNA released by the bacteria travelled to the brain, where it was incorporated and expressed. This was further confirmed by the intraperitoneal injection of ^{3}H-DNA that was found in the nuclei of the brain cells. Importantly, this also demonstrated that the DNA circulating within an organism could cross the blood–brain barrier.

Similarly, bacterial RNA was shown to be synthesized in frog auricles after the intraperitoneal injection of either *E. coli* or *A tumefaciens* into frogs.[24]

Thus, bacterial DNA (*a*) can be released from the bacteria in the host tissues and (*b*) is able to circulate, integrate into the host genome, and be expressed in the host cells, a process which has been termed *transcession*[16] The question arises as to the generality of the release of DNA from eukaryote cells in a similar manner to that of its release from prokaryotes. Can it enter other cells, be integrated into the genome, and be expressed in the host cells?

TABLE 2. Bacteria used and amount and type of antibiotic employed prior to assessing bacterial RNA production in frog brain

Bacterium	Antibiotic concentration	^{3}H-uridine
Escherichia coli	2,000 μg colimycin	+
Agrobacterium tumefaciens	2,000 μg colimycin	+
Bacillus subtilis	2,000 μg amicilin + 2,000 μg cloxacilin	+

Release of DNA from Living Eukaryote Cells

Release from Frog Auricles

Frog auricles can be kept alive for more than 2 days in Ringer's solution as evidenced by the continued pulsation of the auricles. When the medium was removed after various periods of incubation and, after centrifugation to remove any debris, DNA could be isolated from the supernatant.[25] The DNA was identified by its UV absorption, its diphenylamine and dopamine reactions, its sensitivity to DNase, and resistance to pronase and RNase.[25] It was also shown that the DNA released from the living auricles was double-stranded. No repeat sequences were found, indicating that the DNA was unique, and the AT/GC ratios were similar to those of nuclear DNA, therefore showing that it was unlikely to be of mitochondrial origin. Heating the auricles at 50°C prior to incubation resulted in no DNA being released, thus demonstrating that the DNA was released only from living auricles. If the auricles were sliced into three parts to induce damage through cut surfaces, then no extra DNA was released.

When the auricles were incubated for a period and then transferred to a fresh medium for a similar period, additional DNA was released. However, if after the same initial incubation period the auricles were removed from the incubation medium and returned to the same medium, no additional DNA was released, indicating that the DNA release occurred in a homeostatic manner.[26]

Furthermore, if the auricles were fed ^{3}H-thymidine for periods of time and then transferred to Ringer's solution without ^{3}H-thymidine, they released radioactive ^{3}H-DNA that was more heavily labeled than the nuclear DNA.[25]

DNA Release from Nonstimulated and Stimulated Lymphocytes

Similar results to those reported for frog auricles were found for human blood lymphocyte cultures[27] as well as rabbit and rat spleen cell preparations that contained primarily a lymphocyte cell population.[28,29]

Newly synthesized DNA is also released from living, stimulated lymphocytes.[30] The DNA released from stimulated lymphocytes was found to be of higher molecular weight than that released from nonstimulated cells, as determined by separation by agarose gel chromatography.[31]

DNA Release from Chick Embryo Fibroblasts

Adams and McIntosh[32] found newly synthesized DNA released from living chick embryo fibroblasts but not from dead and dying cells.

DNA Release from Plant Protoplasts

Preliminary results with isolated protoplasts from two cultivars (xanthi and frisson) of *Pisum sativum* showed that newly synthesized DNA is maximally released from viable protoplasts only at 7 h of incubation after pulsing the preparations with ^{3}H-thymidine (S. Ochatt and P.B. Gahan, unpublished data).

Thus, newly synthesized DNA is released in a homeostatic fashion from amphibian, avian, mammalian, and plant cells. Moreover, the data from the frog auricles imply that the released DNA from the cells must also be able to pass through the auricular tissue in order to be released into the surrounding medium.

Can This Released Eukaryote DNA Enter Other Cells?

Adams and McIntosh[33] studied the uptake of released DNA by chick embryo fibroblasts. The newly synthesized DNA released from a batch of chick embryo fibroblasts was isolated by agarose gel chromatography, concentrated, and placed in fresh incubation medium together with a new batch of fibroblasts for 18 h. The ^{3}H-DNA was recovered from the recipient fibroblasts. A similar set of experiments were performed with the ^{3}H-DNA released from the chick embryo fibroblasts placed in the culture medium together with a rat spleen cell preparation. The fibroblast DNA was recovered from the rat spleen cells after incubating for 18 h. The DNA was seen to move to the nuclei as measured both biochemically and autoradiographically. By 20 min of incubation, there were equal amounts of DNA in both the nuclei and the cytoplasm.

Thus, the newly synthesized DNA appears to move freely into either recipient amphibian, avian, mammalian, and plant cells *in vitro* or into whole plant tissues.

Can These Recipient Cells Integrate and Express This DNA?

The uptake and integration of prokaryote DNA into plant cells in the intact plant, and its subsequent expression, demonstrates that this is possible.[22] The uptake and expression of the DNA shown for eukaryote cells may be demonstrated by the expression of nontumor cell DNA in tumor cells and of tumor cell DNA in nontumor cells.

Adams *et al.*[34] ran experiments in which the DNA released from J774 cells (leukemia), P497 cells (glial tumor), and nonstimulated lymphocytes together with a cytosolic DNA fraction (equivalent to the released DNA) from hepatocytes were purified and isolated by agarose gel chromatography.

The tumor cell lines were each incubated in the presence of either of the two tumor cell line DNAs, the hepatocyte DNA, or the lymphocyte DNA, and the lymphocytes were incubated in the presence of either of the tumor cell line DNAs or the lymphocyte DNA. ^{3}H-thymidine was added to the cultures and the amount of nuclear incorporation of the ^{3}H-thymidine into DNA (the index of DNA synthesis) was determined. The levels of DNA synthesis in the tumor cell lines were the same in the presence of either of the DNAs released by the two tumor cell lines, but was reduced by ca. 60% in the presence of either the hepatocyte or the lymphocyte DNAs. Conversely, the incubation of the lymphocytes in the presence of either of the tumor cell line DNAs or the lymphocyte DNA showed an increased DNA synthesis (ca. 60%) in the presence of the tumor DNAs but not in the presence of the lymphocyte DNA. Thus, in these experiments, the DNAs have been taken up by the cells and expressed.

CONCLUSIONS

The foregoing experiments indicate that living cells release into the environment a newly synthesized DNA, complexed with glycolipoprotein and RNA, from where it can circulate within the organism and be taken up by other cells. Here, the DNA can be incorporated genomically and expressed. Thus, one role for this DNA can be as a messenger that could be involved with signaling between cells and tissues within the organism.

The DNA fraction would appear to be nuclear in origin[35] since pulse labeling with ^{3}H-thymidine showed the appearance of labeled DNA in the nucleus, followed by its appearance in the cytoplasm prior to being found in the cell environment. In addition, reports show a DNA fraction associated with leukocyte plasma membranes.[36–38] It is possible that this represents the DNA leaving the cell, and where it may receive at least part of the glycolipoprotein complex. This DNA does not represent a loss of genomic DNA since the cells from which it is released are living cells and not in the process of degenerating. Pelc and collaborators[6] found that the DNA in many nuclei in nondividing cell populations became labeled on feeding with ^{3}H-thymidine. This was termed *metabolic DNA*, where extra copies of genes were being synthesized. Subsequently, the labeling was explained by some workers as being due to the presence of DNA repair. However, further studies by[21,39,40] revealed that the metabolic DNA in nuclei of cells in tomato stems was not repaired out of the system. Moreover, although the labeling was initially in the cells at the base of the stem, with no labeling of shoot apices, 2–3 weeks later autoradiographic studies showed that some nuclei in the shoot apex were labeled. The labeling of the nuclear DNA could not be attributed to the breakdown products of the labeled DNA at the base of the stem being sent to the shoot apex since such molecules would be dispersed around the plant and would be present in too small an amount in any one nucleus to be detected by autoradiography. One explanation was that the labeled DNA had moved from cells at the base of the shoot to the apex. Thus, one explanation of the so-called metabolic DNA of Pelc would be that of copying of the genome DNA to form a messenger DNA.

There are a number of different fractions of nucleic acids present in the blood of patients suffering from cancer and other disorders (see articles in this volume). While it is clear that some of the DNA may be derived by apoptosis from, for example, blood and tumor cells, this does not explain the origin of all the fractions. A DNA complex similar to the one described above, which is released from living cells, has already been identified in the blood of cancer patients.[41] It is possible that such a DNA fraction can act as a messenger as already demonstrated for the tumor cell lines J774 and P497 and in so doing, may play a part in the formation of metastases. Garcia-Olmo[42] has already proposed that a fraction of the circulating DNA may be involved in metastasis. Thus, it is considered that the DNA fraction complexed with glycolipoprotein and released from living cells would be a strong candidate for initiating this process.

REFERENCES

1. STROUN, M. *et al.* 1963. Modifications transmitted to the offspring, provoked by heterograft in *Solanum melongena*. Arch. Sci. Genève **16**: 1–21.
2. CHAYEN, J. 1949. U-V Absorbance by nucleic acids in the cytoplasm of root tip cells. Nature **164**: 930–932.
3. CHAYEN, J. 1958. Cytoplasmic localization of DNA in root tip cells. Exp. Cell Res. Suppl. **6**: 115–121.
4. GAHAN, P.B. *et al.* 1962. Cytoplasmic localization of deoxyribonucleic acid in *Allium cepa*. Nature **195**: 1115–1116.
5. GAHAN, P.B. *et al.* 1962. Cytoplasmic localization of deoxyribonucleic acid in *Allium cepa*. Nature **195**: 1115–1116.
6. GAHAN, P.B. & J. CHAYEN. 1965. Cytoplasmic deoxyribonucleic acid. Int. Rev. Cytol. **18**: 223–248.
7. PELC, S.R. 1968. Turnover of DNA and function. Nature **219**: 162–163.
8. BELL, E. 1969. I-DNA: its packaging and its relation to protein synthesis during differentiation. Nature **224**: 326–328.
9. ADAMS, D.H. 1985. The problem of cytoplasmic DNA: its extrusion/uptake by cultured cells and its possible role in cell-cell information transfer. Int. J. Biochem. **17**: 1133–1141.
10. GAHAN, P.B. 2003. Messenger DNA in higher plants. Cell Biochem. Func. **21**: 207–209.
11. SENARATNA, T. *et al.* 1991. Direct DNA uptake during the imbibition of dry seeds. Plant Sci. **79**: 223–228.
12. STROUN, M. & P. ANKER. 1971. Bacterial nucleic acid synthesis in plants following bacterial contact. Molec. Gen. Genetics **113**: 92–98.
13. STROUN, M. *et al.* 1971. Effect of the extent of transcription of plant cells and bacteria on the transcription in plant cells of DNA released from bacteria. FEBS Lett. **13**: 161–164.
14. STROUN, M. *et al.* 1970. Natural release of nucleic acids from bacteria into plant cells. Nature **227**: 607–608.

15. STROUN, M.P. *et al.* 1971. *Agrobacterium tumefaciens* ribonucleic acid synthesis in tomato cells and crown gall induction. J. Bacteriol. **106**: 634–639.
16. STROUN, M.P. *et al.* 1977. Circulating nucleic acids in higher organisms. Int. J. Cytol. **51**: 1–48.
17. STROUN, M. & L. LEDOUX. 1967. Apparition de DNA de densités diffèrentes chez "Solanum lycopersicon esc." Après absorption de DNA bactérien ou de bactéries. Arch. Int. Physiol. Bioch. Soc. Bélge Biochim. **75**: 371–373.
18. STROUN, M. *et al.* 1967. Translocation of DNA of bacterial origin in *Lycopersicum esculentum* by ultracentrifugation in caesium chloride gradient. Nature **215**: 975–976.
19. GAHAN, P.B. *et al.* 1973. An autoradiographic study of bacterial DNA in *Lycopersicum esculentum*. Ann. Bot. **37**: 681–685.
20. GAHAN, P.B. *et al.* 1974. Effect of exogenous DNA on acid deoxyribonuclease activity in intact roots of *Vicia faba* L. Ann. Bot. **38**: 222–226.
21. HURST, P.R. & P.B. GAHAN. 1975. Turnover of DNA in ageing tissues of *Lycopersicon esculentum*. Ann. Bot. **39**: 71–76.
22. GAHAN, P.B. *et al.* 2003. Evidence that direct DNA uptake through cut shoots leads to genetic transformation of *Solanum aviculare* Forst. Cell Biochem. Func. **21**: 11–17.
23. ANKER, P. & M. STROUN. 1972. Bacterial ribonucleic acid in frog brain after a bacterial peritoneal infection. Science **178**: 621–623.
24. ANKER, P. *et al.* 1972. Bacterial RNA synthesis in frog auricles after intraperitoneal injection of bacteria. Experientia **28**: 488–489.
25. STROUN, M. *et al.* 1977. Spontaneous release of newly synthesized DNA from frog auricles. Arch. Sci. Genève **30**: 229–241.
26. STROUN, M. & P. ANKER. 1972. "In vitro" synthesis of DNA spontaneously released by bacteria or frog auricles. Biochimie **54**: 1443–1452.
27. ANKER, P. *et al.* 1975. Spontaneous release of DNA by human blood lymphocytes as shown in an in vitro system. Cancer Res. **35**: 2375–2382.
28. OLSEN I. & G. HARRIS. 1974. Uptake and release of DNA by lymphoid tissue and cells. Immunology **27**: 973–987.
29. ADAMS, D.H. & P.B. GAHAN. 1983. The DNA extruded by rat spleen cells in culture. Int. J. Biochem. **15**: 547–552.
30. ADAMS, D.H. & P.B. GAHAN. 1982. Stimulated and non-stimulated rat spleen cells release different DNA complexes. Differentiation **22**: 47–52.
31. ROGERS, J.C.D. *et al.* 1972. Excretion of deoxyribonucleic acid by lymphocytes stimulated with phytohaemagglutinin or antigen. Proc. Natl. Acad. Sci. USA **69**: 1685–1691.
32. ADAMS, D.H. & A.A.G. MCINTOSH. 1984. The cytosol origin of macromolecules extruded by cultured chick embryo fibroblast cells. Int. J. Biochem. **16**: 721–726.
33. ADAMS, D.H. & A.A.G. MCINTOSH. 1985. Studies on the cytosolic DNA of chick embryo fibroblasts and its uptake by recipient cultured cells. Int. J. Biochem. **17**: 1041–1051.
34. ADAMS, D.H. *et al.* 1997. In vitro stimulation by tumour cell media of [^{3}H] thymidine incorporation by mouse spleen lymphocytes. Cell Biochem. Func. **15**: 119–126.
35. ADAMS, D.H. & C. CHALLEN. 1988. The chick embryo fibroblast cytosolic DNA complex: a possible cell-cell messenger. Int. J. Biochem. **20**: 921–928.

36. BENNETT, R.M. *et al.* 1985. DNA binding to human leukocytes. J. Clin. Invest. **76**: 2182–2190.
37. JUCKETT, D.A. & B. ROSENBERG. 1982. Actions of cis-diamminedichloro-platimnum on cell surface nucleic acids in cancer cells as determined by cell electrophoresis techniques. Cancer Res. **42**: 3565–3573.
38. REID, B.L. & A.J. CHARLSON 1979. Cytoplasmic and cell surface deoxyribonucleic acids with consideration to their origins. Int. Rev. Cytol. **60**: 27–42.
39. HURST, P.R. *et al.* 1973. Turnover of labelled DNA in differentiated collenchyma. Differentiation **1**: 261–266.
40. GAHAN, P.B. 1976. DNA turnover in nuclei of cells from higher plants [abstract]. Riv. Istochim. Norm. Pat. **20**: 108–109.
41. STROUN, M. *et al.* 1987. Isolation and characterization of DNA from the plasma of cancer patients. Eur. J. Cancer Clin. Oncol. **23**: 707–712.
42. GARCIA-OLMO, D.C. *et al.* 2004. Circulating nucleic acids in plasma and serum (CNAPS) and its relation to stem cells and cancer metastasis: state of the issue. Histol. Histopathol. **19**: 575–583.

Immunological Aspects of Circulating DNA

PHILIPPE ANKER AND MAURICE STROUN

OncoXL, 1228 Plan les Ouates/Geneva, Switzerland

ABSTRACT: Nude mice were injected with DNA released by T lymphocytes previously exposed to inactivated herpes symplex type 1 or polio viruses. The serum of these mice was tested for its neutralizing activity. Injected nude mice synthesized antiherpetic or antipolio antibodies, depending on the antigen used to sensitize the T lymphocytes. Mice injected with DNA released by human T cells produced antibodies carrying human allotypes as they could be neutralized by antiallotype sera. However, mice that were injected with DNA released by antigen-stimulated murine T lymphocytes produced antiviral antibodies, which were not neutralized by anti-human allotype sera.

KEYWORDS: lymphocytes; antibodies; DNA

INTRODUCTION

What can be the relation between circulating nucleic acids and immunology? We have worked for many years on the spontaneous release and uptake of DNA by living cells, particularly in human lymphocytes.[1] We found that lymphocytes spontaneously release DNA within a homeostatic mechanism.[2] We and others had found that mitogen-stimulated lymphocytes excreted more DNA than did unstimulated lymphocytes.[3,4] This led us to hypothesize that the DNA released by lymphocytes might contain an immunological message.

Thus we incubated lymphocytes obtained from different tuberculin (PPD) or hepatitis B (HB)-positive or -negative donors in presence of one of these antigens. The DNA released by these lymphocytes in the culture medium was then tested for its information content using successively two cell-free systems. The ability of the resulting protein product to bind specifically to the stimulating antigen was examined by immunoadsorption chromatography. We observed that the DNA excreted by stimulated lymphocytes was transcribed into an RNA that coded for an antigen-binding protein, whereas DNA released by unstimulated lymphocytes did not. The antigen-binding protein

Address for correspondence: Philippe Anker, OncoXL, 14 chemin de Aulx, 1228 Plan les Ouates/Geneva, Switzerland. Voice: +41-22-3458507; fax: + 41-22-7368421.
e-mail: panker@worldcom.ch

Ann. N.Y. Acad. Sci. 1075: 34–39 (2006). © 2006 New York Academy of Sciences.
doi: 10.1196/annals.1368.004

bound specifically to tuberculin- or HB Sepharose-coated columns. Finally after elution from the column, the protein sedimented at 19S in a linear sucrose gradient.[5]

We then used an allogenic T–B lymphocyte cooperation involving lymphocyte subsets from donors with different allotypes and were able to show that B lymphocytes cultured in presence of a supernatant of T cells previously exposed to UV-inactivated herpes simplex virus (HSV) synthesized an antiherpetic antibody, carrying some allotypic markers of the T cell donor. Moreover, DNA purified from the T cell supernatant had the same effect on B lymphocytes as the crude supernatants.[6] Of course, we wanted to know whether the same transfer of immunological information could occur also *in vivo*. Therefore we[7] injected athymic nude mice with DNA released by human T lymphocytes that had previously been exposed to inactivated herpes simplex type 1 (HSV) or inactivated polio virus (Sabin type 1). We then tested the serum of these mice for their antiviral activity.

MATERIALS AND METHODS

The viruses were inactivated by UV prior to the addition to the T lymphocyte culture.

The T lymphocytes were isolated by passaging through nylon wool columns, which gave 98% rosetting lymphocytes as shown by acid esterase coloration.[8] They were cultured for 5 days in TC199 medium supplemented with autologous serum in the presence of antigen (1 plaque-forming unit/cell). After incubation, the lymphocytes were centrifuged at 200 × *g* for 10 min. The supernatant was then centrifuged at 48,000 *g* and the DNA extracted by protease K, followed by phenol treatment, passage through a hydroxyapatite column and a Cs_2SO_4 gradient for 72 h at 160,000 *g*.

The DNA (0.02 or 0.2 μg/mouse) was injected intraperitoneally into 6-week-old Swiss nude mice. After 5 days, the mice were killed and the antiviral activity of their serum was tested. Several dilutions of the mouse serum (1/10–1/800) were preincubated for 30 min with $1–2 \times 10^{3}$ HSV or polio plaque-forming units per mL; 0.2-mL samples of this mixture were then added to a monolayer of monkey kidney cells (Vero CCL 81, ATCC 232–253). After a 30-min adsorption, the monolayer was then washed with medium, and incubated for 1 day in the case of polio neutralization and 2 days when HSV was neutralized. The number of lysed cell plaques was then determined.

Mice serum samples presenting antiviral activity were incubated with anti-human allotypic sera and subsequently added to Vero cell cultures. The allotype specificity of antiviral activity inhibition was confirmed by a further incubation with anti-D antibody of corresponding allotype, which restored antiviral activity of the antibody presenting a certain allotypic specificity. The human T cell donor was positive for all allotypes determined.

All tests were performed by one of us on samples provided under code.

RESULTS AND DISCUSSION

FIGURE 1 shows that injected nude mice synthesized specifically antiherpetic or antipolio antibodies, depending on the antigen used to sensitize the T lymphocytes *in vitro*. The numbers in the ordinate show the highest dilution of the mice sera that showed a neutralizing activity higher than 3Σ. Some of the mice that had been injected with very small quantities of DNA released by human T cells exposed either to HSV or polio viruses, either 0.2 or 2 μg of DNA/mouse, showed specific antiviral activity at rather high dilutions of their serum (up to 1/800 in the case of polio). These antibodies were not found in either the serum of uninjected control mice or mice injected with inactivated herpes or polio viruses,

As can be seen in TABLE 1, mice injected with DNA released by human T cells produced antibodies carrying human allotypes because they could be neutralized by antiallotype sera. In this case the human T cell donor was GM a+, b+, f+, x+, and Km (1+). Almost all allotypes are found in the antibodies made by the nude mice.

However, in another experiment the nude mice that were injected with DNA released by antigen-stimulated murine T lymphocytes isolated from normal mice produced antiviral antibodies that were not neutralized by anti-human allotype sera. They just made mice antibodies (TABLE 2).

It appears from our results that the DNA purified from the nucleoprotein complex secreted by T cells which had been exposed to a viral antigen can, after intraperitoneal injection, circulate in the mouse and transmit information triggering a specific antiviral activity by the mouse B cells.

It is difficult to define the exact role of this translocation of DNA between T and B lymphocytes in addition to well-known immunological mechanisms operating during T–B cooperation. Considering the results of our plaque assays, it appears that the antibody-induced reduction in viral plaque numbers is always on the order of 50%, remaining at this level up to the highest active serum dilutions. It could be due to the fact that as in a primary response, the antibodies formed are of low affinity. The DNA released by antigen-exposed T cells might offer similarities with processed pseudogenes, which have been reported in human lymphocytes and could be the result of DNA transcription, RNA splicing, and retrotranscription into DNA.

Because many people have doubted these results, we decided to produce a hybridoma. As antigen we chose to stay in our HSV system.

We had the opportunity to repeat our experiment in Rockville at Biotech Research, a firm directed by Dr. R.C. Ting. In order to check our results by another way, the team in his laboratory succeeded in making a hybridoma by providing DNA excreted by T cells that had been previously exposed to our HSV antigen. This hybridoma also produced human antiherpetic antibodies measured by ELISA tests; however, these appeared to be transient and lost antibody-synthesizing capacity after 3 days. Back in Geneva, we also obtained a transient hybridoma with the same characteristics.

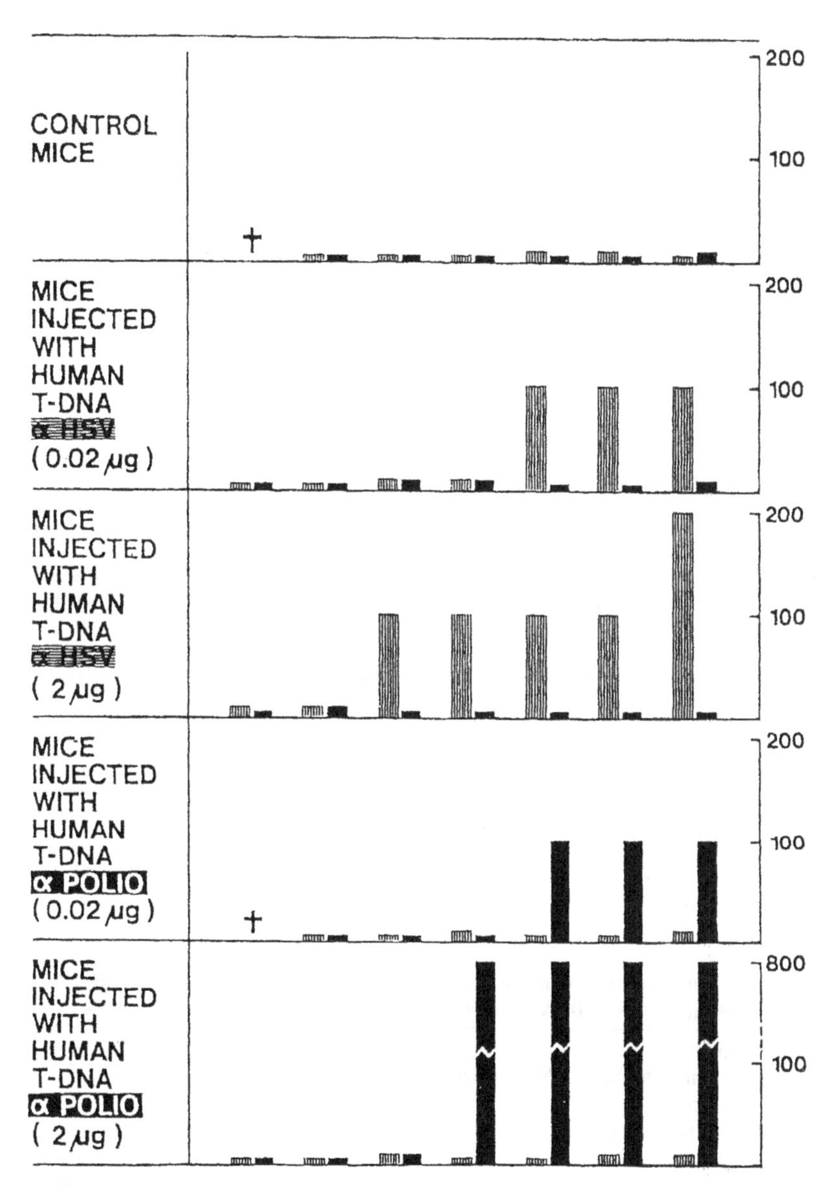

FIGURE 1. Neutralizing activity of the serum of nude mice injected with DNA released by human T cells previously exposed to UV-inactivated HSV or polio virus. Mice were injected with either 0.02 or 2 μg of human DNA. Control mice remained uninjected. After 5 days the mice were killed and their serum collected and frozen until tested. Numbers in the ordinate represent highest dilutions of serum still presenting neutralizing activity (more than 3σ below virus control). Neutralization of HSV (*hatched area*); neutralization of polio (*solid area*). Human T-DNA α (HSV) or T-DNA α (polio) = DNA released by HSV or polio virus–exposed human T lymphocytes. +, mouse that died in the course of the experiment.

TABLE 1. Detection of human allotypes (Gm and Km) on antiherpetic antibodies formed by nude mice injected with DNA released by HSV-exposed human T lymphocytes

Pretreatment of virus	Number of plaques (mean $\pm$ 2 σ_x)	Value of control without serum (%)	Allotype of the antibody
(a) No pretreatment	95 $\pm$ 5.2	100	—
(b) Serum of mouse injected with DNA released by human T cells	53 $\pm$ 6.4	55	—
(c) Serum of b pretreated with			
Anti-Gm a	84 $\pm$ 2.8	88	Gm a+
Anti-Gm a + anti-D Gm a	53 $\pm$ 0.7	55	
Anti-Gm b	83 $\pm$ 2.8	87	Gm b+
Anti-Gm b + anti-D Gm b	48 $\pm$ 2.8	51	
Anti-Gm f	44 $\pm$ 2.1	46	Gm f–
Anti-Gm g	58 $\pm$ 2.1	60	Gm f–
Anti-Gm x	94 $\pm$ 0	99	Gm x+
Anti-Gm x + anti-D Gm x	50 $\pm$ 7.8	52	Gm x+
Anti-Km (1)	82 $\pm$ 0	86	Km (1)+
Anti-Km (1) + anti-D Km (1)	48 $\pm$ 10.6	50	

NOTE: The T lymphocyte donor was Gm (a+, b+, f+, g+, x+) and Km (1)+. Antiherpetic mice serum samples were preincubated in either the presence or absence of antiallotypic sera anti-Gm a, anti-Gm b, anti-Gm f, anti-Gm g, anti-Gm x, and anti-Km (1). The antiallotypic activity competed with anti-D Gm a, anti-D Gm b, anti-D Gm x, and anti-D Km (1). Plaque counting was performed on Vero cells infected with HSV after 2 days of incubation. 2σ_x is 2 standard errors of the mean.

We have dropped the immunological part of our research because we were never able to get any funding to continue this work, but we would be happy if some immunologist working on viruses or an oncologist working on cancer antigens were to become interested in this approach. All this forgotten work might then have some medical importance.

TABLE 2. Detection of human allotypes (Gm and Km) on antiherpetic antibodies formed by nude mice injected with DNA released by HSV-exposed murine T lymphocytes

No pretreatment	65 $\pm$ 2.9	100	
Serum of mice injected with DNA released by mouse T cells	32 $\pm$ 15.5	49	
Serum of e pretreated with:			
Anti-Gm a	34 $\pm$ 0.7	52	Gm a–
Anti-Gm b	25 $\pm$ 0.7	38	Gm b–
Anti-Gm f	37 $\pm$ 2.1	55	Gm f–
Anti-Gm g	35 $\pm$ 0	54	Gm g–
Anti Gm x	37 $\pm$ 7.8	56	Gmx–
Anti-Km (1)	34 $\pm$ 3.5	52	Km (1)–

NOTE: The T lymphocyte donor was Gm (a+, b+, f+, g+, x+) and Km (1)+. Antiherpetic mice serum samples were preincubated either in the presence or absence of antiallotypic sera anti-Gm a, anti-Gm b anti-Gm f, anti-Gm g, anti-Gm x, and anti-Km (1). The antiallotypic activity was put in competition with anti-D Gm a, anti-D Gm b, anti-D Gm x, and anti-D Km (1). Plaque counting was performed on Vero cells infected with HSV after 2 days of incubation.2σ_x is 2 standard errors of the mean.

REFERENCES

1. STROUN, M. *et al.* 1977. Circulating nucleic acids in higher organisms. Int. Rev. Cytol. **51:** 1–48.
2. ANKER, P. *et al.* 1975. Spontaneous release of DNA by human blood lymphoctyes as shown in an in vitro system. Cancer Res. **35:** 2375–2382.
3. STROUN, M. *et al.* 2000. The origin and mechanism of circulating DNA. Ann. N. Y. Acad. Sci. **906:** 161–168.
4. ROGERS, J.C. *et al.* 1972. Excretion of deoxyribonucleic acid by lymphocytes stimulated with phytohemagglutinin or antigen. Proc. Natl. Acad. Sci. USA **69:** 1685–1689.
5. JACHERTZ, D. *et al.* Information carried by the DNA released by antigen-stimulated lymphocytes. Immunology **37:** 7537–7563.
6. JACHERTZ, D. *et al.* 1979. Transfer of genetic information from T to B lymphocytes in the course of an immune response. Biomedicine **31:** 153–44.
7. ANKER, P. *et al.* 1984. Nude mice injected with DNA released by antigen stimulated human T lymphocytes produce specific antibodies expressing human characteristics. Cell Biochem. Funct. **2:** 33–37.
8. MULLER, J.G. *et al.* 1975. Non specific esterase activity: a criterion for differentiation of T and B lymphocytes in mouse lymphnodes. Eur. J. Immunol. **5:** 270–272.

Biology of Circulating mRNA

Still More Questions Than Answers?

MICHAEL FLEISCHHACKER

Charité-Universitätsmedizin Berlin, Medizinische Klinik m.S. Onkologie u. Hämatologie, Molekularbiologisches Labor, Alte Apotheke, Schumannstr. 20-21, 10117 Berlin, Germany

ABSTRACT: A few years after the first description of free-circulating DNA in plasma and serum, the detection of tumor-associated overexpressed mRNA in plasma was also reported. This observation has been confirmed and it seems to be clear that the presence of free-circulating RNA is a ubiquitous phenomenon. In this short review I will discuss some basic aspects of the release mechanisms for the RNA, its biological meaning, and clinical value.

KEYWORDS: extracellular RNA; review; plasma; free-circulating RNA; cancer

BASIC ASPECTS

The existence of extracellular RNA has been known for a long time. One of the first papers demonstrating the presence of extracellular RNA was published by Stroun and co-workers.[1] This group showed that nonstimulated human lymphocytes as well as frog auricles release a complex into culture medium which not only contains DNA but RNA as well. When the culture medium was renewed, a constant amount of RNA was shedded and could be obtained in the cell culture supernatant. The RNA release was dependent on intact and viable cells, as a mild heat treatment and killing of the cells did not increase the amount of extracellular RNA, and slicing of frog auricles did not lead to more RNA in the medium. The release of RNA seems to be a homeostatic mechanism, comparable to the one which regulates the shedding of DNA into the environment.[2] The extracellular RNA appeared to be heavily methylated, associated with DNA, single-stranded, of low molecular weight (2.5 to 4S), and was not tRNA. Stroun *et al.* also showed that the RNA from this complex has a stimulating effect on DNA synthesis *in vitro*. The total amount of nucleic acids

Address for correspondence: Michael Fleischhacker, Charité-Universitätsmedizin Berlin, Medizinische Klinik m.S. Onkologie u. Hämatologie, Molekularbiologisches Labor, Alte Apotheke, Schumannstr. 20-21, 10117 Berlin, Germany. Voice: +4930 450 51 33 04; fax: +4930 450 51 39 64. e-mail: michael.fleischhacker@charite.de

Ann. N.Y. Acad. Sci. 1075: 40–49 (2006). © 2006 New York Academy of Sciences.
doi: 10.1196/annals.1368.005

in normal human plasma had been determined by Kamm and Smith and found to be 144 ng/mL for plasma RNA.[3] When Hamilton *et al.* critically evaluated this and other data published so far for the quantity of RNA in plasma, they found that in most papers the RNA concentrations given were much higher (up to 58-fold) than they were able to detect.[4]

Given the fact that the concentration of RNase in normal persons is high and even higher in cancer patients, and that RNA-degrading enzymes are extremely stable, it is surprising that researchers are able to detect any RNA in plasma and serum samples at all.[5]

Therefore it was reasoned that the RNA released from the cells into the environment must be complexed and in a form resistant against the action of the ubiquitous presence of RNases. When prelabeled Chinese hamster ovary cells were treated with trypsin under conditions in which cells remain fully viable, the release of a macromolecular substance containing ^{32}P and ^{3}H was described.[6] In contrast, a ribonuclease treatment affected neither the ^{32}P nor the ^{3}H radioactivity, while a pronase treatment had only little effect on the release of macromolecular 32P. The authors concluded from these experiments, that RNA together with glycoproteins is released from the external cell surface. The presence of nucleic acid–containing structures on the surface of *in vitro* cultivated mouse tumor cells was also reported by Juckett and Rosenberg.[7] They incubated mouse sarcoma cells with nucleases and observed a decrease of 10% to 20% in the electrophoretic mobility of whole cells. A similar effect was seen when tumor-bearing animals were treated with a therapeutic dose of cisplatin. It was found that both DNase and RNase had a similar effect on the cells and that both enzymes act on the same site. The inhibition of nucleic acid or protein synthesis produced a loss of the cell surface nucleic acids. Digestion of the cells with trypsin or pronase removes the nuclease susceptibility and might indicate that the nucleic acids are anchored by interaction with proteins and sugar moieties. The fact that high concentrations of nucleases are needed to digest the surface nucleic acids led the authors to suggest that they might exist in a nuclease-resistant form, like DNA:RNA hybrids. A few years later, Carr and co-workers demonstrated the presence of membrane vesicles by means of ultramicroscopy.[8] They were able to demonstrate the presence of these vesicles in the plasma of patients with certain types of leukemia but not in the plasma from healthy people, but from patients with solid tumors and from patients with a nonmalignant hematologic disease. A spontaneous release of RNA–lipid complexes from *in vitro* cultivated human colon adenocarcinoma cells was also observed by Rosi *et al.*[9] Later, this group confirmed and extended their observation and showed that these vesicles are shed by healthy and viable cells only, but not by cells that were kept at 4°C.[10] These vesicles contain a polyA+ RNA with a major 5-kb fraction, which is active in an *in vitro* protein synthesis system. A shedding of extracellular plasma membrane vesicles from murine lymphoma cells was also described by Barz *et al.*, who analyzed two different cell lines with variable metastatic capacity.[11]

The shedding capacity of the highly metastatic cell line was much higher than that of the cell line with low metastatic capacity. In addition, the lipid composition of the shed vesicles reflected the membrane composition of the cell line from which they were derived. That apoptotic bodies might be the structures in which the free-circulating RNA is actually floating in plasma and serum, was suggested when Halicka *et al.* demonstrated that HL-60 cells undergoing apoptosis package DNA and RNA into different apoptotic bodies.[12] Experiments with an *in vitro* cultivated human melanoma cell line, in which apoptosis was induced by the addition of an anti-CD95 monoclonal antibody confirmed the assumption that these apoptotic particles harbor RNA.[13] When these apoptotic bodies were incubated with human serum for 30 min, it was still possible to isolate and amplify RNA from these bodies. In contrast, when purified RNA instead of apoptotic bodies was added to human serum, even after a short incubation of only 1 min, there was no detectable RNA, confirming results by Komeda *et al.*[14] A different approach was taken by Ng *et al.* to determine the nature of free-circulating RNA.[15] Plasma from healthy individuals and patients with hepatocellular carcinoma was filtered through 0.22 μM filter or ultracentrifuged and the concentration of GAPDH mRNA was determined by quantitative real-time PCR. It was shown that the GAPDH mRNA concentration was 9- to 15fold higher in untreated samples in comparison to filtered or ultracentrifuged samples. In contrast to RNA, the concentration of DNA (also measured by quantitative real-time PCR) showed no reduction by filtering. The authors concluded from these experiments that the majority of free-circulating mRNA in healthy people as well as tumor patients is particle-associated.

Thus, it seems that at least part of the free-circulating RNA is complexed with other molecules, rendering them resistant to the action of RNA-degrading enzymes. Whether these complexes are apoptotic bodies or are related to the recently rediscovered exosomes is unknown so far.[16] The same holds true for the question whether the liberated RNA is only a byproduct of cell death or is actively secreted into the environment.

DOES CIRCULATING RNA HAVE A BIOLOGICAL MEANING?

Galand *et al.* were among the first to describe the uptake of RNA by ascites tumor cells.[17,18] The ingestion of RNA seems to be an active process, since the uptake decreases when the metabolic activity of the cells is lowered by lowering the incubation temperature. This group also showed that up to 4% of the normal RNA content of the cells can be taken up. The treatment of cells with heterologous (liver and yeast) or homologous RNA leads to an activation of amino acid incorporation into cellular proteins. The base composition as well as the structure of the ingested RNA seems to be maintained after ingestion, suggesting a specific role of the ingested RNA.

The communication between cells seems to be important for a variety of different functions and the exchange of nucleic acids might be one means of achieving this goal. Evidence for a transfer of RNA between *in vitro* cultivated human cells was described by Kolodny, who cocultivated cells of different weight (the donor cells were incubated in a medium containing tantalum particles) and demonstrated a transfer of intact labeled RNA from heavy-weight donor cells into nonlabeled light recipient cells.[19] The inhibition of RNA synthesis in the recipient cells by actinomycin did not affect the RNA transfer. There was no evidence of a DNA transfer between these cells. The same group also described the ability of normal and transformed cells to shed RNA of low molecular weight (5S) into the culture medium.[20] These RNA molecules are resistant to serum nucleases present in the medium. The RNA appears to be heavily methylated (also shown by Stroun *et al.*) and its methylation pattern is simple in comparison to the pattern observed in tRNA and rRNA. There seems to be a difference between this shed medium RNA of low molecular weight and the RNA that is transferred between cells in contact. In contrast to the former, the latter RNA is representative of all major RNA species, whereas the shed RNA contains low molecular weight species only. Wieczorek reported the existence of a proteolipid complex that is bound to glycosphingolipids from melanoma cells.[21] The RNA that could be isolated from this complex was able to induce a morphologic change and a rise in proliferation rate when transferred into susceptible recipient cells.[22]

Whether any of the RNA species described above has a biological meaning, such as a participation in the immune response or other biological processes, has yet to be seen.[23]

CLINICAL SIGNIFICANCE OF CIRCULATING RNA

The first report on the association of the quantity of a circulating RNA–proteolipid complex in sera of patients with malignant disorders and their clinical status was published by Wieczorek *et al.* [21] They considered this complex a potential tumor marker on the basis of the observation that it was not detectable in patients with various benign disorders. They observed that the complex rapidly disappeared after surgical removal of the tumor (approximately 2 days), that the presence of the complex was associated with the tumor mass, and that a correlation between response to a therapy and the amount of circulating RNA–proteolipid complex could be demonstrated.

More than 10 years later Kopreski and co-workers unequivocally demonstrated the presence of tumor-associated overexpressed mRNA in the plasma of cancer patients.[24] Since then, this observation has been confirmed in more than a dozen papers (summarized in TABLE 1). In most of these reports qualitative assays were used, and only two groups applied a quantitative test, which makes a comparison of the results obtained in different laboratories almost

TABLE 1. Summary of papers in which circulating mRNA was detected in plasma and other body fluids

Tumor patients	Overexpressed genes	Detection method	Compartment	Reference
Melanoma	tyrosinase, c-abl	qualitative RT-PCR	plasma/serum	Kopreski *et al.* 1999[24]
Melanoma	tyrosinase, gp100, MART-1	qualitative RT-PCR	plasma/serum	Hasselmann *et al.* 2001[25]
Follicular lymphoma	hTERT	qualitative RT-PCR	plasma/serum	Dasi *et al.* 2001[16]
Breast	5T4	qualitative RT-PCR	plasma/serum	Kopreski *et al.* 2001[27]
Breast	mammaglobin	qualitative RT-PCR	plasma/serum	Gal *et al.* 2001[28]
Breast	mammaglobin, CK19	qualitative RT-PCR	plasma/serum	Silva *et al.* 2001[29]
Breast	hTR, hTERT	qualitative RT-PCR	plasma/serum	Chen *et al.* 2000[30]
Hepatocellular carcinoma	hTERT	qualitative RT-PCR	plasma/serum	Miura *et al.* 2003[31]
Colorectum	CEA, CK19	qualitative RT-PCR	plasma/serum	Silva *et al.* 2002[32]
Colorectum	ß-catenin	quantitative RT-PCR	plasma/serum	Wong *et al.* 2004[33]
Colorectum	hTERT	quantitative RT-PCR	plasma/serum	Lledo *et al.* 2004[34]
Breast, melanoma, thyroid	hTR, hTERT	qualitative RT-PCR	plasma/serum	Novakovic *et al.* 2004[35]
Prostate	PSMA, CEA	qualitative RT-PCR	plasma/serum	Papadopoulou *et al.* 2004[36]
Lung	Her-2/neu, hnRNP-B1	qualitative RT-PCR	plasma/serum	Fleischhacker *et al.* 2001[37]
Lung	hnRNP-B1	qualitative RT-PCR	plasma/serum	Sueoka *et al.* 2005[38]
Lung	hnRNP-B1, PGP 9.5, MAGE-2, Her-2/neu, aurora, pericentrin, hTR, hTERT	qualitative RT-PCR	cell-free bronchial lavage fluid	Schmidt *et al.* 2004[39]

impossible.[33,34] Wong *et al.* observed a correlation between the increased amount of circulating ß-catenin mRNA and the presence of a tumor of the colorectum.[33] By quantification of the ß-catenin mRNA they were able to show a difference between patients with a tumor (8737 copies/mL), with an adenoma (1218 copies/mL), and healthy controls (291 copies/mL). A statistical analysis demonstrated that plasma ß-catenin mRNA concentration was correlated to the tumor stage only. Moreover, the plasma ß-catenin mRNA concentration decreased significantly not only after tumor removal in 16/19 cancer patients, but also in a few patients whose adenoma was removed. In contrast to ß-catenin mRNA, the concentration of GAPDH mRNA was similar in all three patient groups. This led the authors to conclude that the quantification of the plasma ß-catenin mRNA concentration might be a useful tool as a screening marker not only for colorectal cancer but also for dysplastic alterations of the colorectum. When another group examined patients suffering from the same disease for the presence of circulating human telomerase reverse transcriptase (hTERT) mRNA in their plasma, an increased concentration of hTERT mRNA was found in comparison to healthy controls, but there was no association with the tumor stage.[34] The lack of specificity of the test is demonstrated in a paper by Miura *et al.*, who analyzed patients with hepatocellular carinoma (HCC), liver cirrhosis (LC), and chronic hepatitis (CH) for the expression of hTERT gene in their plasma.[31] The test was positive in 90% of HCC patients, but also in 70% of LC patients and 42% of CH patients. The difficulty of finding a relationship between the presence of tumor-associated RNA in plasma and clinicopathological parameters is also illustrated in other papers. When Silva *et al.* used an assay for the detection of cytokeratin 19 (CK19) and mammaglobin in plasma, they found an association with the tumor size and the proliferation index, but were unable to observe a statistical relationship between the presence of circulating tumor cells and the detection of epithelial mRNAs in plasma.[28] The same group examined colorectal cancer patients for the presence of CEA and CK19 coding mRNA in their plasma samples and found an association between these parameters. The reason for these discordant results is unknown so far.[32]

There are only two reports in which tumor-associated mRNA in plasma was analyzed in patients before and after therapy.[36,37] In both papers two marker genes were used for the analysis (Her-2/neu and hnRNP-B1 for lung cancer and PSMA and CEA for prostate cancer patients). In both examinations the number of positive cases was low when patients after therapy were analyzed. Unfortunately, the patient groups that were analyzed before and after therapy were not the same, and qualitative rather than quantitative PCR assays were used, making these results difficult to interpret.

In theory, the determination of the total amount of free-circulating RNA might be a disease marker, as had been shown already for DNA. There was indeed an increased amount of plasma RNA quantity in patients with benign and malignant lung disease.[40] A more refined analysis using a quantitative real-time PCR assay and hnRNP-B1 mRNA as target confirmed this observation.[38] The aforementioned observations were made in plasma and validated for cell-free bronchial lavage (BL) fluid when it was shown that the total amount of RNA was higher in tumor patients than in patients with benign lung disease.[41] These results from BL supernatant open a new field of research in lung cancer detection, because so far only cells from BL have been used for further diagnostic work. Since the release of RNA into the environment is not restricted to any specific disease condition but seems to be a very nonspecific reaction when cells, organs, or whole organisms are put under stress, it is doubtful whether the RNA quantification as a stand-alone method might be a clinically useful tool.

CONCLUSIONS

From the foregoing it seems to be clear that (i) the free-circulating RNA can be detected in plasma, serum, and other body fluids and also from cell-free supernatants of *in vitro* cultivated cells, and that (ii) the RNA is complexed with other molecules, rendering them resistant against nucleolytic attacks. The answers to the following questions are still unknown: What are the mechanisms by which the RNA is released into circulation? Is it an active process or a nonspecific byproduct of apoptosis and/or necrosis? or Are several different mechanisms involved? Is there a quantitative and/or qualitative difference between "particulate" RNA released into the environment and the RNA transferred directly from cell to cell?

It seems to be clear that normal and tumor cells are able to incorporate free-circulating RNA and "recycle" them. Does this process have a biological meaning? Additionally, is the circulating RNA a true reflection of cellular RNA?

Finally, it had been clearly demonstrated, that at least part of the circulating RNA in cancer patients is tumor-derived; that is, genes overexpressed in tumor cells can also be found in the circulation. But is quantification of the total amount of RNA or of tumor-associated mRNA species useful and could any of these tests become a clinically meaningful tool? Which, if any of the assays could be used for diagnostic or prognostic purposes, for the determination of the efficacy of a given therapy, or for an early detection of a relapse?

The answers to all these questions will very much depend on the development of new research tools and our capability of asking the correct questions.

ACKNOWLEDGMENT

I would like to thank Dr. B. Schmidt for critically reading the manuscript and helpful comments.

REFERENCES

1. STROUN, M. *et al.* 1978. Presence of RNA in the nucleoprotein complex spontaneously released by human lymphocytes and frog auricles in culture. Cancer Res. **38**: 3546–3554.
2. STROUN, M. *et al.* 1977. Circulating nucleic acids in higher organisms. Int. Rev. Cytol. **51**: 1–48.
3. KAMM, R.C. & A.G. SMITH. 1972. Nucleic acid concentrations in normal human plasma. Clin. Chem. **18**: 519–522.
4. HAMILTON, T.C. *et al.* 1979. Ribonucleic acid in plasma from normal adults and multiple myeloma patients. Clin. Chem. **25**: 1774–1779.
5. REDDI, K.K. *et al.* 1976. Elevated serum ribonuclease in patients with pancreatic cancer. Proc. Natl. Acad. Sci. USA **73**: 2308–2310.
6. RIEBER, M. *et al.* 1974. An "external" RNA removable from mammalian cells by mild proteolysis. Proc. Natl. Acad. Sci. USA **71**: 4960–4964.
7. JUCKETT, D.A. *et al.* 1982. Actions of cis-diamminedichloroplatinum on cell surface nucleic acids in cancer cells as determined by cell electrophoresis techniques. Cancer Res. **42**: 3565–3573.
8. CARR, J.M. *et al.* 1985. Circulating membrane vesicles in leukemic blood. Cancer Res. **45**: 5944–5951.
9. ROSI, A. *et al.* 1988. RNA-lipid complexes released from the plasma membrane of human colon carcinoma cells. Cancer Lett. **39**: 153–160.
10. CECCARINI, M. *et al.* Biochemical and NMR studies on structure and release conditions of RNA-containing vesicles shed by human colon adenocarcinoma cells. Int. J. Cancer **44**: 714–721.
11. BARZ, D. *et al.* 1985. Characterization of cellular and extracellular plasma membrane vesicles from a non-metastasizing lymphoma (Eb) and its metastasizing variant (ESb). Biochem. Biophys. Acta **814**: 77–84.
12. HALICKA, H.D. *et al.* 2000. Segregation of RNA and separate packaging of DNA and RNA in apoptotic bodies during apoptosis. Exp. Cell Res. **260**: 248–256.
13. HASSELMANN, D.O. *et al.* 2001. Extracellular tyrosinase mRNA within apoptotic bodies is protected from degradation in human serum. Clin. Chem. **47**: 1488–1489.
14. KOMEDA, T. *et al.* 1995. Sensitive detection of circulating hepatocellular carcinoma cells in peripheral venous blood. Cancer **75**: 2214–2219.
15. NG, E.K. *et al.* 2002. Presence of filterable and nonfilterable mRNA in the plasma of cancer patients and healthy individuals. Clin. Chem. **48**: 1212–1217.
16. JOHNSTONE, R.M. 2005. Revisiting the road to the discovery of exosomes. Blood Cells Mol. Dis. **34**: 214–219.
17. GALAND, P. *et al.* 1966. Uptake of exogenous ribonucleic acid by ascites tumor cells. I. Autoradiographic and chromatographic studies. Exp. Cell Res. **43**: 381–390.
18. GALAND, P. *et al.* 1966. Uptake of exogenous ribonucleic acid by ascites tumor cells. II. Relations between RNA uptake and the cellular metabolism. Exp. Cell Res. **43**: 391–397.
19. KOLODNY, G.M. 1971. Evidence for transfer of macromolecular RNA between mammalian cells in culture. Exp. Cell Res. **65**: 313–324.
20. KOLODNY, G.M. *et al.* 1972. Secretion of RNA by normal and transformed cells. Exp. Cell Res. **73**: 65–72.

21. WIECZOREK, A.J. *et al.* 1987. Diagnostic and prognostic value of RNA-proteolipid in sera of patients with malignant disorders following therapy: first clinical evaluation of a novel tumor marker. Cancer Res. **47**: 6407–6412.
22. WIECZOREK, A.J. 1984. RNA-proteoglycolipid from melanoblastoma serum transforms human bone-marrow reticulum cells. Yale J. Biol. **57**: 432A.
23. MITSUHASHI, S. *et al.* 1978. Ribonucleic acid in the immune response. Mol. Cell. Biochem. **20**: 131–147.
24. KOPRESKI, M.S. *et al.* 1999. Detection of tumor messenger RNA in the serum of patients with malignant melanoma. Clin. Cancer Res. **5**: 1961–1965.
25. HASSELMANN, D.O. *et al.* 2001. Detection of tumor-associated circulating mRNA in serum, plasma and blood cells from patients with disseminated malignant melanoma. Oncol. Rep. **8**: 115–118.
26. DASI, F. *et al.* 2001. Real-time quantification in plasma of human telomerase reverse transcriptase (hTERT) mRNA: a simple blood test to monitor disease in cancer patients. Lab. Invest. **81**: 767–769.
27. KOPRESKI, M.S. *et al.* 2001. Circulating RNA as a tumor marker: detection of 5T4 mRNA in breast and lung cancer patient serum. Ann. N. Y. Acad. Sci. **945**: 172–178.
28. GAL, S. *et al.* 2001. Detection of mammaglobin mRNA in the plasma of breast cancer patients. Ann. N. Y. Acad. Sci. **945**: 192–194.
29. SILVA, J.M. *et al.* 2001. Detection of epithelial messenger RNA in the plasma of breast cancer patients is associated with poor prognosis tumor characteristics. Clin. Cancer Res. **7**: 2821–2825.
30. CHEN, X.Q. *et al.* 2000. Telomerase RNA as a detection marker in the serum of breast cancer patients. Clin. Cancer Res. **6**: 3823–3826.
31. MIURA, N. *et al.* 2003. Sensitive detection of human telomerase reverse transcriptase mRNA in the serum of patients with hepatocellular carcinoma. Oncology **64**: 430–434.
32. SILVA, J.M. *et al.* 2002. Detection of epithelial tumour RNA in the plasma of colon cancer patients is associated with advanced stages and circulating tumour cells. Gut **50**: 530–534.
33. WONG, S.C. *et al.* 2004. Quantification of plasma beta-catenin mRNA in colorectal cancer and adenoma patients. Clin. Cancer Res. **10**: 1613–1617.
34. LLEDO, S.M. *et al.* 2004. Real time quantification in plasma of human telomerase reverse transcriptase (hTERT) mRNA in patients with colorectal cancer. Colorectal Dis. **6**: 236–242.
35. NOVAKOVIC, S. *et al.* 2004. Detection of telomerase RNA in the plasma of patients with breast cancer, malignant melanoma or thyroid cancer. Oncol. Rep. **11**: 245–252.
36. PAPADOPOULOU, E. *et al.* 2004. Cell-free DNA and RNA in plasma as a new molecular marker for prostate cancer. Oncol. Res. **14**: 439–445.
37. FLEISCHHACKER, M. *et al.* 2001. Detection of amplifiable messenger RNA in the serum of patients with lung cancer. Ann. N. Y. Acad. Sci. **945**: 179–188.
38. SUEOKA, E. *et al.* 2005. Detection of plasma hnRNP B1 mRNA, a new cancer biomarker, in lung cancer patients by quantitative real-time polymerase chain reaction. Lung Cancer **48**: 77–83.
39. SCHMIDT, B. *et al.* 2004. Detection of cell-free nucleic acids in bronchial lavage fluid supernatants from patients with lung cancer. Eur. J. Cancer **40**: 452–460.

40. LAKTIONOV, P.P. *et al.* 2004. Extracellular circulating nucleic acids in human plasma in health and disease. Nucleosides Nucleotides Nucleic Acids **23**: 879–883.
41. SCHMIDT, B. *et al.* 2005. Quantification of free RNA in serum and bronchial lavage: a new diagnostic tool in lung cancer detection? Lung Cancer **48**: 145–147.

Investigation of the Origin of Extracellular RNA in Human Cell Culture

KATRIN BÖTTCHER,[a] ALEXANDER WENZEL,[b] AND
JENS M. WARNECKE[a,b]

[a]*Institut für Molekulare Medizin, Universitätsklinikum Schleswig-Holstein, Campus Lübeck, Ratzeburger Allee 160, D-23538 Lübeck, Germany*

[b]*Kompetenzzentrum für Drug Design & Target Monitoring, Maria Goeppert Str. 1, D-23562 Lübeck, Germany*

ABSTRACT: We have used the human ECV 304 cell line to study the origin and fate of extracellular RNA (exRNA) in cell culture. Quantification of different extracellular RNA species using reverse transcription followed by quantitative PCR revealed a prevalent fraction of ribosomal RNAs. Comparison of intracellular and extracellular ribosomal RNA copy numbers allowed the calculation of the number of destroyed cells that would result in the corresponding number of extracellular rRNAs. Interestingly, this number was comparable to the amount of destroyed cells as determined by the measurement of extracellular lactate dehydrogenase activity.

KEYWORDS: extracellular RNA; RT-PCR; cell culture

INTRODUCTION

Ex vivo experiments of Stroun and co-workers[1] with frog auricles as well as cell culture experiments of Kolodny [2] have proven that extracellular RNA (exRNA) is sequestered by living cells, a constant level of RNA is maintained outside cells, and that polymeric RNA can be transferred between cells in human cell culture systems. Wieczorek and co-workers found exRNA to be associated with lipids and proteins in cell culture as well as in serum/plasma of tumor patients.[3] The biological significance of exRNA was emphasized by Kopreski *et al.*[4] when a correlation between the presence of cell-free tyrosinase mRNA and malignant melanoma could be demonstrated. In addition to the detection of tumor-specific RNA in plasma and serum of patients with

Address for correspondence: Jens M. Warnecke, Institut für Molekulare Medizin, Universitätsklinikum Schleswig-Holstein, Campus Lübeck, Ratzeburger Allee 160, D-23538 Lübeck, Germany.
e-mail: warnecke@uni-luebeck.de

Ann. N.Y. Acad. Sci. 1075: 50–56 (2006).
doi: 10.1196/annals.1368.006

different types of malignancies,[5] Ng and co-workers further characterized the plasma RNA as particle-associated in a series of filtration experiments.[6] The detection of exRNA bound to the surface of cells in human circulation[7] offered another possibility of RNA transportation outside cells. However, the mode of release as well as the biological fate of ex RNA remains to be elucidated.

By quantification of rRNA among the exRNA fraction we demonstrate that a significant amount of exRNA is generated by destruction of cellular membranes as measured by the lactate dehydrogenase test.

METHODS

Cell Culture and Preparation of RNA

Human bladder carcinoma ECV 304 cells were seeded at a density of 1×10^6 in 175-cm^2 cell flasks using 15 mL medium 199 with 10% fetal calf serum (FCS), grown at 37°C, 5% CO_2, and split every 48 h. Total cellular RNA was prepared by detaching the cells using 0.1 % trypsin, 1 mM EDTA solution for 5 min at 37°C. Reaction was stopped by adding FCS and cells were spun down for 3 min at $800 \times g$. RNA was extracted using the Qiagen Midi-Kit following the manufacturer's instructions including digestion of DNA on the column.

For preparation of cytosol, cells were resuspended in 10 mM Tris/HCl pH 7.5 and membranes were destroyed by three consecutive freeze-and-thaw cycles. The mixture was spun for 10 min at $16000 \times g$ and the supernatant was used as cytosolic fraction.

For isolation of exRNA used in RT-qPCR reactions, cells were washed twice with pre-warmed PBS followed by incubation in OPTIMEM without FCS. Supernatant was collected and filtered through 5-μM filters by gravity flow to separate cells. Ten microliter flow-through was concentrated by centrifugation using Amicon Ultra concentrator (Millipore) with an exclusion size of 5 Kd to a volume of 400 μL. RNA was prepared by Qiagen Midi kit as described above.

For isolation of total exRNA, cells were treated as just described, except that cells were incubated with medium 199. After separation of cells by filtration through a 5-μM filter and concentration by Amicon Ultra concentrator, the volume was adjusted with H_2O to 400 μL. RNA was extracted by two consecutive phenol/chloroform extractions in the presence of 84 mM sodium acetate and 1 mM EDTA at pH 4.5. The resulting aqueous phase was washed with chloroform, and RNA was precipitated by addition of 2.5 volumes of ethanol and 0.04 volumes of 3M sodium acetate. RNA was quantified by absorption measurement at 260 nm.

cDNA Synthesis and Quantitative PCR

Reverse transcription of RNA was performed using Superscript II reverse transcriptase and 150 ng random hexamer primer (Invitrogen) as described by the manufacturer. Ten microliters of RNA isolate was used as template in 20 μL reaction volume. A reaction without reverse transcriptase was performed as non-RT reaction.

The sequence data for rRNAs and GAPDH are available at the NCBI database. Primers were constructed using primer express v2.0 (Applied Biosystems) and are available from the authors upon request. Quantitative PCR assays were performed using the qPCR core Kit (Eurogentec) in the presence of Sybr-Green. Two microliters of cDNA or non-RT reaction were amplified in a total volume of 25 μL. Fifty cycles of a two-step PCR (annealing 60°C, denaturation 95°C) were performed using a AbiPrism 7900HT sequence detection system (Applied Biosystems). To be able to transform the cycle threshold values (ct-values) into absolute RNA copy numbers, calibration curves were prepared from serial dilution of gel-purified PCR fragments of 5.8S, 18S, and 28S rDNA as well as GAPDH cDNA. PCR reactions were performed in duplicate and mean values were used for further calculation. In the case of ribosomal RNAs, the values of the non-RT reactions were subtracted to avoid miscalculation due to contamination of genomic DNA.

Lactate Dehydrogenase Assay

Enzymatic activity of lactate dehydrogenase (LDH) was determined by the reduction of pyruvate to lactate in the presence of NADH. The reaction rate is proportional to the decrease in absorption of NADH at 340 nm. Briefly, 0.14 mg NADH and 0.09 mg pyruvate were dissolved in 990 μL buffer, containing 81.3 mM Tris/HCL and 200 mM NaCl at pH 7.2. After addition of 10 μL enzymatic solution the time-dependent decrease in absorption at 340 nm was recorded. Enzymatic activity was derived from a calibration curve consisting of samples with known LDH activity (6 points ranging from 100 – 0.625 $\times 10^{-2}$ units). Measurements were performed in duplicate and mean values were used for further calculations. Enzymatic activities were determined by using two independent calibration curves.

RESULTS

To determine the amount of ribosomal RNAs in the supernatant of ECV 304 cell culture we first collected the supernatant after 3 h of incubation with 1×10^6 cells. To separate nonadhering cells a filtration through 5-μM filters by gravity flow was performed. Ten microliter flow-through was concentrated

TABLE 1. Analysis of lactate dehydrogenase activity and 18S rRNA copy numbers in cell culture supernatural

Experiment/ passage	18S rRNA[a] (copies/mL)	No. of cells calculated[a] 18S rRNA ($\times 10^4$)	No. of cells calculated from LDH ($\times 10^4$)
P3	8.2×10^8	2.6 ± 1.1	2.6 ± 0.6
P13	6.0×10^8	2.0 ± 0.8	1.7 ± 0.9

Results are mean ± SD.

18s rRNA copy numbers were determined by cDNA synthesis and quantitative PCR of two independent isolations from supernatant of cell culture. To calculate the amount of ribosomal RNA that can be generated by cell destruction, calibration curves were prepared by plotting the copy number of 18s rRNA versus the amount of cells ($5 \times 10^2 - 1.5 \times 10^4$ ECV 304 cells) used for isolation. To determine the number of destroyed cells in culture, calibration curves of lactate dehydrogenase activity versus the amount of cells used for cytosolic preparation were used.

[a]Calculation is based on two independently generated calibration curves.

to allow an efficient isolation of RNA. After cDNA synthesis and quantitative PCR of two independent isolations from supernatant of cell culture from passage 3 and 13, the amount of 18S rRNA was determined to be 8.2×10^8 and 6.0×10^8 copies per mL of supernatant, respectively (TABLE 1). To determine the amount of ribosomal RNA that can be generated by cell destruction we isolated RNA from $5 \times 10^2 - 1.5 \times 10^4$ ECV 304 cells and determined the amount of 18s rRNA therein. Calibration curves were obtained from two independent experiments by plotting the copy number of 18S rRNA versus the amount of cells used for preparation. Using these calibration curves we calculated the amount of destroyed cells in cell culture from passage 3 and 13 to be $(2.6 \pm 1.1) \times 10^4$ and $(2.0 \pm 0.8) \times 10^4$ cells, respectively (TABLE 1). Using amplicons for the quantification of 5.8S and 28S rRNA similar results were obtained (data not shown).

To be able to calculate the amount of destroyed cells in the cell culture used for isolation of exRNA we performed a lactate dehydrogenase assay. Cytosolic fractions were prepared from increasing amount of ECV 304 cells ranging from $5 \times 10^3 - 5 \times 10^5$ cells. The enzymatic activity was determined using a calibration curve derived from purified lactate dehydrogenase. To determine the number of destroyed cells in culture a calibration curve of lactate dehydrogenase activity versus the amount of cells used for cytosolic preparation was used. Measurement of lactate dehydrogenase activity in the supernatant of 1×10^6 cells resulted in $(2.6 \pm 0.6) \times 10^4$ destroyed cells (TABLE 1). This result was not significantly altered when cells of a higher passage were used (TABLE 1).

The results demonstrate that the amount of cells with disintegrated cellular membranes present in the culture would be sufficient to explain the rRNA copy numbers obtained. We next sought to determine whether the ribosomal RNAs from cells with destroyed membranes persist in the cell culture supernatant. Cytosolic fractions were prepared from 5×10^5 cells and incubated for up to 3

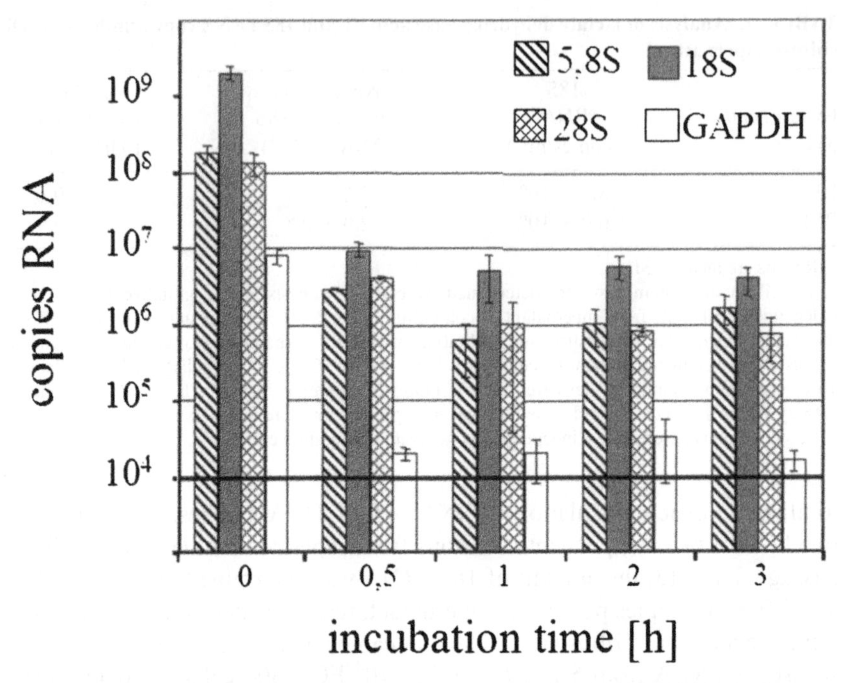

FIGURE 1. Time-dependent decrease of rRNAs as well as GAPDH RNA in the presence of FCS. Cytosolic fractions were prepared from 5×10^5 cells and incubated for up to 3 h in the presence of 10% FCS at 37°C. The ***Y*** axis represents absolute RNA copy number in the experiment, whereas the ***X*** axis represents the time of incubation. The lower limit of detection is marked by the *black line*. Mean values of two independent isolations were used for calculation.

h in the presence of 10% FCS at 37°C. Time-dependent decrease of rRNAs as well as GAPDH was measured by RT-qPCR as described above. As shown in FIGURE 1 in the presence of FCS a rapid decrease of GAPDH RNA as well as rRNAs was seen within 1 h of incubation. Prolonged incubation up to 3 h did not result in further decrease of the ribosomal RNAs, while determination of GAPDH was already at the lower detection limit. These results demonstrate that roughly 0.5% of the starting amount of rRNAs persists for at least 3 h in the presence of FCS. In the presence of medium lacking FCS rRNAs as well as GAPDH RNA were stable for at least 3 h (data not shown).

To determine whether the amount of total exRNA correlates with the extent of cell death we compared lactate dehydrogenase activity and the amount of exRNA in the supernatant in a total of 24 independent experiments under various conditions. ECV 304 cells from different passages in various densities were grown as described in the METHODS section. In addition, some cell cultures were treated with 500 μM staurosporine for 3 h before supernatant was collected. RNA was isolated by phenol/chloroform extraction to increase yield especially by minimizing loss of shorter RNA fragments. FIGURE 2 shows a plot of the number of destroyed cells as calculated by lactate dehydrogenase assay versus the corresponding concentration of exRNA. The relationship between destroyed cells and exRNA was found to be significant ($r = 0.843$; $P < 0.001$) using the Pearson correlation test.

DISCUSSION

Ex vivo experiments as well as cell culture experiments have shown that exRNA is sequestered by living cells and a constant level of RNA is maintained outside cells. The experiments presented in this article provide evidence that a fraction of exRNA, the ribosomal RNA species, can be explained by destruction of cells in culture. The correlation of lactate dehydrogenase activity in the cell culture supernatant with the amount of exRNAs might be a result of apoptosis or unregulated ways of cell death. The amount of cells destroyed in a culture of 10^6 ECV cells (roughly 2.6×10^4) corresponds to approximately 1.5% of cell death, which is quite common in this type of experiment. As the

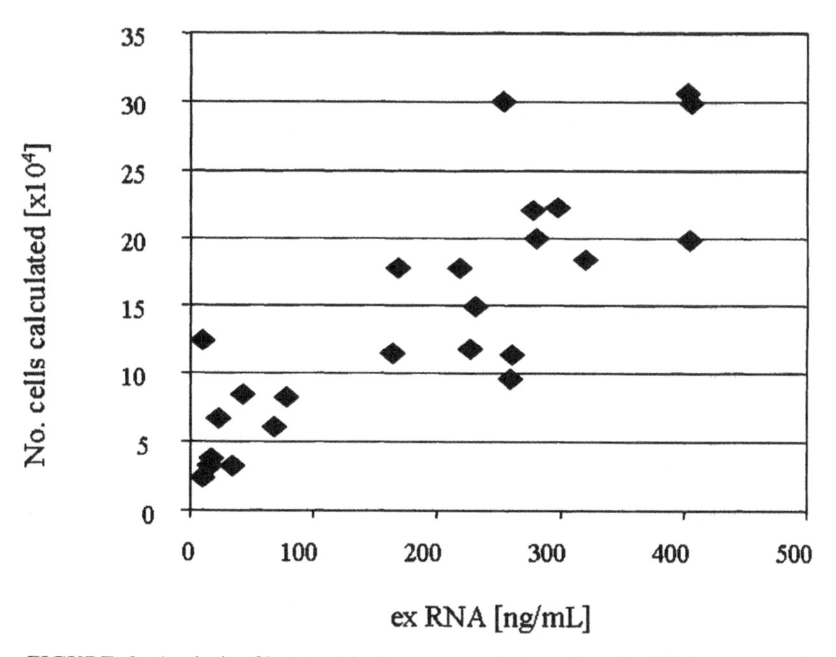

FIGURE 2. Analysis of lactate dehydrogenase activity and total exRNA concentration in cell culture supernatant. ExRNA concentration was determined by absorption measurement at 260 nm. To determine the number of destroyed cells in culture, calibration curves of lactate dehydrogenase activity versus the amount of cells used for cytosolic preparation were used. The *Y* axis represents the number of dead cells in cell culture as calculated by lactate dehydrogenase assay whereas the *X* axis represents the concentration of exRNA.

RNA inside cells largely consists of ribosomal RNAs, the fraction of exRNA generated by unregulated cell death will predominantly consist of rRNA or rRNA fragments. Consequently, analysis of these RNA species among the exRNA defines the background on which a fraction of putative functional RNAs has to be investigated. This idea is supported by the fact that Kolodny has found a high degree of 2′-O-methylation among exRNA.[2] This type of modification is frequently found in ribosomal RNAs. Incubation of cytosolic fractions with FCS clearly demonstrated that rRNA in the ribosomal context is at least partially protected from degradation. Approximately, 0.5% of the initial amount of rRNA was still detectable after 3 h.

A correlation between exRNA and destroyed cells is also found if the amount of total exRNA is investigated. Supernatants containing high lactate dehydrogenase activities generally contain higher amounts of total exRNA as demonstrated by the statistically significant correlation. However, experiments with identical amounts of destroyed cells as calculated by LDH assay may show total RNA amounts that differ significantly. We cannot exclude that exRNA as well as lactate dehydrogenase activity is also derived from apoptotic vesicles or cells in the later stages of apoptosis. Further experiments are needed to elucidate whether a higher amount of apoptotic vesicles and/or functional export of RNA by different mechanisms result in elevated amounts of exRNA.

ACKNOWLEDGMENTS

We thank B. Oschmann for technical assistance.

REFERENCES

1. STROUN, M. *et al.* 1978. Presence of RNA in the nucleoprotein complex spontaneously released by human lymphocytes and frog auricles in culture. Cancer Res. **38**: 3546–3554.
2. KOLODNY, G.M. 1972. Cell to cell transfer of RNA into transformed cells. J. Cell Physiol. **79**: 147–150.
3. WIECZOREK, A.J. *et al.* 1985. Isolation and characterization of an RNA-proteolipid complex associated with the malignant state in humans. Proc. Natl. Acad. Sci. **82**: 3455–3459.
4. KOPRESKI, M.S. *et al.* 1999. Detection of tumour messenger RNA in the serum of patients with malignant melanoma. Clin. Cancer Res. **5**: 1961–1965.
5. GOESSL, C. 2003. Non-invasive molecular detection of cancer—the bench and the bedside. Curr. Med. Chem. **8**: 691–706.
6. NG, E.K. 2002. Presence of filterable and nonfilterable mRNA in the plasma of cancer patients and healthy individuals. Clin. Chem. **48**: 1212–1217.
7. TAMKOVICH, S.N. 2005. Circulating nucleic acids in blood of healthy male and female donors. Clin. Chem. **51**: 1317–1319.

Hypoxia-Induced Membrane-Bound Apoptotic DNA Particles

Potential Mechanism of Fetal DNA in Maternal Plasma

AARON F. OROZCO,[a] FARIDEH Z. BISCHOFF,[b,d] CASSANDRA HORNE,[a] EDWINA POPEK,[c] JOE LEIGH SIMPSON,[b,d] AND DOROTHY E. LEWIS[a]

[a]*Department of Immunology, Baylor College of Medicine, Houston, Texas, USA*

[b]*Department of OB-GYN, Baylor College of Medicine, Houston, Texas, USA*

[c]*Department of Pathology, Baylor College of Medicine, Houston, Texas, USA*

[d]*Department of Human and Molecular Genetics, Baylor College of Medicine, Houston, Texas, USA*

ABSTRACT: Fetal DNA is found in the plasma of pregnant women that appears to be stable for PCR amplification. Although the underlying mechanism giving rise to this DNA in plasma remains unclear, the source of these fragments may be from apoptotic bodies (Apo-Bodies) created from dying cells. Trophoblast apoptosis is essential for normal placental development, given the enormous amount of proliferation, differentiation, and migration during pregnancy. Through flow cytometric analysis coupled with real-time PCR, our lab has shown that aggregates of acridine orange (AO)-stained material (apoptotic particles) are resistant to DNase treatment, disrupted by sodium dodecyl sulfate (SDS), and contain fetal DNA. Because the placenta continuously remodels in an hypoxic environment, our hypothesis is that fetal DNA in maternal plasma comes from hypoxia-induced dying trophoblasts and that this DNA circulates predominately in the form of Apo-Bodies. We have developed a model culture system for analysis of Apo-Bodies derived from JEG-3 cells, an extravillous trophoblastic cell line, undergoing various methods of cell death: hypoxia-induced, etoposide-induced, and heat stress (necrosis like)–induced cell death. Under conditions of similar propidium iodide (PI) uptake, suggesting comparable levels of death, both hypoxia- and etoposide-induced Apo-Bodies increase in concentration over time, whereas heat-induced levels of particles remain fairly constant, indicating that production of DNA-associated Apo-Bodies is a continuous process. Hypoxia, which is likely to be responsible for trophoblast cell death *in vivo*, produced membrane-bound Apo-Bodies containing DNA. Our

Address for correspondence: Farideh Z. Bischoff, Department of OB-GYN, Baylor College of Medicine, One Baylor Plaza, Houston, Texas 77030. Voice: 713-798-8885; fax: 713-798-5575. e-mail: bischoff@bcm.tmc.edu

Ann. N.Y. Acad. Sci. 1075: 57–62 (2006). © 2006 New York Academy of Sciences.
doi: 10.1196/annals.1368.007

results are consistent with the characteristics of membrane-bound particles containing fetal DNA found in maternal plasma.

KEYWORDS: **apoptotic bodies; fetal DNA**

INTRODUCTION

Fetal DNA concentration increases in maternal plasma as gestation progresses.[1] Similarly, elevated levels of DNA have been found in patients with systemic lupus erythematosus (SLE) and cancer. Whether the DNA is derived from apoptotic cells is unclear; however, cell-free tumor-specific DNA fragments between 180–200 bp in length, which is the hallmark of apoptotic DNA, have been found in the plasma of cancer patients with solid tumors.[2–4] Furthermore, pregnancies complicated with placenta previa, hyperemesis gravidarum, and pre-eclampsia (PE) are associated with abnormal levels of fetal cell death[5] and increased cell-free fetal DNA.[6,7] Although the literature suggests fetal DNA may be derived from apoptotic fetal cells, the source of the DNA and the mechanisms of release remain unknown.

Two potential sources of fetal DNA detected in maternal plasma during placental development are syncytiotrophoblasts and extravillous cytotrophoblasts (EVCT), both of which are derived from villous cytotrophoblast stem cells. The role of syncytiotrophoblasts is to mediate exchange of gas, nutrients, and waste products between the fetus and mother.[8] Three to 4 weeks after syncytiotrophoblast formation, aging syncytiotrophoblasts undergo apoptosis, followed by DNA degradation, packaging of apoptotic nuclei into apical membranes known as syncytial knots, followed by release into the maternal circulation.[9] Because it is thought that apoptotic syncytiotrophoblasts and shed syncytial knots are cleared in the maternal lung capillaries, there is likely another major source of fetal DNA.

Such a source of fetal DNA in maternal circulation may come from the subpopulation of trophoblasts called EVCTs. EVCTs initiate physical contact between fetus and mother[8] and give rise to nonproliferative invasive daughter cells. These invasive trophoblasts undergo "rolling adhesion" across the uterine luminal epithelium, invade the uterine tissues (interstitial trophoblasts), and anchor the placenta to the uterus. After successful adhesion, interstitial trophoblasts further differentiate into single invasive trophoblasts, which remodel the maternal uterine spiral arteries (endovascular trophoblast) in a manner similar to tumor invasion. Because continuously generated EVCTs differentiate and migrate into the uterine wall and spiral arteries, apoptosis is likely required to maintain tissue homeostasis. Smaller single cell endovascular trophoblasts are one of a few types of cells, including nRBCs, detected in the maternal blood,[10] suggesting fetal DNA may be released from circulating apoptotic EVCTs.

Researchers have used fetal cell detection to screen for pregnancies at risk for genetic abnormalities. Because trophoblasts and nRBCs in the maternal circulation are rare, accounting for only 2–4 of 10^6 to 10^9 maternal cells, extensive protocols and costly equipment are needed for detection.[11] Although many have been focused on fetal cell enrichment and detection, others have concentrated on maternal plasma fractions containing fetal (trophoblast and/or nRBC-derived) DNA and RNA in early pregnancy (9–13 weeks).[12]

Lo *et al.*, demonstrated cell-free fetal DNA in the plasma of healthy pregnant women as early as 7 weeks of gestation.[13] Circulating fetal DNA concentration is almost 1000-fold greater in the maternal plasma than fetal cells in maternal cellular fractions.[1] In addition, plasma fetal DNA levels increase as the embryo develops,[14] with increase up to 30% during the late third trimester.[15] Fetal DNA in the maternal circulation is also dramatically reduced within a few days of birth,[1] suggesting a continuous *in vivo* supply and rapid clearance during pregnancy.

Our preliminary data using flow-sorted acridine orange (AO)-stained maternal plasma suggest that detected fetal DNA may be housed and protected in membrane-bound particles similar to apoptotic bodies (Apo-Bodies).[16] Because the placenta continuously remodels in a hypoxic environment, our hypothesis is that fetal DNA in maternal plasma is derived from hypoxia-induced dying trophoblasts and that this DNA circulates predominately in the form of Apo-Bodies.

Hypoxia-Induced Apoptosis Produces Apo-Bodies

We tested our hypothesis by developing a model culture system for analysis of apoptotic DNA and Bodies derived from hypoxia-induced apoptotic JEG-3 cells, an established extravillous trophoblastic cell line. To allow for comparison of Apo-Body concentration between hypoxia- and mitochondria-induced apoptosis, similar levels of cell death were attained. To determine whether particles were also derived from necrotic cells, which involves osmotic swelling and leakage, JEG-3 cells were also heat stressed. Two million JEG-3 cells were exposed to hypoxia (50 μM rotenone)-, mitochondria (30 μM etoposide), or heat-stressed (1 h at 60°C) cell death, examined at 24 and 48 h for propidium iodide (PI) uptake compared to untreated cells. At these concentrations, both hypoxia- and mitochondria-induced apoptosis show similar PI uptake at 24 h (39% and 41%) and 48 h (81% and 79%), whereas heat-stressed cells were 100% PI (+) at both 24 h and 48 h.

To determine Apo-Body concentration, we established a quantitative assay using 10- μm fluorescent beads and flow cytometric analysis. A standard concentration (as determined by hemocytometer counting) of fluorescent beads was used. Twenty-microliter beads were added to either 480 μL PBS or to 430 μL PBS plus 50 μL Apo-Bodies for a total volume of 500 μL. Ten thousand beads were counted per 500 μL sample on an EPICS XL-MCL Flow Cytometer (Coulter). Apo-Body concentration was determined by the following equation:

$$\text{Concentration of Apo-Bodies} = \text{Specific Apo-Body Count} \times \frac{\text{Concentration of Beads}}{\text{Bead Count}}$$

$$\text{Specific Apo-Body Count} = \text{Total Apo-Body Count} - \text{Beads In Sample}$$

Apoptotic/necrotic supernatant from 2 million JEG-3 cells was collected and centrifuged at 19,900 rpm (75,000 *g*) for 30 min at 4°C (SW 40 Ti Swinging-Bucket Rotor in a Beckman L8-M, Class H, machine) and Apo-Bodies were quantified using an EPICS XL-MCL Flow Cytometer (Coulter). At 48 h, quantitative analysis show hypoxia-, mitochondria-, and necrosis-like particles at a range of $4–50 \times 10^6$ bodies, suggesting that hypoxia- and mitochondria-induced cell death produced 20–50 Apo-Bodies per cell, whereas heat stress resulted in only 4 bodies per cell compared to 1 body per 5 cells in untreated samples.

Membrane-Bound Apo-Bodies Contain DNA

To determine whether Apo-Bodies are membrane-bound and contain DNA, two independent labeling techniques were used. First, Apo-Bodies were labeled with PKH26, a fluorescent dye with long aliphatic tails, which inserts into the lipid regions of the cell membrane, and PicoGreen, which preferentially binds dsDNA. Secondly, Apo-Bodies were stained with the B subunit of cholera toxin (CTB), which binds to ganglioside G_{M1} lipids found on the cell surface (lipid rafts), and PI, which binds nucleic acids. After 48 h, 65% of hypoxia-induced Apo-Bodies were double positive for PKH26 (membrane) and PicoGreen (DNA) and 46% were double positive for CTB (lipid rafts) and PI (nucleic acids). Similarly, mitochondria-induced Apo-Bodies were 80% double positive for membrane (PKH26) and DNA (PicoGreen), with less staining for both lipid rafts (CTB) and DNA (PI) (50%). Our data show that Apo-Bodies and more specifically, hypoxia-induced Apo-Bodies contain membrane-bound, DNA- containing particles.

Apoptotic Particles Contain Stable and Amplifiable DNA

To determine whether the DNA contained in Apo-Bodies is stable for PCR amplification, we performed PCR for the Y-sequences. DNA was extracted from hypoxia- and mitochondria-induced Apo-Bodies at 24 h and 48 h and PCR amplified for Y sequences. Our data show hypoxic apoptotic DNA that is stable and suitable for specific Y sequence amplification by PCR.

CONCLUSION

Past research shows increasing levels of circulating fetal DNA through gestational progression. A greater level is also found in complications of pregnancy, such as pre-eclampsia. Similarly, trophoblast apoptosis, which is necessary for normal placental development, increases with gestation and is even greater in complicated pregnancies. Currently, the source and mechanisms of circulating fetal DNA is speculative. Our lab has shown that flow-sorted particles from maternal plasma contain particles with male fetal DNA, which is resistant to DNase and SDS treatment.[16] Therefore, we hypothesized that circulating fetal DNA may be protected from nucleases because of containment within Apo-Bodies.

Our *in vitro* studies show that dying cells package apoptotic DNA into membrane-bound vesicles, which we call Apo-Bodies. Furthermore, kinetic analysis demonstrates hypoxia- and mitochondria-induced Apo-Bodies increase over time, whereas necrotic-induced particles remain constant. Recently, it was shown that cell contraction and the formation of membrane blebs, which occurs prior to the release of Apo-Bodies, is dependent on myosin ATPase.[17] Because necrotic cells are depleted of ATP, perhaps they cannot form membrane blebs and thereby release only a few bodies (Necro-Bodies). Our data also suggest that production of Bodies is a process that may provide a continuous source of fetal DNA compared to heat-stressed necrosis-like particles.

We also demonstrated that CTB binds to sphingolipids which are commonly associated with lipid rafts. Interestingly, HLA-G has been shown to co-localize in lipid rafts of JEG-3 cells.[18] Whether HLA-G or other trophoblast-specific markers are expressed on Apo-Bodies will be examined.

Finally, we show that DNA packaged into Apo-Bodies can be amplified for specific Y sequences. Whether apoptotic particles contain both DNA/RNA and whether they are resistant to nuclease treatment are currently being investigated.

Although these results were obtained from an *in vitro* culture system, they support the theory that fetal DNA in maternal plasma may come from dying trophoblasts in an hypoxic environment.

REFERENCES

1. Lo, Y.M. *et al.* 1998. Quantitative analysis of fetal DNA in maternal plasma and serum: implications for noninvasive prenatal diagnosis. Am. J. Hum. Genet. **62:** 768–775.
2. Chen, X.Q. *et al.* 1996. Microsatellite alterations in plasma DNA of small cell lung cancer patients. Nat. Med. **2:** 1033–1035.
3. Nawroz, H. *et al.* 1996. Microsatellite alterations in serum DNA of head and neck cancer patients. Nat. Med. **2:** 1035–1037.

4. JAHR, S. *et al.* 2001. DNA fragments in the blood plasma of cancer patients: quantitations and evidence for their origin from apoptotic and necrotic cells. Cancer Res. **61:** 1659–1665.
5. ISHIHARA, N. *et al.* 2002. Increased apoptosis in the syncytiotrophoblast in human term placentas complicated by either preeclampsia or intrauterine growth retardation. Am. J. Obstet. Gynecol. **186:** 158–166.
6. SMID, M. *et al.* 2001. Quantitative analysis of fetal DNA in maternal plasma in pathological conditions associated with placental abnormalities. Ann. N. Y. Acad. Sci. **945:** 132–137.
7. LO, Y.M. *et al.* 1999. Redman quantitative abnormalities of fetal DNA in maternal serum in preeclampsia. Clin. Chem. **45:** 184–188.
8. ZHOU, Y. *et al.* 2003. The human placenta remodels the uterus by using a combination of molecules that govern vasculogenesis or leukocyte extravasation. Ann. N. Y. Acad. Sci. **995:** 73–83.
9. HUPPERTZ, B. & J.C. KINGDOM. 2004. Apoptosis in the trophoblast-role of apoptosis in placental morphogenesis. J. Soc. Gynecol. Investig. **11:** 353–362.
10. OUDEJANS, C.B. *et al.* 2003. Circulating trophoblast in maternal blood. Prenat. Diagn. **23:** 111–116.
11. VONA, G. *et al.* 2002. Enrichment, immunomorphological and genetic characterization of fetal cells circulating in maternal blood. Am. J. Pathol. **160:** 51–58.
12. VAN WIJK, I.J. *et al.* 1996. Enrichment of fetal trophoblast cells from the maternal peripheral blood followed by detection of fetal deoxyribonucleic acid with a nested X/Y polymerase chain reaction. Am. J. Obstet. Gynecol. **174:** 871–878.
13. LO, Y.M. *et al.* 1997. Presence of fetal DNA in maternal plasma and serum. Lancet **350:** 485–487.
14. ARIGA, H. *et al.* 2001. Kinetics of fetal cellular and cell-free DNA in the maternal circulation during and after pregnancy: implications for noninvasive prenatal diagnosis. Transfusion **41:** 1524–1530.
15. CHAN, L.Y. *et al.* 2003. Serial analysis of fetal DNA concentrations in maternal plasma in late pregnancy. Clin. Chem. **49:** 678–680.
16. BISCHOFF, F.Z. *et al.* 2005. Cell-free fetal DNA in maternal blood: kinetics, source and structure. Hum. Reprod. Update **11:** 59–67.
17. CROFT, R.D. *et al.* 2005. Actin-myosin-based contraction is responsible for apoptotic nuclear disintegration. J. Cell. Biol. **168:** 245–255.
18. COMISKEY, M. *et al.* 2003. Evidence that HLA-G is the functional homolog of mouse Qa-2, the *ped* gene product. Hum. Immunol. **64:** 999–1004.

Fetal Nucleic Acids in Maternal Body Fluids

An Update

DIANA W. BIANCHI, TUANGSIT WATAGANARA, OLAV LAPAIRE, MAY LEE TJOA, JILL L. MARON, PAIGE B. LARRABEE, AND KIRBY L. JOHNSON

Division of Genetics, Departments of Pediatrics, Obstetrics, and Gynecology, Tufts-New England Medical Center and Tufts University School of Medicine, Boston, Massachusetts, USA

ABSTRACT: Our laboratory continues to be actively involved in the development of new biomarkers for prenatal diagnosis using maternal blood and amniotic fluid. We have also developed a mouse model that demonstrates that cell-free fetal (cff) DNA is detectable in the pregnant maternal mouse. In human maternal plasma and serum we have analyzed factors that are important in the clinical interpretation of cff DNA levels. Maternal race, parity, and type of conception (natural or assisted) do not affect cff DNA levels, but maternal weight does. We have also analyzed the relationship between placental volume, using a three-dimensionsal ultrasound examination, and cff DNA levels. Surprisingly, there is no association between these values. Finally, we are using specific disease models (such as congenital diaphragmatic hernia and twin-to-twin transfusion) to understand the effects of gestational age and specific pathology on fetal gene expression by analyzing cell-free mRNA levels in maternal plasma. In the amniotic fluid we have focused on improvements in recovery of cff DNA and mRNA. By optimizing recovery we have made some interesting observations about differences in fetal DNA between blood and amniotic fluid. In addition, we have successfully hybridized cff DNA in amniotic fluid to DNA microarrays, permitting assessment of fetal molecular karyotype. We also have preliminary data on fetal gene expression in amniotic fluid. Finally, we remain actively involved in promoting noninvasive prenatal testing in the United States, such as encouraging the use of fetal DNA for fetal rhesus D assessment. On the other hand, we are cautious and concerned about the accuracy of "at-home" kits for fetal gender detection.

KEYWORDS: cell-free fetal DNA; prenatal diagnosis; amniotic fluid; fetomaternal trafficking; microarray; comparative genomic hybridization

Address for correspondence: Diana W. Bianchi, Tufts-New England Medical Center, Box 394, 750 Washington Street, Boston, MA 02111 USA. Voice: 617-636-1468; fax: 617-636-1469.
e-mail: Dbianchi@tufts-nemc.org

Ann. N.Y. Acad. Sci. 1075: 63–73 (2006). © 2006 New York Academy of Sciences.
doi: 10.1196/annals.1368.008

INTRODUCTION

The biennial conference on circulating nucleic acids in plasma and serum (CNAPS) provide us with a milestone opportunity to assess scientific progress occurring in the field at local and global levels. While the focus of this article will be on the advances emanating from our laboratory since CNAPS III, in general the field has evolved significantly. Several trends are notable: first of all, there has been a significant increase in the number of scientific publications in the field. Using the Pub Med search engine and the key words "cell-free fetal DNA," the year 2004 saw a dramatic rise in published papers on this topic. Another trend is the movement from technical advances occurring in individual laboratories to multicenter studies examining diagnostic issues related to quality assurance,[1] and development of international groups, such as the European Union's Special Advances in Fetal Evaluation (SAFE) network, which are exploring larger-scale issues related to accuracy, efficiency, and costs of testing. Noninvasive prenatal diagnosis of fetal Rhesus D is transitioning from research to clinical care in Europe. Regrettably, in the United States, there are not yet organized efforts to bring noninvasive prenatal diagnosis of Rhesus D to clinical practice.[2] Instead, pregnant women can purchase an at-home kit to determine the gender of their fetus (see "The Baby Gender Mentor" at www.pregnancystore.com) using fetal DNA in a maternal blood sample captured on filter paper. The scientific accuracy of this at-home test is completely unknown and its use unregulated, yet the test has received enormous positive publicity in the American print and television media.

In this manuscript we will summarize our recent findings with regard to the prenatal diagnostic applications for cell-free fetal (cff) and maternal DNA in maternal blood and amniotic fluid. In addition, we will discuss some of our work in the murine model, and speculate as to the future directions for our research.

Factors that Influence Cell-Free DNA Levels in Normal Pregnancies

Multiple studies have suggested that fetal cell-free DNA levels in maternal plasma and/or serum are elevated in certain clinical scenarios, such as a fetus with trisomy 21 or with clinical symptoms of pre-eclampsia. These studies have compared DNA values in affected pregnancies with so-called "normal controls." We decided to question what is "normal," and to ask what factors influence "normal levels" of fetal and total cell-free DNA.

Since it is well known that several demographic variables, such as maternal weight, ethnic background, and smoking status affect levels of maternal serum proteins in serum screening assays, we hypothesized that DNA levels might be similarly affected. We therefore investigated the association between cff DNA levels during the first and second trimesters of pregnancy, and multiple demographic factors, including maternal age.[3] The first-trimester study group consisted of fresh plasma samples from 91 women with confirmed male fetuses undergoing elective termination of pregnancy for personal indications (no known fetal pathology). The second-trimester study group consisted of 126 archived residual serum screening samples from women with confirmed male fetuses and no known fetal or neonatal pathology. Fetal DNA was isolated and amplified using real-time quantitative polymerase chain reaction for the Y chromosome–specific sequence *DYS1*. As has been demonstrated previously, gestational age is significantly associated with fetal DNA levels, although much more dramatically in the first trimester as compared with the second. Fetal DNA levels increase, on average, 21% per week in the first trimester. Maternal age, however, does not affect fetal DNA levels. Similarly, ethnic background (identified as Caucasian, African-American, Hispanic, or Asian) does not affect DNA levels. We observed no differences between smokers and nonsmokers, although information regarding the number of cigarettes smoked was unavailable. The most significant finding was that there was a significant inverse correlation between maternal weight and fetal DNA levels (after adjustment for gestational age) in second-trimester samples only. According to our study, a weight-correction factor of 26.5% may be needed when fetal DNA analysis is performed in the second trimester, especially when the pregnant woman weighs more than 170 pounds.[3]

Following the theme of examining factors that are adjusted for the interpretation of maternal serum screening tests, we decided to investigate whether type of conception affects cff DNA levels.[4] In *in vitro* fertilization (IVF) conceptions, increased second-trimester levels of maternal serum human chorionic gonadotropin (hCG) lead to increased false positive rates for Down's syndrome. In some studies, increased levels of cff DNA have been correlated with increased hCG levels. We hypothesized that cff DNA, if derived entirely from the placenta, might also be elevated in IVF conceptions. We therefore conducted a blinded case–control study to compare fetal DNA levels (using *DYS1*) as well as quadruple screen markers in archived maternal serum samples from women carrying male fetuses who conceived by IVF with women who conceived naturally. No statistically significant differences in mean cff DNA levels were observed between cases and controls. Furthermore, there were no significant correlations observed between cff DNA and each of the four standard maternal serum screening markers (hCG, alpha-fetoprotein, unconjugated estriol, and inhibin A) ($P = 0.18$, 0.22, 0.57, and 0.25, respectively). This provides additional evidence that cff DNA functions as an independent serum marker, as has been shown previously in a separate study performed in our laboratory.[5]

We continue to be interested in the tissue source of the circulating cff DNA. Although many reports associate increased circulating fetal DNA levels with either abnormal or invasive placenta, it is unlikely that the placenta is the exclusive contributor to the pool of fetal nucleic acids in the maternal circulation.

Potential alternative sources of fetal sequences include fetal hematopoietic cells and direct transfer of DNA fragments from organ-derived cells. We tested the hypothesis that if the placenta is the major contributor of cff DNA, then an increased placental volume, as measured by three-dimensional ultrasonography, should be associated with higher maternal plasma cff DNA values.[6] In a blinded analysis, we obtained frozen plasma samples from 143 pregnant women in the first trimester who presented for placental volume measurement at the Ludwig Boltzmann Institute for Clinical Gynecology and Obstetrics in Vienna, Austria. Cff DNA was measured using real-time amplification of the *DYS1* sequence, and raw values were adjusted for both gestational age and maternal body mass index as per our previous study.[3] Using linear regression analysis, we found that neither placental volume (expressed in mL) nor placental quotient (a ratio between placental volume and fetal crown–rump length, expressed in mm^2) showed a statistically significant association with maternal plasma cff DNA levels ($P = 0.43$ for each). Thus, our hypothesis was disproved, which was surprising, since a previous study of cff DNA levels in women carrying twins (with a presumed increase in placental mass) showed increased cff DNA levels.[7] An alternative explanation for our finding might be that it is the extent of placental apoptosis, and not placental volume, that determines the amount of nucleic acids released. This alternative explanation is given credence by studies performed in Professor Susan Fisher's laboratory, in which terminal deoxynucleotidyl transferase-mediated dUTP nick end labeling (TUNEL) staining was used to demonstrate the extent of placental apoptosis.

In one study, pre-eclampsia was shown to be consistently associated with widespread apoptosis of cytotrophoblasts that have invaded the uterus.[8] Pre-eclampsia is associated with a fivefold elevation of cff DNA levels in symptomatic women.[9] In a more recent study, there was an unexpectedly variable pattern of TUNEL reactivity in cytotrophoblasts obtained from trisomy 21 placentas.[10] Although there is on average a twofold elevation of cff DNA levels in women carrying fetuses with Down syndrome, on a per-case basis, some women have normal levels and others have elevated levels.[11,12] Thus, the variability in DNA levels agrees with the variability in TUNEL staining results.

Fetal Nucleic Acid Measurement and Specific Clinical Applications in Maternal Plasma and Serum

While progress ensues toward development of a gender-independent universal fetal marker, we are primarily using *DYS1* as a marker for male fetuses to further explore potential noninvasive prenatal diagnostic clinical applications. In collaboration with the Perinatology Research Branch of the National Institutes of Health, we measured cff DNA levels in a group of women at high risk for preterm delivery because of either preterm labor or preterm premature rupture of the membranes.[13] We used Kaplan–Meier and Cox regression analyses to investigate the relationship between cff DNA concentrations and likelihood of preterm (<36 weeks 6 days) delivery. Using raw cff DNA measurements, we calculated multiples of the median (MoM) values by means of a weighted regression, and used the study group that was delivered at term as a reference. A cut-off value of 1.82 MoM was chosen for further analysis. Our results indicated that the cumulative rate of early preterm delivery (<30 weeks) was significantly higher in this group of women at high risk for early delivery when their cff DNA values were greater than or equal to 1.82 MoM (45% vs. 18%) ($P = 0.008$).

Another clinical scenario in which there is a unique opportunity to study fetomaternal trafficking of nucleic acids is in fetoscopic treatment for twin–twin transfusion syndrome (TTS). TTS is a placental malformation that occurs in monochorionic twin pregnancies. Abnormal vascular anastomoses in the shared placenta give rise to an imbalance in feto-fetal transfusion, resulting in oligohydramnios in the donor twin and polyhydramnios in the recipient twin. Fetoscopic laser coagulation of shared vessels is one treatment option. In collaboration with the Eurofoetus trial, we obtained sequential plasma samples (preprocedure, 30 min, 60 min, 24 h, and 48 h after laser impact) from 34 women with twin pregnancies undergoing laser ablation therapy for TTS.[14] The results showed that, compared with baseline, median cff DNA elevations were 0.8% at 30 min, 15.8% at 60 min, 179.5% at 24 h, and 172.9% at 48 h. Factors associated with higher cff DNA levels at 24 h postprocedure included a longer operation time, a higher number of vessels ablated, and subsequent *in utero* fetal death of at least one twin. The tissue source of the cff DNA was not determined in this study. We speculated that it could derive from coagulation necrosis from thermal injury at the vessel impact site, focal subchorionic hemorrhage, or a confined hematoma.

To further explore the tissue source of the nucleic acids released by the thermal ablation technique we decided to study expressed levels of placental and hematopoietic genes in sequential maternal plasma samples. In this next study we compared women undergoing fetoscopic procedures for either TTS ($n = 12$) or fetal congenital diaphragmatic hernia (CDH) ($n = 10$). These invasive fetoscopic procedures contrast well with each other, since laser ablation directly involves the placenta and fetoscopic endoluminal tracheal occlusion (FETO), the treatment for CDH, does not. We hypothesized that in a manner similar to our previous observation that cff DNA levels increase following laser ablation, we would also see an increase in placental mRNA levels. We also hypothesized that since FETO does not directly involve the placenta, we should not observe an elevation in placenta-derived nucleic acids.[15] To test our hypotheses, we performed quantitative real-time reverse-transcriptase polymerase chain reaction amplification on sequential maternal plasma samples to determine levels of total mRNA (using the housekeeping sequence glyceraldehyde-3-phosphate dehydrogenase, *GAPDH*), placental mRNA (human placental lactogen, *hPL*), and fetal hematopoietic mRNA (*gamma globin*). Surprisingly, the maternal plasma median mRNA levels of *GAPDH*, *hPL*, and *gamma globin* remained unchanged following fetal intervention in both study groups. However, using multivariate regression analysis median levels of *GAPDH* mRNA were elevated at all time points in the CDH study group compared to the TTS study group ($P = 0.05$). There were three differences between the CDH and TTS study groups: (1) in CDH there was only one fetus and in TTS there were two; (2) the TTS procedure was performed in the second trimester, while the CDH procedure was performed in the third trimester; (3) all fetuses with CDH received at least one intramuscular injection for pain control and paralysis 5–30 min before the first maternal blood sample was drawn. None of the fetuses with TTS received any injections. These results suggest two important conclusions: one, the physiology of mRNA release into the maternal circulation likely involves a different process than the physiology of DNA release. It is possible that the mRNA that enters the maternal circulation following laser ablation may not be packaged in apoptotic bodies. If the mRNA is packaged in necrotic bodies or is more fragmented as a result of the laser treatment, it may be unstable and may rapidly disintegrate upon entry into the maternal circulation. Unexpectedly, we also observed the consistently higher *GAPDH* levels when FETO was performed. In adults, the ubiquitously expressed DNA sequence, *beta globin*, increases rapidly in the circulation of trauma patients; levels depend on the severity of the trauma as well as associated organ failure.[16] We speculate that *GAPDH* may have similar clinical utility as a universal marker for fetal trauma. Further study is clearly needed to understand the differences in the packaging and kinetics of fetal RNA versus fetal DNA in the maternal circulation.

Widespread prenatal diagnostic application for fetal nucleic acids in maternal blood still awaits the identification and validation of a gender-independent fetal marker. We are exploring differences in gene expression microarrays using extracted mRNA from the blood of pregnant women before and after delivery, and comparing the results to mRNA extracted from their newborn's umbilical cord blood. Results of this study are ongoing but appear promising.

Fetal Cell-Free Nucleic Acids in Amniotic Fluid

Amniotic fluid can only be obtained by performing amniocentesis, a safe and widely practiced prenatal diagnostic technique with a 0.5% chance of associated miscarriage. Thus, although the procedure is not noninvasive, it does provide an opportunity to analyze another maternal body fluid. We have shown that amniotic fluid is a rich source of predominantly, or possibly exclusively, fetal nucleic acids.[17–19] Initially, we developed a method of extraction of cff DNA from the fraction of amniotic fluid supernatant that is normally discarded after cells are cultured and the fluid is assayed for alpha-fetoprotein and acetyl cholinesterase. We wished to determine whether the amniotic fluid cff DNA could be hybridized to DNA microarrays for comparative genomic hybridization (CGH) analysis. We rationalized that by using the discarded amniotic fluid supernatant we would not interfere with the current standard of care (i.e., metaphase analysis of cultured amniocytes), yet we could potentially enhance fetal diagnosis by using array CGH analysis. Microarray CGH analysis has higher sensitivity and higher resolution for the detection of small genomic changes not observed with traditional microscopic methods, such as microdeletions or microduplications. Furthermore, it reduces the inherent subjectivity in Giemsa-banded metaphase chromosome analysis and allows statistical proof of the presence of an abnormality.

In an initial proof-of-principle study we performed CGH microarray analysis on cff DNA isolated from 17 amniotic fluid samples; four were from euploid female fetuses, nine were from euploid male fetuses, three were from fetuses with trisomy 21 (two male, one female), and one was from a fetus with Turner syndrome.[18] All nine samples from the euploid male fetuses showed increased hybridization signal intensity for *SRY* and reduced signal intensity for X chromosome markers compared with female reference DNA ($P < 0.01$). Samples from all four euploid female fetuses had significantly decreased hybridization signals for *SRY* and increased hybridization signals for X chromosome markers compared with male reference DNA ($P < 0.01$). When samples were compared with reference DNA from the same gender, there were no differences in the hybridization signal intensity for either the X or the Y chromosome, which was expected. For all of the 13 euploid fetal samples, there were no differences in hybridization signals for chromosome 21 sequences compared to reference DNA. However, the three samples from trisomy 21 fetuses had increased target-to-reference intensities on most chromosome 21 sequences. The Turner syndrome fetal sample had decreased hybridization signals on seven of nine X chromosome markers compared with normal fetal reference DNA. Our results showed that cff DNA from amniotic fluid supernatant can be used for clinical diagnosis in CGH microarrays to correctly identify fetal gender and whole chromosome losses (45,X) or gains (trisomy 21). We also compared cff DNA from amniotic fluid with DNA isolated from cultured amniocytes in the same fetuses, and found that the arrays performed similarly, albeit with more clone–clone variability in the cff DNA, presumably on account of degradation.

Now that we have demonstrated that cff DNA in amniotic fluid hybridizes to DNA microarrays, we have directed our efforts toward improving DNA extraction. We have significantly improved yield by replacing the previously overloaded Qiagen mini-spin columns with maxi-spin columns, and changing buffers from AL to AVL (Qiagen, Valencia, CA). The new protocol has resulted in a median seven-fold improvement in quantity of cff DNA available for analysis.[20]

Another goal of the laboratory is to better understand the biophysical properties of fetal nucleic acids in amniotic fluid, and whether they differ from those in the maternal circulation. In a manner similar to what has been shown in adult blood[21] we wished to explore whether cell-free RNA in amniotic fluid is associated with particles, and is thus filterable, and whether cff DNA in amniotic fluid is not associated with particles, and thus not affected by filtration.[22] In this group of experiments, the ubiquitous sequences, *GAPDH* and *beta globin* were used for mRNA and DNA measurement, respectively. In seven amniotic fluid samples, the results showed a significant decrease in *GAPDH* mRNA after filtration through a 0.22-μ filter, but filtration did not reduce the amount of *beta globin* DNA sequence detected. Thus, amniotic fluid nucleic acids have similar properties to those found in plasma, which suggests that a universal mechanism exists to process them.

Having shown that fetal mRNA is present in amniotic fluid, and establishing that it is stable enough to be amplified and detected using reverse-transcriptase PCR, we hypothesized that the fetal mRNA in amniotic fluid might permit the study of fetal gene expression *in vivo*.[19] If successful, this would have unique clinical relevance, as there are currently no molecular biological techniques available to monitor the ongoing development of the human fetus. Initially we did not know how much mRNA would be needed for hybridization to gene expression arrays, so we obtained large-volume amniotic fluid samples from four pregnant women undergoing therapeutic amniocentesis for polyhydramnios. Two women were carrying a fetus with hydrops (gestational ages 29 and 32 weeks) and two women had twin fetuses with TTS (gestational ages 20 weeks; the second woman had two samples at 21 and 24 weeks). For comparative purposes we also created pooled controls, consisting of amniotic fluid from euploid male or female fetuses at 17–18 weeks of gestation. RNA was extracted, amplified twice, labeled, and analyzed using Affymetrix U133A microarrays. Data were analyzed using the Affymetrix Microarray Suite 5.0, including the Data Mining software.

The results showed that 36% of 22,283 probe sets represented on the arrays were present in the amniotic fluid. A median value of 20% of all probe sets differed between the hydrops or TTS cases and the pooled control.[19] The mRNA appears to be of fetal, not placental origin, as no placental gene transcripts were detected in any amniotic fluid sample. The absence of placental transcripts in amniotic fluid suggests that fetomaternal trafficking of placental sequences is primarily unidirectional toward the mother. Although we were unable to confirm our results with real-time quantitative reverse-transcriptase PCR for key sequences due to limited amounts of RNA available, the presence of Y chromosome and prostate-related transcripts in the male samples but not the female sample provided some physiologic validation. The expression of some developmental transcripts, such as surfactant proteins, salivary mucin, and keratins, changed with gestational age by up to 64-fold. We also identified a unique transcript, aquaporin-1, a water transporter protein, that was upregulated in cases of TTS. Although this was a pilot study, it is the first *in vivo* study of global gene expression in the developing human fetus. More work needs to be done to improve methods of extraction and amplification of RNA, but we have shown that it is feasible, and more importantly, the gene expression patterns we observed correlated with known variables, such as fetal gender, gestational age, and disease status.[20]

Fetal Nucleic Acids in Animal Models

In a related area of the laboratory we are studying fetomaternal trafficking and the persistence of fetal cells postpartum. To facilitate such work, we use a unique paternally inherited transgene to track fetal cells in the mother. One of the matings we use involves a wild-type female mouse to a male that is transgenic for the enhanced green fluorescent protein (*gfp*). We performed real-time PCR using previously described primers and probes to quantify the number of fetal cells present in maternal organs to better understand fetal cell microchimerism and its health consequences.[23] Knowing that cff DNA is present in human maternal blood during pregnancy, we wished to determine whether the same phenomenon was true in the mouse. Uterine anatomy and placentation significantly differs between mouse and human; thus, we had no idea whether the trafficking of nucleic acids would be similar. Using real-time quantitative PCR for *gfp*, we consistently detected cff DNA in the plasma of wild-type pregnant mice.[24] In all cases, the *gfp* sequence was undetectable following pregnancy. This model may prove to be useful to further study the pathogenesis of pre-eclampsia, as mouse models of pre-eclampsia already exist.

CONCLUSIONS

During the past 2 years we have significantly increased our understanding of demographic and physiologic factors that affect levels of circulating cff DNA in maternal blood. All of this work was performed using Y chromosome sequences, such as *SRY* or *DYS1*, to measure fetal DNA. Large-scale clinical application of fetal DNA measurement awaits further development of a robust and reproducible gender-independent marker, which may be *maspin*.[25] By using clinical samples from women undergoing specific invasive fetal procedures, such as placental laser ablation, we can learn more about the pathophysiology of circulating fetal nucleic acids. Amniotic fluid is a rich and somewhat underappreciated source of fetal nucleic acids. The cff DNA and mRNA in amniotic fluid have significant potential to greatly expand the amount of information we can retrieve from the normal and abnormal developing fetus. At a local level, one of the main challenges for the future in our laboratory is how to manage and analyze all of the data that are acquired from the gene expression microarrays. On a global level, we need to work together to ensure that measurement and analysis of circulating fetal nucleic acids is performed in a responsible, ethical, and scientifically accurate manner to provide the best and most advanced care to the pregnant woman.

ACKNOWLEDGMENTS

We would like to thank our collaborators from the Eurofoetus trial (J. Deprest, J. Jani, L. Lewi, and E. Gratacos) for their help in obtaining plasma samples; M. Metzenbaum and E. Haffner from the Ludwig Boltzmann Institute in Vienna; and U. Tantravahi and J. Cowan for their help in obtaining amniotic fluid samples. Most of the work described here was supported by NIH Grant R01 HD42053.

REFERENCES

1. JOHNSON, K.L. *et al.* 2004. Inter-laboratory comparison of fetal male DNA detection from common maternal plasma samples by real-time PCR. Clin. Chem. **50**: 516–521.
2. BIANCHI, D.W. *et al.* 2005. Noninvasive prenatal diagnosis of fetal Rhesus D: ready for prime(r) time. Obstet. Gynecol. **106**: 841–844.
3. WATAGANARA, T. *et al.* 2004. Inverse correlation between maternal weight and second trimester circulating cell-free fetal DNA levels. Obstet. Gynecol. **104**: 545–550.
4. PAN, P.D. *et al.* 2005. Cell-free fetal DNA levels in pregnancies conceived by IVF. Hum. Reprod. **20**: 3152–3256; e-pub 2005 July 8.
5. FARINA, A. *et al.* 2003. Evaluation of cell-free fetal DNA as a second-trimester maternal serum marker of Down's syndrome pregnancy. Clin. Chem. **49**: 239–242.
6. WATAGANARA, T. *et al.* 2005. Placental volume, as measured by 3-dimensional sonography and levels of maternal plasma cell-free fetal DNA. Am. J. Obstet. Gynecol. **193**: 496–500.
7. SMID, M. *et al.* 2003. Fetal DNA in maternal plasma in twin pregnancies. Clin. Chem. **49**: 1526–1528.
8. DIFEDERICO, E. *et al.* 1999. Preeclampsia is associated with widespread apoptosis of placental cytotrophoblasts within the uterine wall. Am. J. Pathol. **155**: 293–301.
9. LO, Y.M. *et al.* 1999. Quantitative abnormalities of fetal DNA in maternal serum in preeclampsia. Clin. Chem. **45**: 184–188.
10. WRIGHT, A. *et al.* 2004. Trisomy 21 is associated with variable defects in cytotrophoblast differentiation along the invasive pathway. Am. J. Med. Genet. **130**: 354–364.
11. LO, Y.M. *et al.* 1999. Increased fetal DNA concentrations in the plasma of pregnant women carrying fetuses with trisomy 21. Clin. Chem. **45**: 1747–1751.
12. LEE, T. *et al.* 2002. Down's syndrome and cell-free fetal DNA in archived maternal serum. Am. J. Obstet. Gynecol. **187**: 1217–1221.
13. FARINA, A. *et al.* 2005. High levels of fetal cell-free DNA in maternal serum: a risk factor for spontaneous preterm delivery. Am. J. Obstet. Gynecol. **193**: 421–425.

14. WATAGANARA, T. *et al.* 2005. Persistent elevation of cell-free fetal DNA levels in maternal plasma after selective laser coagulation of chorionic plate anastomoses in several mid gestational twin-twin transfusion syndrome. Am. J. Obstet. Gynecol. **192**: 604–609.
15. TJOA, M.L. *et al.* 2006. Circulating cell-free fetal messengerRNA levels after fetoscopic interventions of complicated pregnancies. Am. J. Obstet. Gynecol. e-pub April 18.
16. LAM, N.Y. *et al.* 2003. Time course of early and late changes in plasma DNA in trauma patients. Clin. Chem. **49**: 1286–1291.
17. BIANCHI, D.W. *et al.* 2001. Large amounts of cell-free fetal DNA are present in amniotic fluid. Clin. Chem. **47**:1867–1869.
18. LARRABEE, P.B. *et al.* 2004. Microarray analysis of cell-free fetal DNA in amniotic fluid: a prenatal molecular karyotype. Am. J. Hum. Genet. **75**: 485–491.
19. LARRABEE, P.B. *et al.* 2005. Global gene expression analysis of the living human fetus using cell-free messenger RNA in amniotic fluid. JAMA **293**: 836–842.
20. LAPAIRE, O. *et al.* 2006. Larger columns and change of lysis buffer significantly improve the quantity of extracted cell-free DNA from amniotic fluid. Clin. Chem. **52**: 156–157.
21. NG, E.K.O. *et al.* 2002. Presence of filterable and nonfilterable mRNA in the plasma of cancer patients and healthy individuals. Clin. Chem. **48**: 1212–1217.
22. LARRABEE, P.B. *et al.* 2005. Presence of filterable and nonfilterable cell-free mRNA in amniotic fluid. Clin. Chem. **51**: 1024–1026.
23. KHOSROTEHRANI, K. *et al.* 2005. Natural history of fetal cell microchimerism during and following murine pregnancy. J. Reprod. Immunol. **66**: 1–12.
24. KHOSROTEHRANI, K. *et al.* 2004. Fetal cell-free DNA circulates in the plasma of pregnant mice: relevance for animal models of fetomaternal trafficking. Hum. Reprod. **19**: 2460–2464.
25. CHIM, S. *et al.* 2005. Detection of the placental epigenetic signature of the *maspin* gene in maternal plasma. Proc. Natl. Acad. Sci. USA **102**: 14753–14758.

Fetal DNA in Maternal Plasma

Progress through Epigenetics

Y.M. DENNIS LO

Department of Chemical Pathology, The Chinese University of Hong Kong, Prince of Wales Hospital, Shatin, New Territories, Hong Kong SAR, China

ABSTRACT: The discovery of cell-free fetal DNA in maternal plasma has opened up new possibilities for noninvasive prenatal diagnosis. Most of the work in this field has focused on the detection of fetal genetic markers that are distinguishable from the background maternal DNA. The feasibility of detecting fetal epigenetic markers in maternal plasma using an imprinted locus was demonstrated in 2002. This work has recently led to the development of the first universal fetal epigenetic marker, hypomethylated *maspin*, which can be used for fetal DNA detection in maternal plasma, irrespective of fetal gender and genetic polymorphisms. It is expected that many more fetal epigenetic markers that can be used in this manner will be developed over the next few years. These markers may catalyze the eventual clinical use of circulating fetal DNA for noninvasive prenatal diagnosis, such as for the detection of fetal chromosomal aneuploidies.

KEYWORDS: DNA methylation; noninvasive prenatal diagnosis; epigenetics

INTRODUCTION

The discovery of cell-free fetal DNA in maternal plasma in 1997 has opened up new possibilities for noninvasive prenatal diagnosis.[1] Over the last few years, this noninvasive source of fetal DNA has been used for the prenatal diagnosis of sex-linked disorders,[2] fetal RhD status,[3,4] β-thalassemia,[5–7] congenital adrenal hyperplasia,[8] and other diseases. Furthermore, quantitative aberrations of circulating fetal DNA have been described in a variety of pregnancy-associated disorders, including pre-eclampsia,[9,10] trisomy 21,[11,12] preterm labor,[13,14] hyperemesis gravidarum,[15] and invasive placentation.[16] Despite these promising results, a number of issues have remained unsolved.

Address for correspondence: Y.M. Dennis Lo, Department of Chemical Pathology, The Chinese University of Hong Kong, Prince of Wales Hospital, 30-32 Ngan Shing Street, Shatin, New Territories, Hong Kong SAR, China. Voice: +852 2632 2963; fax: +852 2636 5090.
e-mail: loym@cuhk.edu.hk

Ann. N.Y. Acad. Sci. 1075: 74–80 (2006). © 2006 New York Academy of Sciences.
doi: 10.1196/annals.1368.009

First, for qualitative applications, for example, for fetal gender determination for sex-linked disease, the lack of a Y chromosomal signal on plasma analysis could be interpreted in more than one way. For example, the lack of a signal could imply that the fetus is female; or alternatively, it could be the result of a very low level of circulating fetal DNA that is below the detection limit of the assay; or that the DNA extraction process has been suboptimally performed. To resolve these possibilities, the development of a universal fetal DNA marker that can be used to demonstrate the presence of amplifiable fetal DNA in a particular maternal plasma sample is very important. Several groups have used multiple insertion–deletion polymorphisms for ascertaining the presence of fetal DNA in a particular specimen,[17,18] but this approach would greatly increase the complexity of the resultant assays.

Second, for quantitative applications, the lack of a universal fetal DNA marker makes the situation even worse. Most workers in the field use the Y chromosome from male fetuses as a fetal-specific marker.[10,12,19] However, this approach is only applicable to the 50% of pregnancies involving male fetuses. The development of a universal fetal DNA marker would enhance the practicality of using circulating fetal DNA measurement as a way for predicting pregnancies at risk of complications.

Third, fetal DNA only represents a minor proportion of DNA in the plasma of pregnant women.[19] Thus, most workers in the field have attempted the detection of markers that the fetus has inherited from the father, and which are distinguishable from those of the mother. The detection of DNA fragments that the fetus has inherited from the mother would therefore be of considerable interest conceptually.

Fourth, trisomy 21 remains one of the most important reasons why pregnant women opt for prenatal diagnosis. Since the discovery of circulating fetal DNA in maternal plasma, many workers have debated the best way to use this new source of fetal DNA for the noninvasive prenatal diagnosis of trisomy 21. While the absolute concentration of circulating fetal DNA has been reported by a number of groups to be elevated in pregnant women carrying fetuses affected by trisomy 21,[11,12,20] this approach will only result in a relatively nonspecific screening method for this fetal chromosomal aneuploidy, as many other conditions, such as pre-eclampsia are also associated with an elevated level in circulating fetal DNA.[10,21] The exploration of new approaches for the noninvasive prenatal diagnosis of fetal chromosomal aneuploidies using circulating fetal DNA is therefore important.

In this review, the potential of fetal epigenetic marker detection in maternal plasma to address each of the foregoing issues will be discussed.

Plasma Epigenetic Markers

Epigenetics refers to alternate phenotypic states that are not based on differences in genotype, and are potentially reversible, but are generally stably maintained during cell division.[22] One important epigenetic change is DNA methylation. Shortly after the recent interest in plasma DNA as a molecular diagnostic tool was developed,[23] researchers have demonstrated that aberrant DNA methylation changes that are present in tumor cells are also detectable in the plasma of cancer patients.[24] Soon after this development, Poon *et al.* were the first to investigate whether epigenetic changes that are present in the fetal genome might be detectable in the plasma of a pregnant woman.[25] Poon *et al.* made use of an imprinted region that is methylated if inherited from one's father and unmethylated if inherited from one's mother. These investigators further studied a single nucleotide polymorphism (SNP) in this region and traced the transmission of such a stretch of DNA that a pregnant woman has inherited from her father (thus methylated in the pregnant woman's cells) onto her unborn fetus (thus unmethylated in the fetal cells). Because this stretch of DNA has a different methylation status between the unborn fetus and her mother, Poon *et al.* showed that such fetal DNA, when present in maternal plasma, is distinguishable from the background maternal DNA. The significance of this work is twofold: first, it demonstrates, for the first time, that fetal epigenetic markers can be successfully detected in maternal plasma. Second, it directly confronts the third issue discussed in the INTRODUCTION, namely, that with a careful choice of markers, it is possible to detect a stretch of DNA that a fetus has inherited from its mother in maternal plasma.

The demonstration of the feasibility of fetal epigenetic marker detection in maternal plasma has opened up the quest for more epigenetic markers, especially those with less exacting requirements for an informative pedigree. Because different tissues may exhibit different DNA methylation patterns,[26] it is important to decide which fetal and maternal tissues would be the most appropriate for the search of fetal epigenetic markers that can be detected in maternal plasma. In this regard, many investigators have suggested that the placenta may be the most likely organ for releasing fetal nucleic acids into maternal plasma.[27,28] Furthermore, in pregnancies with chromosomal abnormalities confined to the placenta, the same chromosomal aberrations are also detectable in maternal plasma.[29,30] Thus, the placenta might be the best tissue type for the search of new fetal epigenetic markers. However, the placental epigenetic pattern would need to be compared to a maternal tissue type, namely, the one that is responsible for a large proportion of the maternal DNA in a pregnant woman's plasma. Previous data generated from a bone marrow transplantation (BMT) model indicate that hematopoietic cells are the predominant source of plasma DNA in BMT recipients.[31] If these findings could be extrapolated to the case of a pregnant woman, then hematopoietic cells are probably the best maternal cell types to be analyzed for a representative maternal epigenetic signature for comparison with the placental epigenetic profile. Using this approach, Chim *et al.* discovered that the *maspin* gene is hypomethylated in the placenta and hypermethylated in maternal blood cells.[32] These investigators were prompted to study this gene because recent data indicate that the expression of the *maspin* gene is epigenetically controlled.[33] However, the methylation status of the *maspin* gene in the placenta was not known until now. Chim *et al.* showed that hypomethylated *maspin* sequences could be detected in the plasma of pregnant women and were promptly cleared following delivery.[32] Furthermore, genotyping of a SNP within the *maspin* gene showed that the hypomethylated *maspin* sequences in maternal plasma bore the fetal genotype.[32] To demonstrate the utility of hypomethylated *maspin* sequences as a fetal DNA marker, Chim *et al.* further showed that the median concentration of such sequences was elevated 5.7 times in the plasma of women suffering from pre-eclampsia when compared with controls. These latter results thus directly address the second issue described in the INTRODUCTION.

Future Directions

The demonstration of the feasibility of fetal epigenetic markers for noninvasive prenatal diagnosis has opened up a new area of investigation. It is expected that many more fetal epigenetic markers will be described over the next few years. It is to be expected that the pace of discovery of such markers will hasten with the Human Epigenome Project.[34] It is likely that these markers may exhibit different levels of differential DNA methylation between the placenta and maternal blood cells. The more divergent the pattern of DNA methylation, the more desirable a particular marker would be. It is further proposed that some of these markers may demonstrate disease-associated methylation changes. For example, it is possible that certain DNA methylation changes may be specific for trisomy 21,[35] in which case the fourth problem discussed in the INTRODUCTION could potentially be solved.

Another area that would be a fruitful avenue of research is in the methodology for DNA methylation analysis. Much of the current literature on epigenetics rests upon methods based on bisulfite conversion, such as bisulfite DNA sequencing,[36] and methylation-specific polymerase chain reaction (MSP), or its variants.[37,38] However, the bisulfite conversion step is time-consuming and requires much optimization.[39] It is hoped that future development of new methods for DNA methylation analysis will allow such methods to be more adapted to routine diagnostic use. This would allow one to address the first issue described in the INTRODUCTION, namely, to introduce a straightforward positive control for the presence of fetal DNA in a particular maternal plasma sample.

CONCLUSIONS

In conclusion, the development of epigenetic markers for detecting fetal DNA in maternal plasma has opened up new lines of research for noninvasive prenatal diagnosis. This development has allowed us to address each of the four challenges to the field described in the INTRODUCTION. In addition to diagnostic implications, this area of research may also yield new insights into the biology of development, especially those concerning the placenta.[40,41] It is expected that future technical advances in epigenetic analysis will further enhance the ease with which these developments could be used in a clinical setting.

ACKNOWLEDGMENTS

This work is supported by an Earmarked Research Grant from the Hong Kong Research Grants Council (CUHK 4279/04M).

REFERENCES

1. LO, Y.M.D. *et al.* 1997. Presence of fetal DNA in maternal plasma and serum. Lancet **350**: 485–487.
2. COSTA, J.M. *et al.* 2002. New strategy for prenatal diagnosis of X-linked disorders. N. Engl. J. Med. **346**: 1502.
3. LO, Y.M.D. *et al.* 1998. Prenatal diagnosis of fetal RhD status by molecular analysis of maternal plasma. N. Engl. J. Med. **339**: 1734–1738.
4. FINNING, K.M. *et al.* 2002. Prediction of fetal D status from maternal plasma: introduction of a new noninvasive fetal RHD genotyping service. Transfusion **42**: 1079–1085.
5. CHIU, R.W.K. *et al.* 2002. Prenatal exclusion of beta-thalassaemia major by examination of maternal plasma. Lancet **360**: 998–1000.
6. DING, C. *et al.* 2004. MS analysis of single-nucleotide differences in circulating nucleic acids: application to noninvasive prenatal diagnosis. Proc. Natl. Acad. Sci. USA **101**: 10762–10767.
7. LI, Y. *et al.* 2005. Detection of paternally inherited fetal point mutations for beta-thalassemia using size-fractionated cell-free DNA in maternal plasma. JAMA **293**: 843–849.
8. CHIU, R.W.K. *et al.* 2002. Noninvasive prenatal exclusion of congenital adrenal hyperplasia by maternal plasma analysis: a feasibility study. Clin. Chem. **48**: 778–780.
9. LO, Y.M.D. *et al.* 1999. Quantitative abnormalities of fetal DNA in maternal serum in preeclampsia. Clin. Chem. **45**: 184–188.
10. LEVINE, R.J. *et al.* 2004. Two-stage elevation of cell-free fetal DNA in maternal sera before onset of preeclampsia. Am. J. Obstet. Gynecol. **190**: 707–713.
11. LO, Y.M.D. *et al.* 1999. Increased fetal DNA concentrations in the plasma of pregnant women carrying fetuses with trisomy 21. Clin. Chem. **45**: 1747–1751.
12. ZHONG, X.Y. *et al.* 2000. Fetal DNA in maternal plasma is elevated in pregnancies with aneuploid fetuses. Prenat. Diagn. **20**: 795–798.
13. LEUNG, T.N. *et al.* 1998. Maternal plasma fetal DNA as a marker for preterm labour. Lancet **352**: 1904–1905.
14. FARINA, A. *et al.* 2005. High levels of fetal cell-free DNA in maternal serum: a risk factor for spontaneous preterm delivery. Am. J. Obstet. Gynecol. **193**: 421–425.

15. SEKIZAWA, A. *et al.* 2001. Cell-free fetal DNA is increased in plasma of women with hyperemesis gravidarum. Clin. Chem. **47**: 2164–2165.
16. SEKIZAWA, A. *et al.* 2002. Increased cell-free fetal DNA in plasma of two women with invasive placenta. Clin. Chem. **48**: 353–354.
17. VAN DER SCHOOT, C.E. *et al.* 2003. Real time PCR of bi-allelic insertion/deletion polymorphisms can serve as a reliable positive control for cell-free fetal DNA in non-invasive prenatal genotyping. Blood **102**: 93A.
18. BROJER, E. *et al.* 2005. Noninvasive determination of fetal RHD status by examination of cell-free DNA in maternal plasma. Transfusion **45**: 1473–1480.
19. LO, Y.M.D. *et al.* 1998. Quantitative analysis of fetal DNA in maternal plasma and serum: implications for noninvasive prenatal diagnosis. Am. J. Hum. Genet. **62**: 768–775.
20. FARINA, A. *et al.* 2003. Evaluation of cell-free fetal DNA as a second-trimester maternal serum marker of Down syndrome pregnancy. Clin. Chem. **49**: 239–242.
21. LEUNG, T.N. *et al.* 2001. Increased maternal plasma fetal DNA concentrations in women who eventually develop preeclampsia. Clin. Chem. **47**: 137–139.
22. LAIRD, P.W. 2005. Cancer epigenetics. Hum. Mol. Genet. **14**: **Spec No 1**: R65–R76.
23. CHEN, X.Q. *et al.* 1996. Microsatellite alterations in plasma DNA of small cell lung cancer patients. Nat. Med. **2**: 1033–1035.
24. WONG, I.H.N. *et al.* 1999. Detection of aberrant p16 methylation in the plasma and serum of liver cancer patients. Cancer Res. **59**: 71–73.
25. POON, L.L.M. *et al.* 2002. Differential DNA methylation between fetus and mother as a strategy for detecting fetal DNA in maternal plasma. Clin. Chem. **48**: 35–41.
26. AZHIKINA, T.L. & E.D. SVERDLOV. 2005. Study of tissue-specific CpG methylation of DNA in extended genomic loci. Biochemistry (Mosc). **70**: 596–603.
27. BIANCHI, D.W. 1998. Fetal DNA in maternal plasma: the plot thickens and the placental barrier thins. Am. J. Hum. Genet. **62**: 763–764.
28. NG, E.K.O. *et al.* 2003. mRNA of placental origin is readily detectable in maternal plasma. Proc. Natl. Acad. Sci. USA **100**: 4748–4753.
29. FLORI, E. *et al.* 2004. Circulating cell-free fetal DNA in maternal serum appears to originate from cyto- and syncytio-trophoblastic cells. Case report. Hum. Reprod. **19**: 723–724.
30. MASUZAKI, H. *et al.* 2004. Detection of cell free placental DNA in maternal plasma: direct evidence from three cases of confined placental mosaicism. J. Med. Genet. **41**: 289–292.
31. LUI, Y.Y.N. *et al.* 2002. Predominant hematopoietic origin of cell-free DNA in plasma and serum after sex-mismatched bone marrow transplantation. Clin. Chem. **48**: 421–427.
32. CHIM, S.S.C. *et al.* 2005. Detection of the placental epigenetic signature of the maspin gene in maternal plasma. Proc. Natl. Acad. Sci. USA **102**: 14753–14758.
33. FUTSCHER, B.W. *et al.* 2002. Role for DNA methylation in the control of cell type specific maspin expression. Nat. Genet. **31**: 175–179.
34. RAKYAN, V.K. *et al.* 2004. DNA methylation profiling of the human major histocompatibility complex: a pilot study for the human epigenome project. PLoS Biol. **2**: 2170–2182.
35. KUROMITSU, J. *et al.* 1997. A unique down regulation of h2-calponin gene expression in Down syndrome: a possible attenuation mechanism for fetal survival by methylation at the CpG island in the trisomic chromosome 21. Mol. Cell. Biol. **17**: 707–712.

36. FROMMER, M. *et al.* 1992. A genomic sequencing protocol that yields a positive display of 5-methylcytosine residues in individual DNA strands. Proc. Natl. Acad. Sci. USA **89**: 1827–1831.
37. HERMAN, J.G. *et al.* 1996. Methylation-specific PCR: a novel PCR assay for methylation status of CpG islands. Proc. Natl. Acad. Sci. USA **93**: 9821–9826.
38. LO, Y.M.D. *et al.* 1999. Quantitative analysis of aberrant p16 methylation using real-time quantitative methylation-specific polymerase chain reaction. Cancer Res. **59**: 3899–3903.
39. LIU, L. *et al.* 2004. Profiling DNA methylation by bisulfite genomic sequencing: problems and solutions. Methods Mol. Biol. **287**: 169–179.
40. FULKA, H. *et al.* 2004. DNA methylation pattern in human zygotes and developing embryos. Reproduction **128**: 703–708.
41. MIGEON, B.R. *et al.* 2005. Differential X reactivation in human placental cells: implications for reversal of X inactivation. Am. J. Hum. Genet. **77**: 355–364.

Cell-Free DNA in Maternal Plasma

Is It All a Question of Size?

YING LI,[a] WOLFGANG HOLZGREVE,[a] EDOARDO DI NARO,[b] ANGELOANTONIO VITUCCI,[c] AND SINUHE HAHN[a]

[a]*University Women's Hospital/Department of Research, Basel, Switzerland*

[b]*Department of Obstetrics and Gynecology, University of Bari, Bari, Italy*

[c]*Division of Haematology II, University of Bari, Bari, Italy*

ABSTRACT: Fetal cell-free DNA (cf-DNA) represents only a small fraction of the total cf-DNA in maternal plasma. This feature has rendered it difficult to reliably distinguish fetal alleles which are not very disparate from maternal ones, such as those involving point mutations, by conventional polymerase chain reaction (PCR)-based approaches. It has recently been shown that cell-free fetal DNA molecules have a smaller size than comparable cf-DNA molecules of maternal origin, and that this feature can be exploited for the selective enrichment of fetal DNA sequences, thereby permitting the detection of otherwise masked fetal genetic traits. By the use of this approach, we have shown that it is possible to detect fetal genetic loci for microsatellite markers, as well as point mutations involved in disorders such as achondroplasia and β-thalassemia.

KEYWORDS: circulatory DNA; prenatal diagnosis; electrophoresis; size fractionation

INTRODUCTION

The discovery of fetal cell-free DNA (cf-DNA) in maternal plasma and serum in 1997 opened up a new avenue for the noninvasive analysis of fetal genetic traits, especially for those fetal loci completely absent from the maternal genome, such as Y-chromosome-specific sequences, or the fetal RHD gene in rhesus-D-negative pregnant women.[1-3] These analyses have proven to be so reliable that they are already implemented clinically in several European centers.[4,5]

The analysis of fetal genetic loci less disparate from maternal ones (e.g., point mutations) has been less successful, and has been limited to a small

Address for correspondence: Sinuhe Hahn, Laboratory for Prenatal Medicine, Department of Research, University Women's Hospital, Spitalstrasse 21, CH 4031 Basel, Switzerland. Voice: +41-61-265-9224; fax: +41-61-265-9399.

e-mail: shahn@uhbs.ch

Ann. N.Y. Acad. Sci. 1075: 81–87 (2006).
doi: 10.1196/annals.1368.010

number of case reports.[6] This is largely due to the preponderance of cf-DNA sequences of maternal origin in the maternal circulation, rendering the detection of more subtle fetal differences difficult when using conventional or real-time polymerase chain reaction (PCR)-based approaches.

To date few studies have attempted to examine the biophysical properties of cell-free fetal DNA, but rather have focused on the immediate clinical applications. In order to address this issue, we decided to examine this aspect in greater detail, in the hope that this would lead to the discovery of the differences between fetal and maternal cf-DNA molecules, which could be used to distinguish between them. Previous examinations using quantitative real-time PCR had shown that cell-free fetal DNA was more abundant than that of circulating fetal cells by a factor of almost 1,000.[7] It was also demonstrated that it is unlikely that cell-free fetal DNA stems from the demise of fetal cells that have crossed the placental barrier, such as erythroblasts, in that elevations in cell-free fetal DNA were found to occur in the absence of any elevated transplacental fetal cell trafficking.[8] Furthermore, cell-free fetal DNA was shown to have a very short half-life (<15 min), disappearing very rapidly following delivery, conditions under which fetal cells can persist for months or even decades.[7]

These various findings have led to the suggestion that cell-free fetal DNA may be of placental origin.[9] This suspicion has been confirmed by three lines of separate evidence:

(i) The absence of cell-free fetal DNA sequences in cases of placental mosaicism.
(ii) The presence of cell-free fetal DNA in instances where part of the placenta was retained post partum.
(iii) The quantitative detection of cell-free fetal DNA sequences with placenta-specific methylation patterns.

On the other hand, the vast majority of normal cf-DNA in plasma has been proposed to be derived from the hemopoietic system, presumably from the red-cell lineage as erythroblasts enucleate to mature into erythrocytes.[3] Although this process usually involves a tight interaction with macrophages which engulf and degrade the erythroblast nuclei, enucleation can occur in the absence of such interactions. The release of cf-DNA by the placenta is likely to involve apoptotic shedding of syncytiotrophoblast debris as this monolayer is constantly replenished.[9] Hence, because fundamental differences may exist between cf-DNA released into the circulation by these two different mechanisms, it is possible that physical differences may exist between fetal and maternal cf-DNA species that may aid in distinguishing between them.

Fetal cf-DNA is Smaller than Maternal cf-DNA

In our approach to look for biophysical differences between fetal and maternal cf-DNA, we were curious to determine whether these possessed any apoptotic characteristics.[10] For this reason we separated cf-DNA isolated from maternal plasma by standard gel electrophoresis and examined it by Southern blotting using high-copy Alu and DNA Y-chromosome segment (DYS) 14 sequences. Although the Alu sequence probe was able to pick up characteristic apoptotic fragments, differing in size by approximately 180 bp (base pairs), we were not able to detect any male fetal DNA with the DYS 14 probe, indicating that the level of cell-free fetal DNA was too low to be detectable by such rather archaic approaches. Of interest is that the Alu probe indicated that cf-DNA sequences were present having a size in excess of 24 kb. Because our recent analysis of nuclear DNA in terminal differentiating erythroblasts indicates that this is cleaved into megabase fragments (>40 Mb), it is possible that the large molecular weight cf-DNA we have detected is derived from enucleating erythroblasts (Hristoskova *et al.*, unpublished observations).

Because our approach using Southern blotting was not sensitive enough to detect cell-free fetal DNA, we next looked for the presence of such DNA molecules using real-time PCR. For this analysis, total cf-DNA from maternal plasma was again subjected to gel electrophoresis, following which each lane of the gel-containing DNA was carefully sliced into discrete fragments, having sizes of less than 0.3 kb, 0.3–0.5 kb, 0.5–1.0 kb, 1.0–1.5 kb, 1.5–23 kb, and larger than 23 kb. The cf-DNA in each agarose fragment was eluted and the quantity of total cf-DNA (indicative of that of maternal origin) was quantified by an assay for the GAPDH gene, whereas the amount of fetal cf-DNA was determined using an assay specific for the SRY locus on the Y-chromosome. This examination indicated that the vast proportion of fetal cf-DNA (approximately 70%) had a molecular size of less than 300 bp, whereas most of the maternal cf-DNA (approximately 75%) was greater than 300 bp.

Of interest is that our data obtained using gel electrophoresis is remarkably similar to those obtained independently by Dennis Lo and his coworkers, who used PCR amplicons of different sizes for their analyses.[11] These new facets concerning the biophysical properties of fetal and maternal cf-DNA implied that fetal cf-DNA molecules in maternal plasma can be enriched for on the basis of a smaller size than comparable maternal cf-DNA molecules.

Detection of Highly Polymorphic Microsatellite Sequences

To test this hypothesis we chose to examine a series of highly polymorphic microsatellite loci, also termed STRs (short tandem repeats).[10] The reason for choosing these markers is that they can be used to readily distinguish between fetus and mother in a gender-independent manner. Once we had genotyped both maternal and fetal DNA samples for informative markers which could aid in distinguishing fetal and maternal cf-DNA, we examined a series of maternal plasma samples taken early and late in pregnancy. The analysis of total cf-DNA in maternal plasma indicated that only the maternal STR alleles could

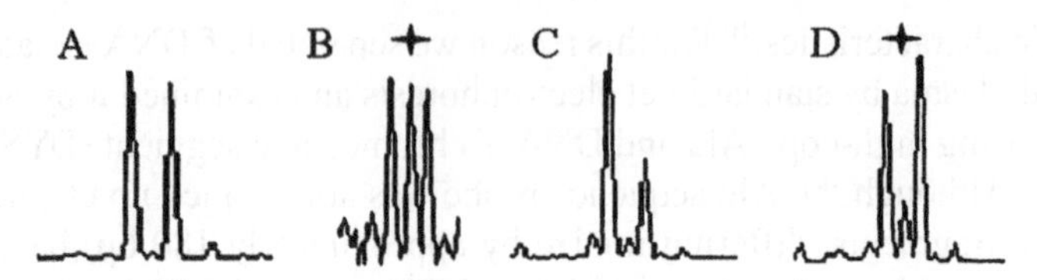

FIGURE 1. Size fraction of cf-DNA in maternal plasma permits the detection of a paternally inherited STR allele (indicated by a *star*) in a fetus with trisomy 21. (**A**) Maternal genomic DNA. (**B**) Fetal genomic DNA, with presence of an additional allele (*star*). (**C**) Total circulatory DNA: inability to detect additional fetal allele. (**D**) Size-fractionated circulatory DNA: ready detection of additional fetal allele (*star*).

be reliably detected, whereas in most cases the paternally inherited fetal STR allele is totally absent (FIG. 1). This is not unexpected, as the same sequences are used to PCR amplify both the maternal and fetal STR alleles. As there is a greater abundance of the maternal STR alleles, the fetal ones are effectively competed out of the PCR reaction. However, when the cf-DNA fraction, which had a size of less than 300 bp was taken, the paternally inherited fetal STR allele was readily detected in all instances (FIG. 1). Indeed, in many cases it appeared as if the maternally inherited fetal STR allele was present in greater abundance than the noninherited maternal allele (FIG. 1).

Therefore, this experiment conducted on a total of 7 samples using a number of different STR markers on chromosome 21 indicated that the size fractionation of cf-DNA from maternal plasma did permit the ready detection of fetal genetic loci which were not apparent in the analysis of the total cf-DNA sample.[10]

Detection of Fetal Point Mutations: Achrondroplasia and β-Thalassemia

Confident that our approach may permit the detection of even more subtle differences between the maternal and fetal genome, we next set out to determine whether we could detect paternally inherited fetal point mutations. In the first instance we examined a sample obtained from a pregnancy at risk for achondroplasia. In this case the father had a dominant mutation in the FGFR3 (fibroblast growth factor receptor 3) gene, which can be detected by Scfl restriction analysis of the PCR amplification product. Our analysis of total cf-DNA indicated that the paternal mutant allele could be faintly detected.[12] The presence of this mutant allele could, however, be much more robustly demonstrated in the cf-DNA fraction having a size of less than 300 bp.[12]

For our next examination we focused our attention on the hemoglobinopathies, probably the largest kind of hereditary genetic disorder in man, with several hundred thousand lethally affected pregnancies occurring yearly.[13]

We have a long-standing interest in the prenatal diagnosis of β-thalassemia, having previously collaborated with investigators at the University in Bari, Italy, on this topic.[14] Because a large number of mutations are involved in this disorder, pregnancies which are not at risk of being affected can be excluded from further invasive prenatal diagnostic analysis if the paternal mutant allele is not inherited by the fetus. For this reason we chose a study cohort of pregnancies at risk for β-thalassemia where the parents do not share the same β-globin mutation.[15]

Although we had only analyzed a single case of achondroplasia, this served to illustrate the weakness of conventional PCR-based assays for the accurate detection of fetal point mutations, even when using size-fractionated cf-DNA. Furthermore, as the hemoglobinopathies are a major health concern in many developing countries, we were interested in a detection system that was not overly expensive, and yet was amenable to high throughput analysis. For this reason we chose to use an allele-specific real-time PCR-based approach, which has been shown to be reliable and robust for the analysis of SNPs on pure genomic DNA. Because we were, however, concerned that the quantity of fetal cf-DNA may still be less than that of maternal origin in our size-fractionated cf-DNA sample, we decided to introduce a further precautionary step to enrich for the mutant fetal allele. This was accomplished by first performing a round of PCR using a PNA (peptide nucleic acid) probe specific for the wild-type maternal allele. As this PNA probe has very high affinity for the wild-type maternal allele, its addition to the PCR reaction would effectively inhibit the efficient amplification of this allele, and permit the amplification of the mutant fetal allele. To detect the presence of the mutant fetal allele we used an allele-specific real-time PCR assay with Cybergreen for the detection. In order to ensure that we were truly detecting the mutant fetal allele and not some erroneous amplification of the wild-type allele, we employed a ΔC_T system, whereby the extent of amplification of the wild-type allele was subtracted from that of the mutant allele. In experimental conditions, wherein the mutant allele was diluted into a wild-type background, this system was shown to permit a clear discrimination between the two alleles, and furthermore yield clear cut-off values that could be used for the assessment of their presence. In our analysis of 32 clinical samples taken early in gestation immediately prior to chorionic villus sampling (median gestational age 10.7 weeks), we looked for the presence of the four most common β-globin point mutations in the population group in Bari, southern Italy: IVSI-1, IVSI-6, IVSI-110, and codon 39.[15] One of the samples in our analysis had to be excluded because we were not able to ascertain the fetal genotype from the invasive prenatal diagnosis. A further two samples had to be excluded since the amount of cf-DNA in these samples was very low. It is possible that this facet may have contributed to the one false analysis we recorded. In all other 29 instances, the absence or presence of the paternal mutant allele was correctly determined, yielding a sensitivity of 100% with a concomitant specificity of almost 94%.[15]

Current Drawbacks and Future Directions

The major current bottleneck and technical deficit of the approach we have established is the actual size fractionation of the cf-DNA, as this currently relies on conventional gel electrophoresis, the cutting of a gel slice with the desired molecular weight DNA, and the elution of the cf-DNA from this agarose fragment. These steps are tedious, time-consuming, and prone to contamination. As such, it should be noted that the approach we have developed is currently more an issue of testing a "proof-of-principle," than being a straightforward application that is so secure that it is ready for the clinic.

It is patently clear that an effective method is needed for the efficacious and robust size fractionation of cf-DNA, which is also cost-effective and amenable to high throughput. Unfortunately such a system does not exist at the moment. Currently, we are exploring the use of centrifugal filtration devices (Amicon Milian SA, Geneva, Switzerland), capillary electrophoresis, and microfluidics systems. The latter is being conducted as part of the exploration or "new technologies" in the EU-funded Special Non-Invasive Advances in Fetal and Neonatal Evaluation (SAFE) project.

A further issue is that current analytical approaches do not permit a reliable quantitation of the amount of mutant fetal allele present. Nor is it possible to determine whether the fetus has inherited the mutant allele from the mother as well. Consequently, the assays as they currently stand can only be used for the exclusion of pregnancies at risk, and not for an actual prenatal diagnosis.

CONCLUSIONS

Two independent analyses showed that fetal cf-DNA molecules in maternal plasma are smaller than comparable maternal cf-DNA molecules. This finding suggested that fetal cf-DNA molecules could be enriched on account of a smaller size than maternal ones. In a series of proof-of-principle experiments we showed that this hypothesis was indeed true, and that this approach could be used to detect a number of fetal genetic loci, including STRs and point mutations, which cannot be reliably interrogated on total cf-DNA in maternal plasma.

Despite this exciting development, several hurdles still need to be cleared before this system can be applied to clinical settings, including mechanisms permitting the efficacious size fractionation of cf-DNA and the robust quantitative analysis of the fetal alleles present in this sample. This latter step is necessary if we are to ascertain whether the fetus has inherited both the paternal and maternal mutant alleles, thereby changing the analysis from one of simply excluding pregnancies at risk on the basis of presence or absence of the mutant paternal allele, to one of an actual prenatal diagnosis. This development will also assist in determining whether cf-DNA can be used for the detection of fetal aneuploidies, the long sought-after holy grail of noninvasive prenatal diagnosis.

REFERENCES

1. Lo, Y.M. *et al.* 1997. Presence of fetal DNA in maternal plasma and serum. Lancet **350:** 485–487.
2. Hahn, S. & W. Holzgreve. 2002. Prenatal diagnosis using fetal cells and cell-free fetal DNA in maternal blood: what is currently feasible? Clin. Obstet. Gynecol. **45:** 649–656.
3. Chiu, R.W. & Y.M. Lo. 2004. The biology and diagnostic applications of fetal DNA and RNA in maternal plasma. Curr. Top. Dev. Biol. **61:** 81–111.
4. Finning, K.M. *et al.* 2002. Prediction of fetal D status from maternal plasma: introduction of a new non-invasive fetal RHD genotyping service. Transfusion **42:** 1079–1085.
5. Rijnders, R.J. *et al.* 2004. Clinical applications of cell-free fetal DNA from maternal plasma. Obstet. Gynecol. **103:** 157–164.
6. Saito, H. *et al.* 2000. Prenatal DNA diagnosis of a single-gene disorder from maternal plasma. Lancet **356:** 1170.
7. Lo, Y.M. *et al.* 1998. Quantitative analysis of fetal DNA in maternal plasma and serum: implications for non-invasive prenatal diagnosis. Am. J. Hum. Genet. **62:** 768–775.
8. Zhong, X.Y. *et al.* 2002. Cell-free fetal DNA in the maternal circulation does not stem from the transplacental passage of fetal erythroblasts. Mol. Hum. Reprod. **8:** 864–870.
9. Hahn, S. *et al.* 2005. Fetal cells and cell free fetal nucleic acids in maternal blood: new tools to study abnormal placentation? Placenta **26:** 515–526.
10. Li, Y. *et al.* 2004. Size separation of circulatory DNA in maternal plasma permits ready detection of fetal DNA polymorphisms. Clin. Chem. **50:** 1002–1011.
11. Chan, K.C. *et al.* 2004. Size distributions of maternal and fetal DNA in maternal plasma. Clin. Chem. **50:** 88–92.
12. Li, Y. *et al.* 2004. Improved prenatal detection of a fetal point mutation for achondroplasia by the use of size-fractionated circulatory DNA in maternal plasma-case report. Prenat. Diagn. **24:** 896–898.
13. Weatherall, D.J. 2004. Thalassaemia: the long road from bedside to genome. Nat. Rev. Genet. **5:** 625–631.
14. Di Naro, E. *et al.* 2000. Prenatal diagnosis of beta-thalassaemia using fetal erythroblasts enriched from maternal blood by a novel gradient. Mol. Hum. Reprod. **6:** 571–574.
15. Li, Y. *et al.* 2005. Detection of paternally inherited fetal point mutations for beta-thalassemia using size-fractionated cell-free DNA in maternal plasma. JAMA **293:** 843–849.

Fetal Blood Group Genotyping

Present and Future

GEOFF DANIELS, KIRSTIN FINNING, PETE MARTIN, AND JO SUMMERS

International Blood Group Reference Laboratory, National Blood Service, Bristol, United Kingdom

ABSTRACT: Prediction of fetal blood group from DNA is usually performed when the mother has antibodies to RhD, to assess whether the fetus is at risk from hemolytic disease of the fetus and newborn (HDFN). Over the last five years RhD testing on fetal DNA in maternal plasma has been introduced. At the International Blood Group Reference Laboratory (IBGRL) we employ real-time quantitative polymerase chain reaction (RQ-PCR) to detect *RHD* exons 4, 5, and 10, which also reveals *RHDψ*. *SRY* and, in RhD-negative (RhD–) females, eight biallelic polymorphisms are incorporated in an attempt to provide an internal positive control. Since 2000 we have tested 533 pregnancies for RhD. In 327 pregnancies where the RhD of the infant is known, we had one false-positive and one false-negative result. In 2004 we introduced fetal typing from DNA in maternal plasma for K, Rhc, and RhE, which represent single nucleotide polymorphisms (SNPs) on the *KEL* and *RHCE* genes.

We have begun trials on an automated method for fetal RhD typing from DNA in maternal plasma. This is designed to test fetal RhD in all pregnant RhD– women, to identify the 40% with an RhD– fetus so that antenatal RhD immunoglobulin (Ig) prophylaxis can be avoided. Similar trials have already been reported by Sanquin Research Laboratories in Amsterdam.

KEYWORDS: blood groups; Rh; RhD; Kell; hemolytic disease of the fetus and newborn; free fetal DNA in maternal plasma

INTRODUCTION

Despite the success of anti-RhD Ig prophylaxis in substantially reducing the prevalence of hemolytic disease of the fetus and newborn (HDFN) caused by anti-RhD, some RhD-negative (RhD–) women still produce anti-RhD. In the United Kingdom alone about 500 fetuses develop HDFN each year, leading to 25–30 deaths.[1] After anti-RhD, the antibodies that cause

Author for correspondence: Dr. Geoff Daniels, International Blood Group of Reference Laboratory, National Blood Service, Southmead Road, Bristol, BS10 5ND, United Kingdom. Voice: +44-(0)117-991-2116; fax: +44-(0)117-959-1660.

e-mail: geoff.daniels@nbs.nhs.uk

Ann. N.Y. Acad. Sci. 1075: 88–95 (2006). © 2006 New York Academy of Sciences.
doi: 10.1196/annals.1368.011

the most cases of severe HDFN are anti-K (KEL1) of the Kell system and anti-Rhc.[2]

The most common use of molecular blood group genotyping is for predicting the RhD type of the fetus of a woman with anti-RhD to assist in assessing the risk of HDFN. If the fetus is RhD-positive (RhD+), it is at risk from HDFN and the appropriate management of the pregnancy can be arranged; if it is RhD–, there is no risk and unnecessary invasive procedures can be avoided. Although this sort of test is usually done for RhD, it is occasionally required for Rhc, RhE, and K, and rarely for other groups, such as Fy^{a}. Prior to 2001, fetal blood group genotyping was usually performed on DNA isolated from amniocytes and occasionally from chorionic villi (CV). Now we are able to predict RhD, Rhc, RhE, and K phenotypes from free fetal DNA in maternal plasma, avoiding the invasive procedures of amniocentesis and CV sampling.

Genetics of the Rh Blood Group System

Rh phenotypes are controlled by *RHD* and *RHCE*, a pair of homologous genes on chromosome 1, which have 10 exons each and share about 94% identity.[2] In people of European origin, the RhD– phenotype almost always results from homozygosity for a complete deletion of *RHD*. Most RhD+ people are either homozygous or hemizygous for *RHD* (FIG 1). The basic test for predicting RhD phenotype from DNA is polymerase chain reaction (PCR) amplification of a fragment of *RHD* to determine whether the gene is present. In the design of PCR primers care must be taken to ensure that only *RHD*, and not *RHCE*, is amplified.

Rh is a very complex blood group system with numerous variants.[2] Variant haplotypes exist in which all or part of *RHD* is present, but no RhD antigen is expressed on the red cells. In partial RhD phenotypes some or most of *RHD* is missing, yet some RhD epitopes are expressed. Fortunately, most of these *RHD* variants are rare, but there is one, *RHDΨ*, that is particularly important because it is common in people of African origin; approximately 66% of RhD– Black Africans and 24% of RhD– African Americans have *RHDΨ*.[3] *RHDΨ* contains all 10 exons, but is inactive because of two inactivating mutations, a 37-bp duplication in exon 4 and a nonsense mutation (Y269X) in exon 6 (FIG 1). There are also four characteristic single nucleotide polymorphisms (SNPs) in exons 4 and 5. Another abnormal gene that is relatively common in Africans and produces no RhD antigen, despite the presence of some *RHD* exons, is *RHD–CE–D^{S}*. This hybrid gene comprises exons 1 and 2, the 5′ end of exon 3, and exons 9 and 10 from *RHD*, and the 3′ end of exon 3 and exons 4–8 from *RHCE*.[4]

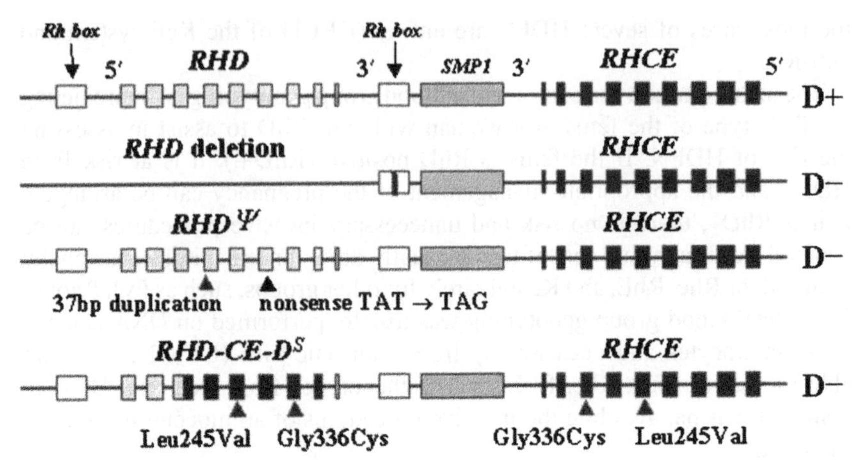

FIGURE 1. Four Rh haplotypes, showing *RHD* and *RHCE* in opposite orientation, the *Rh boxes*, regions of high identity, flanking *RHD*, and the gene *SMP1* between the two Rh genes. At the top is the common haplotype that produces RhD; the other three produce no RhD.

Prediction of RhD Phenotype from Fetal DNA in Maternal Plasma

The method we currently use routinely at the International Blood Group Reference Laboratory (IBGRL) for detecting fetal *RHD* in DNA isolated from plasma of pregnant RhD– women is a modification of the method described by Finning *et al.*[5,6] It involves amplification of *RHD* exons 4, 5, and 10, with exons 4 and 10 as a multiplex. The reactions for exons 4 and 5 do not amplify *RHD*Ψ, whereas those for exon 10 do, signaling its presence. Like most laboratories carrying out this sort of test, we employ real-time quantitative polymerase chain reaction (RQ-PCR), with Taqman technology, in which the quantity of product is measured at every cycle. Its advantages are that it is highly sensitive, it is quantitative, so we can be sure we are amplifying fetal and not maternal DNA, and the reaction takes place and is analyzed in a closed system, reducing the risk of contamination.

As only 3–6% of the DNA isolated from maternal plasma is of fetal origin, a complication of tests for detecting fetal *RHD* in plasma of RhD– pregnant women is to provide a suitable internal control reaction that amplifies fetal DNA, but not maternal DNA. A test for the Y-borne gene *SRY* is included in the initial test and provides an internal control if the fetus is male. When a fetus appears to be RhD– and female, eight biallelic insertion or deletion polymorphisms[7] are used in an attempt to detect a fetal gene of paternal origin that is not present in the mother. Initially, DNA isolated from maternal buffy coat is tested for all eight polymorphisms, and then the DNA isolated from the maternal plasma is tested for those sequences that are not present in the maternal genome. In 80% of cases at least one of these will provide a suitable control. Having carried out all of these tests, only about 4% of samples would be expected to have given an uncontrolled RhD– result. These are then reported to the clinician with an explanation that the test could not be effectively controlled.

Since IBGRL introduced noninvasive fetal RhD testing in 2001, we have tested DNA from 533 RhD– pregnant women. In 327 of these we have been able to obtain information on the RhD type of the baby after delivery. The correct result was obtained in all but two cases, one false-positive and one false-negative, and changes have subsequently been made to improve the accuracy. Although we cannot guarantee a 100% accurate result, most clinicians consider the small chance of obtaining a false result as preferable to the risks involved in amniocentesis. Referrals for RhD testing on fetal DNA in maternal plasma have now almost completely replaced those for tests on fetal DNA obtained by amniocentesis or CV sampling.

Fetal Genotyping for Rhc, RhE, and K

After RhD, the two Rh antibodies that most commonly cause HDFN are anti-Rhc and anti-RhE. The c/C Rh polymorphism is primarily governed by a 307C>T (P103S) SNP in *RHCE* exon 2 and the e/E polymorphism by 676G>C (A226P) in *RHCE* exon 5, though expression of the antigens may be affected to some degree by other SNPs. Anti-K of the Kell blood group system is also a major cause of HDFN. The k/K polymorphism is encoded by a 698C>T (T193M) SNP in exon 6 of *KEL*. Finning *et al.*[8] have developed methods for Rhc, RhE, and K fetal testing on fetal DNA from maternal plasma. Development of these tests was potentially more challenging than RhD testing, because Rhc, RhE, and K represent SNPs rather than deletion of a whole gene. These tests are carried out by RQ-PCR and Taqman technology, each test employing an allele-specific primer.

TABLES 1–3 show the results of fetal Rhc, RhE, and K testing in a clinical setting. Of the 35 results of Rhc testing in which we have been informed of the true phenotype obtained by serological tests on red cells of the baby after delivery, 29 were Rhc+, 5 Rhc–, and there was 1 false-positive result. In one case no result was obtained. For RhE, 39 results have been confirmed,18 RhE+, 20 RhE–, and 1 incorrect result, a false-negative. For K, 68 results have been confirmed: 20 K+, 46 K–, and 2 incorrect results, both false-negatives, although a subsequent modification to the technique would have eliminated those errors. In six cases no result was obtained.

The specificity of the K test proved to be a problem, with mispriming on the *k* allele. This was alleviated by incorporating locked nucleic acids (LNAs) into the allele-specific forward primer. LNAs are nucleic acid analogues with a 2′O–4′C methylene bridge, which restricts the flexibility of the ribofuranose ring and locks the structure into a rigid bicyclic formation. Oligonucleotides

TABLE 1. Results of Rhc testing on fetal DNA in maternal plasma

	No. tested	
Result	Total	Result confirmed
Rhc+	59	29
Rhc–	9	5
False-positive	1	1
False-negative	0	0
No result	1	
Total	70	35

containing LNAs have an exceptionally high affinity to complementary DNA strands and excellent mismatch discrimination.[9] Both the 3′allele-specific nucleotide and a mismatch nucleotide at the third position from the 3′end contained LNAs. This reduced the sensitivity of the test, but enhanced its specificity.

Although we are still validating the noninvasive c, E, and K testing at IBGRL, we carry out this testing as a service and clinicians use our results as additional information to that obtained by other means as to the likelihood that the fetus has HDFN. During the period April 2004–April 2005 we carried out tests on 35 pregnancies for Rhc, 28 for RhE, and 53 for K. About 80% of these were just from the United Kingdom, demonstrating that there is a clear demand for these services.

Workshops and External Quality Assurance

As fetal blood group genotyping is becoming more widely used for clinical diagnosis, it is important that it is properly regulated. One important aspect of regulation must be through external quality assurance schemes. A European Network of Excellence on Special Noninvasive Advances in Fetal and Neonatal Evaluation (SAFE), funded by the European Union, provides money for research and for workshops, so that there is more communication among workers in the field (http://www.safenoe.org).[10]

TABLE 2. Results of RhE testing on fetal DNA in maternal plasma

	No. tested	
Result	Total	Result confirmed
RhE+	28	18
RhE–	26	20
False-positive	0	0
False-negative	1	1
No result	0	
Total	65	39

TABLE 3. Results of K testing on fetal DNA in maternal plasma

	No. tested	
Result	Total	Result confirmed
K+	29	20
K–	105	46
False-positive	0	0
False-negative	2	2
No result	6	
Total	142	68

In 2004, the International Society of Blood Transfusion (ISBT) and the International Council for Standardization in Hematology (ICSH) organized a workshop on molecular blood group genotyping, which culminated in a feedback meeting and was the precursor of an international quality assurance scheme.[11] This workshop included an exercise in which samples representing plasma from two pregnant women were sent to 24 laboratories. For ethical and logistic reasons mixtures of adult plasma were used to represent one hemizygous RhD+ male fetus and one RhD– female fetus. Fifteen laboratories returned results (13 from Europe, 1 from Canada, and 1 from Brazil) and all obtained the correct result for RhD type. Eight laboratories tested for *SRY* as an internal control and all obtained the correct result; the other seven did not include an internal control. Eleven of the laboratories employed RQ-PCR and four used conventional PCR. The organizers felt that the testing might not have been so successful had real plasma samples from pregnant women been used. A matter of some concern is that six of the laboratories declared that they would be prepared to report the RhD– result to the clinic despite the absence of any positive internal control. Consequently, a recommendation from the workshop was that when no paternally derived fetal marker has been detected in tests on maternal plasma, a fetal RhD– result should be reported with a caveat that it had not been possible to include all appropriate controls.

Another workshop will take place in 2006 with a feedback meeting at the ISBT Congress in Cape Town in September 2006 (http://isbt-web.org/capetown/). This time the organizers plan to distribute plasma samples from RhD– pregnant women. Information about the workshops can be found at http://www.blood.co.uk/research.

Future Developments

Because the provision of Anti-RhD Ig prophylaxis to RhD– women who deliver an RhD+ baby has greatly reduced the prevalence of anti-RhD immunization, a major cause of RhD immunization has become feto-maternal hemorrhage during the pregnancy. To prevent this, it is now common policy to give anti-RhD Ig to RhD– pregnant women antenatally. In the United Kingdom, two doses are given, at 28 and 34 weeks.[1] This treatment must be given to all RhD– pregnant women as the RhD group of their fetus is not known, yet in a European population about 40% of RhD– pregnant women have an RhD– fetus. Consequently, a simple, noninvasive, fully automated method of predicting fetal RhD phenotype from maternal plasma would reduce wastage of anti-RhD Ig, making the test cost-effective. Van der Schoot (personal communication) has estimated that the total cost of this assay could be around one half of the cost of antenatal immunoprophylaxis. More importantly, however, it would save patients from receiving unnecessary therapy with blood products.

At IBGRL, we have begun trials of screening for fetal *RHD* in plasma of RhD– pregnant women (in collaboration with the Birmingham Blood Centre, United Kingdom), using robotic DNA preparation and RQ-PCR. The test involves detection of *RHD* exons 5 and 7, with exon 5 discriminating against detection of *RHD*Ψ. Sanquin Research at CLB, Amsterdam, also using RQ-PCR, has already carried out trials on more than 1,200 pregnancies and reported greater than 99% concordance with the serological results of the baby.[12]

We predict that by 2010 in many countries it will be the standard of care to genotype the fetus for the appropriate blood group in all women with anti-RhD, anti-Rhc, or anti-K; or at least in those who have a heterozygous partner. In addition, at least in a few countries, we predict that by 2010 fetal RhD type will be determined for all pregnant RhD– women to assess whether they require antenatal Ig prophylaxis.

RQ-PCR is currently the method of choice for detecting blood group polymorphisms in free fetal DNA in maternal plasma. This technology, however, is not ideal and will probably be replaced. At the moment, the most likely candidate appears to be PCR analysis by mass spectrometry with time-of-flight measurement.

REFERENCES

1. National Institute for Clinical Excellence. 2002. Technology appraisal guidance 41. Guidance on the use of routine antenatal anti-D prophylaxis for RhD-negative women. NICE, London.
2. Daniels, G. 2002. Human blood groups, 2nd edition. Blackwell Science. Oxford.
3. Singleton, B. K. *et al.* 2000. The presence of an *RHD* pseudogene containing a 37 base pair duplication and a nonsense mutation in most Africans with the Rh D-negative blood group phenotype. Blood **95**: 12–18.
4. Faas, B. H. W. *et al.* 1997. Molecular background of VS and weak C expression in Blacks. Transfusion **37**: 38–44.
5. Finning, K. M. *et al.* 2002. Prediction of fetal D status from maternal plasma: introduction of a new non-invasive fetal *RHD* genotyping service. Transfusion **42**: 1079–1085.
6. Finning, K. *et al.* 2004. A clinical service in the UK to predict fetal Rh (Rhesus) D blood group using free fetal DNA in maternal plasma. Ann. N. Y. Acad. Sci. **1022**: 119–123.

7. ALIZADEH, M. *et al.* 2002. Quantitative assessment of hematopoietic chimerism after bone marrow transplantation by real-time quantitative polymerase chain reaction. Blood **99**: 4618–4625.
8. FINNING, K. *et al.* 2005. Prediction of fetal K, c and E blood groups from free fetal DNA in maternal plasma [abstract]. Transfus. Med. **15** (Suppl. 1): 31.
9. PETERSEN, M. & J. WENGEL. 2003. LNA: a versatile tool for therapeutics and genomics. Trends Biotechnol. **21**: 74–81.
10. HULTEN, M. 2005. The SAFE network of excellence.(abstract). Clin. Chem. **51**: 7.
11. DANIELS, G., & C.E. VAN DER SCHOOT. *et al.* 2005. Report of the first international workshop on molecular blood group genotyping. Vox. Sang. **88**: 136–142.
12. VAN DER SCHOOT, E. *et al.* 2005. Antenatal prophylaxis can be given based on non-invasive fetal RHD-genotyping in all D-negative pregnant women. [abstract]. Clin. Chem. **51**: 8.

Placental RNA in Maternal Plasma

Toward Noninvasive Fetal Gene Expression Profiling

NANCY B. Y. TSUI AND Y. M. DENNIS LO

Department of Chemical Pathology, The Chinese University of Hong Kong, Prince of Wales Hospital, Hong Kong, China

ABSTRACT: The recent demonstration of the presence of placenta-derived fetal RNA in maternal plasma has opened up new opportunities for noninvasive prenatal investigation. Circulating fetal RNA analysis could in principle be applied to all pregnancies without the limitations by fetal gender or polymorphisms between the mother and fetus. The use of fetus- or disease-specific circulating RNA markers would greatly increase the number of markers that can be used for prenatal monitoring. With the recent advances in microarray technology, new placenta-derived plasma RNA markers could be rapidly identified. These newly identified placental transcripts were robustly detectable in maternal plasma and were pregnancy-specific. Remarkably, the relative concentrations of the placental mRNA in maternal plasma directly reflect the gene expression patterns in the placenta, an observation which suggests that placental gene expression level plays a predominant role in determining the placental mRNA concentrations in maternal plasma. Thus, fetal mRNA measurement in maternal plasma may be a useful tool for noninvasive prenatal placental gene expressing profiling.

KEYWORDS: plasma RNA; circulating placental RNA; noninvasive prenatal diagnosis

INTRODUCTION

Prompted by the promising development in the use of circulating fetal DNA, cell-free fetal RNA in maternal plasma has recently been considered as another useful fetal genetic material for noninvasive prenatal diagnosis and monitoring. Because of its fetal-specificity, circulating fetal RNA markers could potentially be applied to all pregnancies without the limitations of

Author for correspondence: Y. M. Dennis Lo, Department of Chemical Pathology, The Chinese University of Hong Kong, Prince of Wales Hospital, Room 38023, 1/F Clinical Sciences Building, 30–32 Ngan Shing Street, Shatin, New Territories, Hong Kong Special Administrative Region, China. Voice: +852-2632-2963; fax: +852-2636-5090.
e-mail: loym@cuhk.edu.hk

Ann. N.Y. Acad. Sci. 1075: 96–102 (2006).
doi: 10.1196/annals.1368.012

fetal gender or polymorphisms between the mother and fetus. In addition, circulating fetal RNA could provide valuable information on the gene expression of the fetus. In this review, some recent advancements in this field are discussed.

Detection of Circulating Fetal RNA

Circulating fetal RNA was discovered in 2000, when Poon *et al.* reported the presence of *Y-chromosome-specific zinc finger protein* mRNA in plasma of women carrying male fetuses.[1] Afterward, by the development of a robust and sensitive protocol for extracting and quantifying circulating RNA, Ng *et al.* have successfully quantified two placenta-derived mRNAs, namely mRNA coding for *human placental lactogen (hPL)*, and the β-*subunit of human chorionic gonadotropin (*β-*hCG)*, in maternal plasma.[2] In the study of Ng *et al.*,[2] both *hPL* and β-*hCG* mRNA transcripts were readily detectable in maternal plasma. A correlation between the plasma mRNA levels of *hPL* and β-*hCG* and their corresponding protein levels at various gestational ages was demonstrated. In addition, *hPL* mRNA was rapidly cleared from the maternal plasma after delivery, showing that the placenta is an important organ for releasing fetal RNA into maternal plasma.

Clinical Significance

Plasma *CRH* mRNA has been shown to be a potential marker for studying preeclampsia. Ng *et al.* have shown that in pregnant women with the complication of pre-eclampsia, their median maternal plasma *CRH* mRNA level was 10.5 times higher than nonpre-eclamptic pregnant women.[3] In comparison, a fivefold increase in fetal DNA concentration has been demonstrated in the plasma of pre-eclamptic pregnant women.[4] In a more recent study, the relationship between circulating *CRH* mRNA levels and severity of pre-eclampsia has been investigated.[5] Although no significance difference was reached, the plasma *CRH* mRNA concentration has been shown to be positively related to the severity of pre-eclampsia.[5]

Basing their work on the encouraging finding that circulating β-*hCG* mRNA was readily detectable in first-trimester maternal plasma samples of normal pregnant women,[2] Ng *et al.* have investigated its clinical usefulness in fetal chromosomal aneuploidy.[6] In comparison with normal euoploid pregnancies, the authors have shown that the β-*hCG* mRNA concentration in maternal serum was 2.2-fold elevated in trisomy 21 pregnancies, but 9.4-fold reduced in trisomy 18 pregnancies.[6] The data suggested that the circulating β-*hCG* mRNA could be a possible noninvasive molecular marker for studying aneuploid pregnancies.

Noninvasive Placental Gene Expression Profiling

Recently, Tsui *et al.* have used microarray technology to systematically identify new placenta-derived RNA markers that could be detected in maternal plasma.[7] In the study of Tsui *et al.*, a marker screening strategy was devised based on the previous observations that the placenta is an important source of circulating fetal RNA in maternal plasma[2] and that hematopoietic cells are the predominant source of plasma nucleic acids in bone marrow transplant subjects.[8] Panels of placenta-specific transcript markers, which were predominantly expressed in placenta compared with maternal blood cells, were identified for both early and the late pregnancies.[7] Six transcripts selected from the candidate transcript panels were shown to be present in maternal plasma, suggesting that the release of placental mRNA into the maternal circulation is a phenomenon that could be generalized to other placenta-expressed transcripts. In general, the transcript that had a relatively high microarray signal was more readily detectable in maternal plasma by real-time quantitative reverse transcriptase polymerase chain reaction (QRT-PCR). In addition, the rapid clearance of all six transcripts from maternal plasma after delivery confirmed their placental specificity.

It is of special interest that the relative abundance of the placental mRNA in maternal plasma correlated well with its gene expression levels in the placenta.[7] In contrast, no correlation was observed between the plasma and placental mRNA levels of non–placenta-specific transcripts, namely, *glyceraldehyde-3-phosphate-dehydrogenase (GAPDH)* and β-*globin.*[7] These data demonstrate that placental gene expression level plays an important role in determining the placental RNA concentrations in maternal plasma. Thus, fetal mRNA detection in maternal plasma may indeed be a useful tool for noninvasive prenatal placental gene expression profiling. This finding has implications for the feasibility of identifying new pregnancy-related disease RNA markers by microarray, which could subsequently be used for noninvasive investigation in maternal plasma.

Circulating Fetal RNA and DNA: Comparison and Contrast

Over the past few years, studies on circulating fetal RNA have grown rapidly and a number of biological and analytical issues concerning the analysis of circulating fetal DNA have been revisited for the analysis of circulating fetal RNA. These two types of fetus-derived extracellular nucleic acid species have some features that are similar and others that are different.

First Appearance in Maternal Plasma

With the continuous development of sensitive assays, circulating fetal DNA has been shown to be detectable as early as the third week of gestation, and has been reliably detected at the fifth week.[9,10] Recently, Chiu *et al.* showed that circulating fetal RNA could be detected early in the pregnancy as well.[11] Plasma β-*hCG* mRNA, which is abundant in early pregnancy,[2] first appears in the maternal plasma at the fourth gestational week. The transcript was detected in 100% of the maternal plasma samples at the eighth week of gestation.[11]

Post-Delivery Clearance

Circulating fetal DNA is cleared rapidly from maternal plasma after delivery, with an apparent half-life of 16.3 min.[12] Moreover, most reports have suggested that cell-free fetal DNA is unlikely to persist in maternal plasma.[12–14] Similar to the circulating fetal DNA, circulating fetal-derived *hPL* RNA has recently been shown to be cleared rapidly from the maternal plasma after delivery with the clearance half-life of 14.1 min.[11] The rapid clearance of the fetal nucleic acids suggests that prenatal diagnosis performed using fetal DNA and RNA in maternal plasma is unlikely to be susceptible to false results due to previous pregnancies.

Molecular Integrity

The molecular characteristic of fetal DNA molecules in maternal plasma has been shown to mainly consist of short DNA fragments.[15] Recently, similar characterization has been carried out on circulating RNA molecules. Wong *et al.* studied the integrity of RNA molecules in plasma by using QRT-PCR assays to target multiple regions along a transcript.[16] These authors demonstrated that for both *GAPDH* mRNA and six individual placental transcripts, the 5′ mRNA fragments were generally more abundant than the 3′ mRNA fragments in maternal plasma. They further discussed that instead of intact molecules, most mRNA fragments in maternal plasma showed various degrees of degradation from their 3′ ends. Biologically, this finding provides clues for the degradation mechanisms of RNA in various tissues and in plasma. Technically, this information suggested that the detection of placental RNA in plasma could be improved by designing assays targeting the 5′ regions of RNA transcripts.

Stability in Blood after Venesection

The detection of placental RNA in maternal plasma is perhaps surprising in view of the inherent instability of RNA. *hPL* and β-*hCG* mRNA molecules have been shown to be very stable in maternal plasma. No significant changes in circulating *hPL* and β-*hCG* mRNA concentrations were observed up to 24 h after maternal blood collection.[2] Similar to circulating fetal RNA, circulating fetal DNA concentration remained largely unchanged over 24 h after maternal blood collection.[17] This stability suggests that both fetal DNA and RNA markers are practical molecular markers for clinical use.

Particality

Circulating RNA has been shown to be associated with subcellular particles of different sizes, which are removable by filtration and ultracentrifugation.[18] Approximately 70% of *hPL* and β-*hCG* mRNA can be removed from the maternal plasma by filtering the plasma sample through a 0.45-μm filter. One possible source of these particle-associated placental RNA molecules would be the apoptotic shedding of microparticles from the placenta into the maternal circulation.[19] Recently, Gupta *et al.* have demonstrated that fetal DNA and RNA molecules were present in syncytiotrophoblast microparticles generated from the placenta *in vitro.*[20] Such particle association could possibly contribute to the stability of the plasma RNA molecules by conferring protection against ribonuclease attack.[21] In contrast to circulating RNA, circulating DNA was mostly non–particle-associated. No significant change in plasma β-*globin* DNA concentration was observed after filtration of plasma samples.[18]

Fetomaternal Transfer

Ng *et al.* have found that *hPL* mRNA was easily detectable in maternal plasma but was absent in cord plasma.[2] The authors suggested that the placental barrier was relatively impermeable to the transfer of *hPL* mRNA from the placenta into the fetal circulation, but whether this apparent unidirectionality of RNA transfer is a general phenomenon will have to await similar studies using other placental mRNA species. In contrast to circulating RNA, the fetomaternal DNA transfer has been shown to be a bidirectional process.[22,23] Quantitative comparison of the two trafficking directions has shown that fetal DNA was detectable in all maternal plasma samples at a median fractional concentration of 3×10^{-2}; while maternal DNA was detected in 30% of cord plasma samples at a median fractional concentration of 3×10^{-3}.[22]

CONCLUSIONS

Circulating fetal RNA is an exciting field for noninvasive prenatal investigation. Encouraging findings have been reported on the reliable detection and the possible clinical application of circulating fetal RNA. With the advent of microarray technology, it is envisioned that more fetus- and disease-specific plasma RNA markers will be developed in the near future.

ACKNOWLEDGMENTS

This work is supported by an Earmarked Research Grant (CUHK4474/03M) from the Hong Kong Research Grants Council.

REFERENCES

1. POON, L.L.M. *et al.* 2000. Presence of fetal RNA in maternal plasma. Clin. Chem. **46**: 1832–1834.
2. NG, E.K.O. *et al.* 2003. mRNA of placental origin is readily detectable in maternal plasma. Proc. Natl. Acad. Sci. USA **100**: 4748–4753.
3. NG, E.K.O. *et al.* 2003. The concentration of circulating corticotropin-releasing hormone mRNA in maternal plasma is increased in preeclampsia. Clin. Chem. **49**: 727–731.
4. LO, Y.M.D. *et al.* 1999. Quantitative abnormalities of fetal DNA in maternal serum in preeclampsia. Clin. Chem. **45**: 184–188.
5. FARINA, A. *et al.* 2004. Circulating corticotropin-releasing hormone mRNA in maternal plasma: relationship with gestational age and severity of preeclampsia. Clin. Chem. **50**: 1851–1854.
6. NG, E.K.O. *et al.* 2004. Evaluation of human chorionic gonadotropin beta-subunit mRNA concentrations in maternal serum in aneuploid pregnancies: a feasibility study. Clin. Chem. **50**: 1055–1057.
7. TSUI, N.B.Y. *et al.* 2004. Systematic micro-array based identification of placental mRNA in maternal plasma: towards non-invasive prenatal gene expression profiling. J. Med. Genet. **41**: 461–467.
8. LUI, Y.Y.N. *et al.* 2002. Predominant hematopoietic origin of cell-free DNA in plasma and serum after sex-mismatched bone marrow transplantation. Clin. Chem. **48**: 421–427.
9. HONDA, H. *et al.* 2002. Fetal gender determination in early pregnancy through qualitative and quantitative analysis of fetal DNA in maternal serum. Hum. Genet. **110**: 75–79.
10. GUIBERT, J. *et al.* 2003. Kinetics of SRY gene appearance in maternal serum: detection by real time PCR in early pregnancy after assisted reproductive technique. Hum. Reprod. **18**: 1733–1736.
11. CHIU, R.W.K. *et al.* 2006. Time profile of appearance and disappearance of circulating placenta-derived mRNA in maternal plasma. Clin. Chem. **52**: 313–316.
12. LO, Y.M.D. *et al.* 1999. Rapid clearance of fetal DNA from maternal plasma. Am. J. Hum. Genet. **64**: 218–224.
13. JOHNSON-HOPSON, C.N. & C.M. ARTLETT. 2002. Evidence against the long-term persistence of fetal DNA in maternal plasma after pregnancy. Hum. Genet. **111**: 575.
14. SMID, M. *et al.* 2003. No evidence of fetal DNA persistence in maternal plasma after pregnancy. Hum. Genet. **112**: 617–618.
15. CHAN, K.C.A. *et al.* 2004. Size distributions of maternal and fetal DNA in maternal plasma. Clin. Chem. **50**: 88–92.
16. WONG, B.C.K. *et al.* 2005. Circulating placental RNA in maternal plasma is associated with a preponderance of 5′ mRNA fragments: implications for non-invasive prenatal diagnosis and monitoring. Clin. Chem. **51**: 1786–1795.

17. Angert, R.M. *et al.* 2003. Fetal cell-free plasma DNA concentrations in maternal blood are stable 24 hours after collection: analysis of first- and third-trimester samples. Clin. Chem. **49**: 195–198.
18. Ng, E.K.O. *et al.* 2002. Presence of filterable and nonfilterable mRNA in the plasma of cancer patients and healthy individuals. Clin. Chem. **48**: 1212–1217.
19. Huppertz, B. *et al.* 1998. Villous cytotrophoblast regulation of the syncytial apoptotic cascade in the human placenta. Histochem. Cell Biol. **110**: 495–508.
20. Gupta, A.K. *et al.* 2004. Detection of fetal DNA and RNA in placenta-derived syncytiotrophoblast micro particles generated in vitro. Clin. Chem. **50**: 2187–2190.
21. Hasselmann, D.O. *et al.* 2001. Extracellular tyrosinase mRNA within apoptotic bodies is protected from degradation in human serum. Clin. Chem. **47**: 1488–1489.
22. Lo, Y.M. *et al.* 2000. Quantitative analysis of the bidirectional fetomaternal transfer of nucleated cells and plasma DNA. Clin. Chem. **46**: 1301–1309.
23. Bauer, M. *et al.* 2002. Detection of maternal deoxyribonucleic acid in umbilical cord plasma by using fluorescent polymerase chain reaction amplification of short tandem repeat sequences. Am. J. Obstet. Gynecol. **186**: 117–120.

Development and Application of a Real-Time Quantitative PCR for Prenatal Detection of Fetal α^0-Thalassemia from Maternal Plasma

WARUNEE TUNGWIWAT,[a,b] SUPAN FUCHAROEN,[b] GOONNAPA FUCHAROEN,[b] THAWALWONG RATANASIRI,[c] AND KANOKWAN SANCHAISURIYA[b]

[a]*Biomedical Science Program, Graduate School, Khon Kaen University, Khon Kaen, Thailand*

[b]*Centre for Research and Development of Medical Diagnostic Laboratories, Faculty of Associated Medical Sciences, Khon Kaen University, Khon Kaen, Thailand*

[c]*Department of Obstetrics & Gynecology, Faculty of Medicine, Khon Kaen University, Khon Kaen, Thailand*

ABSTRACT: In order to provide a noninvasive prenatal diagnosis of α^0-Thalassemia (Southeast Asian [SEA] deletion), we have developed a real-time quantitative semi-nested polymerase chain reaction (PCR) method for identifying the fetal α^0-Thalassemia in maternal plasma. Analysis was performed using DNA extracted from 200 μL plasma from 13 pregnant women during 8–20 weeks of gestation who carried fetuses with normal (2), α^0-Thalassemia carrier (8), Hb H disease (1), and homozygous α^0-Thalassemia (Hb Bart's hydrops fetalis (2). The α^0-Thalassemia was detected using a two-step PCR. Plasma DNA was amplified conventionally using α^0-Thalassemia-specific primers and a portion of the first PCR product was subjected to a semi-nested real-time q-PCR using the SYBR green I chemistry for fluorescence detection. Calibration curve for α^0-Thalassemia quantification was prepared by assaying serial dilution of genomic DNA of an α^0-Thalassemia carrier. Differences in the C_T (threshold cycle) values and calculated concentrations of amplified DNA among normal fetus, α^0-Thalassemia carrier, Hb H disease, and homozygous α^0-Thalassemia were clearly observed, which could help in prenatal prediction of the fetal genotype. This noninvasive prenatal detection of α^0-Thalassemia in maternal plasma should enhance prenatal diagnostic options for this common genetic disorder in routine DNA diagnostic setting.

Address for correspondence: Supan Fucharoen, Department of Clinical Chemistry, Faculty of Associated Medical Sciences, Khon Kaen University, Khon Kaen, Thailand 40002. Voice: 66-43-202-083; fax: 66-43-202-083.
e-mail: supan@kku.ac.th

**Ann. N.Y. Acad. Sci. 1075: 103–107 (2006). © 2006 New York Academy of Sciences.
doi: 10.1196/annals.1368.013**

Keywords: fetal DNA; maternal plasma; α^0-Thalassemia; noninvasive prenatal diagnosis

INTRODUCTION

α^0-Thalassemia is a common genetic disorder in Southeast Asia (SEA), with prevalence ranging from 3.5–14% in different regions.[1] The most common form of α^0-Thalassemia is the SEA deletion, which is 20.5 kb in length, deleting both α-globin genes on a chromosome. In a homozygous state, fetuses without α-globin genes would suffer from severe anemia, hypoxia, heart failure, and a fatal condition known as the hemoglobin (Hb) Bart's hydrops fetalis syndrome. Infants with this syndrome almost always either die *in utero* at 23–38 weeks of gestation or soon after birth. The disease also carries an increased risk of maternal complications with severe preeclampsia, antepartum and postpartum hemorrhage, and lethal complications.[2] Whereas screening for an α^0-Thalassemia carrier can be simply performed using a combination of simple blood test and polymerase chain reaction (PCR), most of the prenatal diagnosis relies on the analysis of fetal tissues obtained by invasive procedures.[3–5] The discovery of the presence of fetal DNA in maternal plasma and serum has offered new approaches to noninvasive prenatal diagnosis, especially for those of paternally inherited disorders as well as fetal gender.[6–8] It has also been demonstrated that fetal DNA in maternal circulation increases in pathological pregnancies including those with preterm labor and preeclampsia.[9,10] We have now demonstrated that real-time quantitative (q)-PCR analysis of α^0-specific sequence from maternal plasma could help in prenatal prediction of α^0-Thalassemia genotype of the fetus.

MATERIALS AND METHODS

Subjects and Plasma DNA Preparation

We studied 13 unrelated Thai pregnant women who attended our thalassemia-screening unit at Khon Kaen University because of their risks of having fetuses with severe thalassemia syndromes at 11–20 weeks of gestation. After informed consent, two milliliters of peripheral blood samples were collected with EDTA as anticoagulant. Blood specimens were processed immediately after collection. DNA was extracted from 200 μL of the plasma using the QIAmp Blood Mini Kit (QIAGEN Inc., Hilden, Germany) and was eluted into a final volume of 200 μL.[8,11] A total of 10 μL was used as template for PCR analysis. Chorionic biopsy, amniocentesis, or fetal blood sampling was also necessary for routine prenatal diagnosis of severe thalassemia in fetuses and analysis of these specimens was used as a confirmatory test.

Real-Time q-PCR

Identification of α^0-Thalassemia (SEA deletion) was performed in two steps. In the first step, 10 μL of plasma DNA was subjected to a conventional PCR using α^0-Thalassemia-specific primers A7 (5′ CTCTGTGTTCTCAGTATTG-GAG 3′) and A9 (5′ ATATATGGGTCTGGAAGTGTATC 3′) used routinely in our laboratory.[3] The PCR mixture (100 μL) contained 10 μL plasma DNA, 3 pmol of primers A7 and A9, 200 μM dNTPs, 10 mM Tris-HCl, 1.5 mM $MgCl_2$, and 1 unit of *Taq* DNA polymerase (Promega Co. Madison, WI, USA). After heating to 94°C for 5 min, the PCR process (94°C 1 min, 55°C 1 min, and 72°C 1 min) was carried out on a PCR system 9600 (Perkin Elmer Wellesley, MA, USA) for 30 cycles. A total of 5 μL of the first PCR product was then subjected to a semi-nested real-time q-PCR using primers A7 and A9S (5′ GGCTTACTGCAGCCTTGAACTCCTG 3′), a nested primer to A9 to produce an α^0-Thalassemia-specific fragment of 95 bp (base pairs) in length. The real-time q-PCR was carried out with 3 pmol of each primer in a total volume of 25 μL using a SYBR Green I (Sigma Co. St. Louis, MO, USA) as the fluorescence monitoring chemical on a Rotor Gene 3000 (Corbett Research, Sydney, New South Wales, Australia) with the same cycling conditions. After PCR amplification, a melting curve was generated to verify the specificity and identity of PCR product. All samples were tested in duplicate and both values were presented. Appropriate serial dilutions of quantified genomic DNA of an α^0-Thalassemia carrier was used to construct the standard curve by plotting the C_T (threshold cycle) of each standard against the DNA concentration.

RESULTS AND CONCLUSIONS

FIGURE 1A shows a melt curve of a real-time q-PCR system for the detecting of α^0-Thalassemia (SEA type) in maternal plasma, indicating specificity of the identification system. The system was applied to the identification of α^0-Thalassemia in plasma samples collected from 13 pregnant women. Quantitative data in relation to the concentration of α^0-Thalassemia-specific amplification was determined for each maternal plasma sample. The amounts of α^0-Thalassemia-specific DNA in each case were presented along with the fetal α-globin genotypes obtained by routine PCR analysis of fetal tissue in FIGURE 1B. Among 13 cases examined, routine analysis of fetal tissues indicated two normal fetuses (αα/αα) (FIG. 1B, 1 and 2), eight α^0-Thalassemia carriers ($\alpha\alpha/--^{SEA}$) (FIG. 1B, 3–10), one Hb H disease ($\alpha\alpha/--^{SEA}$) (FIG. 1B, 11), and two Hb Bart's hydrops fetalis ($--^{SEA}/--^{SEA}$) (FIG. 1B, 12–13). As shown in FIGURE 1B, no α^0-Thalassemia-specific amplification was detected from plasma samples of the noncarrier pregnant women carrying normal fetuses, further indicating a specificity of the detection system. The amounts ranging from approximately 250 to 762 ng/μL of α^0-Thalassemia-specific

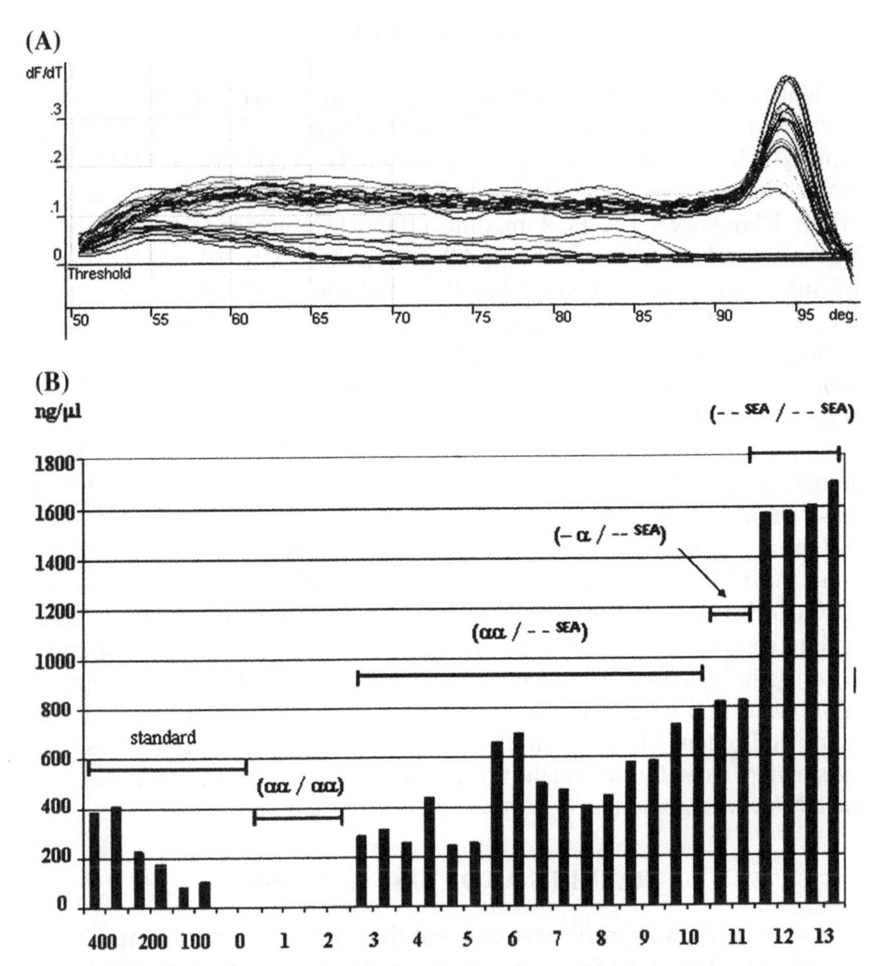

FIGURE 1. (**A**) Melt curve of the real-time q-PCR system for detection of α^0-Thalassemia. (**B**) The amounts (ng/μL) of α^0-Thalassemia-specific DNA determined by real-time q-PCR from maternal plasma. 400, 200, 50, and 0 indicate the concentrations of DNA standards used in the system. 1–13 demonstrate the results of duplicate analysis of 13 maternal plasma samples whose fetal genotypes are indicated.

DNA were detected among 8 maternal plasma samples with α^0-Thalassemia-carrier fetuses. The average level of 830 ng/μL was found for maternal plasma with the Hb H disease ($-\alpha/--^{SEA}$). In contrast, the average amounts of 1,578 and 1,648 ng/μL were detected for samples from two pregnant women carrying fetuses with Hb Bart's hydrops fetalis syndrome ($--^{SEA}/--^{SEA}$). As for other pathological pregnancies,[9,10,12] the amounts of α^0-Thalassemia-specific amplification observed apparently correlated with the disease severity of the fetus. These results demonstrate that the semi-nested real-time q-PCR analysis of maternal plasma DNA described should potentially prove useful in development of a noninvasive prenatal detection of α^0-Thalassemia through maternal plasma. This method would be practical in any laboratory where a real-time PCR is available. The ease, rapidity, and effectiveness shown by this system should be directly applicable to a noninvasive prenatal prediction of a homozygous α^0-Thalassemia by maternal plasma DNA analysis. This should theoretically reduce the number of invasive procedures required for diagnosis of this common genetic disorder in the region.

ACKNOWLEDGMENT

This work was supported by grants from Khon Kaen University and the Thailand Research Fund (TRF), Thiland. W.T. is a Ph.D. candidate receiving a scholarship from the Royal Golden Jubilee Ph.D. program of the TRF (Grant No. PHD/0135/2547).

REFERENCES

1. FUCHAROEN, S. & P. WINICHAGOON. 1987. Hemoglobinopathies in Southeast Asia. Hemoglobin **11:** 65–88.
2. CHUI, D.H.K. & J.S. WAYE. 1998. Hydrops fetalis caused by α^0-Thalassemia: an emerging health care problem. Blood **91:** 2213–2222.
3. PANYASAI, S. *et al.* 2002. A simplified screening for α^0-Thalassemia 1 (SEA type) using a combination of a modified osmotic fragility test and a direct PCR on whole blood cell lysates. Acta. Haematol. **108:** 74–78.
4. FUCHAROEN, G. *et al.* 2004. A simplified screening strategy for thalassaemia and haemoglobin E in rural communities in Southeast Asia. Bull. WHO **82:** 364–372.
5. FUCHAROEN, S. *et al.* 1991. Prenatal diagnosis of thalassemia and haemoglobinopathies in Thailand: experience from 100 pregnancies. Southeast. Asian. J. Trop. Med. Public Health **22:** 16–29.
6. LO, Y.M.D. *et al.* 1997. Presence of fetal DNA in maternal plasma and serum. Lancet **16:** 485–487.
7. LO, Y.M.D. *et al.* 1989. Prenatal sex determination by DNA amplification from maternal peripheral blood. Lancet **2:** 1363–1365.
8. TUNGWIWAT, W. *et al.* 2003. Non-invasive fetal sex determination using a conventional nested PCR analysis of fetal DNA in maternal plasma. Clin. Chim. Acta. **334:** 173–177.
9. LEUNG, T.N. *et al.* 1998. Maternal plasma fetal DNA as a marker for preterm labour. Lancet **352:** 1904–1905.
10. ZHONG, X.Y. *et al.* 2001. Elevation of both maternal and fetal extracellular circulating deoxyribonucleic acid concentrations in the plasma of pregnant women with preeclampsia. Am. J. Obstet. Gynecol. **184:** 414–419.
11. FUCHAROEN, G. *et al.* 2003. Prenatal detection of fetal hemoglobin E gene from maternal plasma. Prenat. Diagn. **23:** 393–396.
12. LEUNG, T.N. *et al.* 2001. Increased maternal plasma fetal DNA concentrations in women who eventually develop preeclampsia. Clin. Chem. **47:** 137–139.

Detection of a Paternally Inherited Fetal Mutation in Maternal Plasma by the Use of Automated Sequencing

ANA BUSTAMANTE-ARAGONES, MARIA GARCIA-HOYOS,
MARTA RODRIGUEZ DE ALBA, CRISTINA GONZALEZ-GONZALEZ,
ISABEL LORDA-SANCHEZ, DAN DIEGO-ALVAREZ,
M. JOSE TRUJILLO-TIEBAS, CARMEN AYUSO, AND CARMEN RAMOS

Department of Genetics, Fundacion Jimenez Diaz, Madrid, Spain

ABSTRACT: The discovery of circulating fetal DNA in maternal blood has been an encouraging step forward in the prenatal diagnostic field. It has opened up the possibility of development of a noninvasive method for the genetic analysis of the fetus. Many techniques have been applied to the study of this fetal DNA, but automated sequencing has been seldom used. The intention of this study was to use the automated sequencing technique for the detection of a paternally inherited fetal mutation in maternal plasma. Maternal plasma samples from a pregnant woman, whose husband had a mutation (Q134X) in the *RP2* gene, which is located in the X-chromosome, were collected at two different gestational ages (10th and 19th week of gestation) in order to determine whether the paternally inherited fetal mutation could be detected by automated sequencing. Restriction analysis was also performed to confirm the results. The fetal mutation was clearly detected in the maternal plasma by the use of automated sequencing. The automated sequencing enables the possibility of analyzing fetal sequences, at a nucleotide level, in order to detect mutations or polymorphisms which are distinguishable from maternal sequences.

KEYWORDS: fetal DNA; maternal blood; mutation; retinitis pigmentosa; automated sequencing

INTRODUCTION

Circulating cell-free fetal DNA in maternal blood was detected for the first time in 1997 by Lo *et al.*[1] This discovery opened up a new perspective for the development of noninvasive methods for prenatal diagnosis, thus avoiding the risk of fetal loss induced by invasive obstetric procedures. A first aim

Address for correspondence: ANA Bustamante, Department of Genetics, Fundacion Jimenez Diaz, Avda. Catolicos 2, 2840 Madrid, Spain. Voice: +34-915504872; fax: +34-915448735.
e-mail: abustamente@fjd.es

**Ann. N.Y. Acad. Sci. 1075: 108–117 (2006). © 2006 New York Academy of Sciences.
doi: 10.1196/annals.1368.014**

has been to look for an appropriate method to isolate the maximum amount of fetal DNA from maternal blood,[2-5] because fetal DNA only represents 3–6% of total DNA.[6] Once the presence of fetal DNA in maternal blood was widely demonstrated by the detection of Y-chromosome sequences, the technique gained momentum and many diverse studies started to appear. The most common approaches that demonstrate the presence of this circulating fetal DNA have been the detection of Y-chromosome sequences in male-bearing pregnancies,[7-9] diagnosis of the fetal Rhesus-D status,[10-12] and the detection of paternally inherited mutations.[13-15] Many other studies are contributing to the better understanding of the biology and behavior of this source of fetal genetic material.[16,17] These studies have demonstrated that there are some factors that increase the quantity of cell-free fetal DNA in maternal circulation, such as gestational age,[18] aneuploidy gestations like trisomy 21[19,20] and trisomy 13[21], and pathological pregnancies associated with placental abnormalities.[22,23] All these reports describe the use of different techniques to detect and quantify fetal DNA in maternal circulation, such as conventional polymerase chain reaction (PCR), real-time PCR, quantitative fluorescence (QF)-PCR, and restriction analysis. However, to our knowledge there is only one report about the use of automated sequencing in fetal DNA from maternal plasma.[24]

Here we present a case of a paternally inherited fetal mutation associated to X-linked retinitis pigmentosa (XLRP) studied in maternal plasma by automated sequencing analysis. Retinitis pigmentosa (RP) is a degenerative disease of the retina characterized by night blindness and visual field constriction. The X-linked form of RP (XLRP; MIM # 268000) is the most severe type of RP because of its early onset and its rapid progression. Five XLRP loci have been mapped: RP_{23}, RP_6, RP_3, RP_2, and RP_{24}. So far, only two XLRP genes have been identified: RPGR (for RP_3 locus)[25] and RP_2.[26]

The pregnant woman underwent a chorionic biopsy because she was a carrier of a chromosomal mosaicism (45,X/46,XX). The cytogenetic results of the biopsy showed that the fetal karyotype was 46,XX, an obligatory carrier of the paternal mutation. The parents gave their consent for the study of the maternal plasma to be performed, knowing that it has only research and not diagnostic purposes.

MATERIALS AND METHODS

Family History and Patient

A pregnant woman at risk of carrying an affected fetus came to our hospital to undergo a chorion biopsy. She was a carrier of a mosaicism 45,X/46,XX. In addition, the father of the fetus was affected by XLRP due to a mutation in

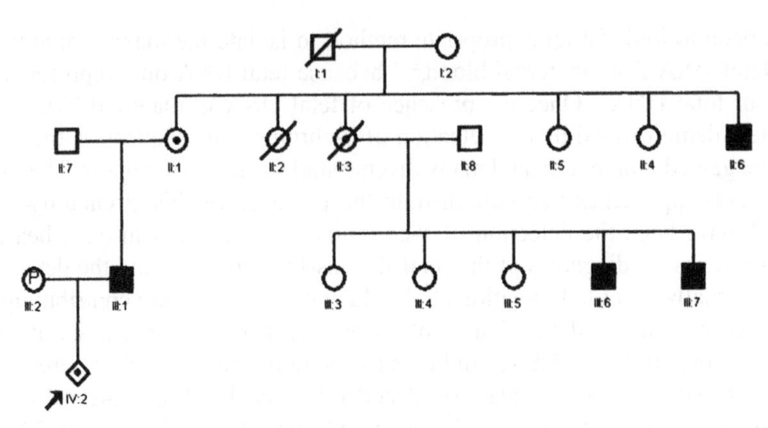

FIGURE 1. Genealogy of the family studied. The father of the fetus (■ III:I) was affected by an XLRP due to a mutation in the RP2 gene. He inherited the mutation from his mother (⊙II:I) who was a healthy carrier. Ⓟ: Pregnant woman. The arrow shows the *probandus* (fetus).

the RP2 gene (Q134X). He inherited this mutation from his mother, who was a healthy carrier (FIG. 1).

Blood Sample Collection

Blood samples were obtained from the pregnant woman with consent according to the Helsinki Declaration (1979). A total of 9 mL of maternal peripheral blood were collected in a tube with EDTA, at two different points of gestation (10th and 19th weeks). The first-trimester sample was taken just before the chorion biopsy. There was no opportunity to collect samples in between. Each tube was centrifuged at 1,500 *g* for 10 min at room temperature. One mL aliquot of plasma was centrifuged at 17,000 *g* for 10 min in an Eppendorf tube at room temperature. Supernatant (about 1 mL) was taken and stored in clean Eppendorf tubes at −20°C.

DNA Extraction

DNA was extracted from 2 mL of maternal plasma following the protocol of the QIAmp DNA Blood Mini Kit (Qiagen, Hilden, Germany) with modifications. The elution process was performed twice using 50 μL of elution buffer. Samples were stored at −20°C.

DNA from the chorion villus sample (CVS) was extracted with the EZ1 DNA Tissue Kit, using the EZ1 Robot (Qiagen) in order to use it as fetal control.

The other DNAs used as controls (from the pregnant woman, the father, and the paternal grandmother) were already available because of a previous familial RP study.

PCR Amplification

RP2 gene comprises 5 exons and the mutation studied (Q134X) lies in the middle of exon 2.[27] A PCR assay was performed to amplify a 300-bp (base pair) region of this exon.

For each gestational age we used two different amounts of DNA (20 μL and 30 μL) as template and 2 μL of DNA from the control samples. The final reaction mix had a volume of 50 μL containing: 30 pmol of each primer (RP2-2b forward: 5′-CAT TGA TGA CTG TAC TAA CTG CAT-3′; RP2-2b reverse: 5′- CTG ACA CAG GTG TAA AGT CAT G-3′) (Applied Biosystems, Foster City, CA), 1× PCR buffer with 2.5nM $MgCl_2$ (Roche, Indianapolis, IN), 200 μM of each deoxynucleotide, and 1U of FastStart Taq DNA Polymerase (Roche). Amplification was performed in a GeneAmp PCR System 2700 thermal cycler (Applied Biosystems). After an initial incubation at 94°C for 7 min, the reaction was cycled for 30 sec at 94°C, 45 sec at 61°C, and 1 min at 72°C, for 40 cycles, followed by a final extension of 7 min at 72°C. A total of 35 μL of amplified PCR product was tested by electrophoresis in a 3% agarose gel and then purified using the QIAgen Purification Kit (Qiagen) with a final elution in 30 μL of elution buffer.

Automated Sequencing

The sequencing reaction was carried out with 5 μL of purified PCR product, 10 pmol of RP2-2b forward primer and dRhodamine Terminator Cycle Sequencing Ready reaction Kit (Applied Biosystems) according to the manufacturer's recommendations. Automated sequencing of the products was performed by the Sequencing Analysis program in the ABI Prism 3100 Genetic Analyzer (Applied Biosystems).

Restriction Analysis

The Q134X (c.400C>T) mutation creates a restriction site for *MseI*. In Q134X mutated sequences *MseI* breaks the 300-bp exon in two fragments of 146 bp and 154 bp each, whereas in wild-type sequences the enzyme does not fragment the exon. The digestion was carried out in a total volume of 37 μL, containing 15 μL amplified PCR product (300 bp), 40 U of *MseI* enzyme (Fermentas), 3.5 μL of Buffer Enzyme R (10 mM Tris-HCl [pH 8.5], 10mM $MgCl_2$, 100mM KCl, 0.1 mg/mL BSA) (MBI Fermentas GMBH, St. Leon-Rot,

Germany), and 14.5 μL of distilled water, and left overnight at 65°C. The digestion product was tested by an electrophoresis in a 9% acrylamide gel. To avoid the possibility of contamination in the loading gel process, plasma samples were separately loaded from control samples.

RESULTS

Sequencing analysis enabled the detection of the paternal mutation (Q134X) in the plasma sample. It was possible to observe at the mutation site the presence of a cytosine and a thymine. The cytosine corresponds to the wild-type X-chromosome whereas the thymine corresponds to the mutated X-chromosome. The presence of both nucleotides was also observed in the biopsy sample, confirming the results. The DNA sample of the paternal grandmother, used as healthy carrier control, revealed the presence of both nucleotides. The paternal sample, as hemizygous control, only showed the thymine at the mutation site (FIG. 2).

The sequencing analysis results were confirmed by restriction analysis. In the plasma sample three fragments of different sizes were observed: fragments of 146 bp and corresponding to the mutated X-chromosome, and a 300-bp fragment corresponding to the wild-type X-chromosome. The three bands were also observed in both the CVS and the grandmother's DNA. The paternal DNA showed the two bands corresponding to the mutated X-chromosome and the maternal DNA only showed the 300-bp band corresponding to the wild-type X-chromosome (FIG. 3).

None of the techniques revealed the presence of the mutation in the plasma sample from the 10th week of gestation. All the results were obtained in the 19th-week sample.

DISCUSSION

The aim of this study was to sequence a paternally inherited fetal mutation from maternal blood using automated sequencing.

The results of the automated sequencing presented here show that this technique allows the detection of the paternally inherited mutation in maternal plasma. The electropherogram showed the presence of a cytosine and a thymine at the same sequence site, showing the cytosine and the thymine corresponding to the wild-type X-chromosome and mutated X-chromosome, respectively. The presence of the thymine in the maternal blood can only be explained by its fetal origin. This clear detection demonstrates that paternally inherited fetal sequences or nucleotides, which are different from maternal ones, can be sequenced in maternal plasma. To our knowledge, there is only one report that has used DNA sequencing to analyze fetal DNA in maternal blood.[24]

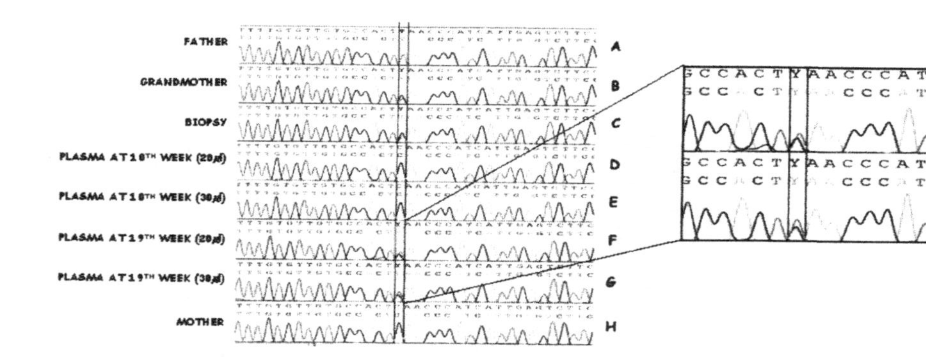

FIGURE 2. Sequencing analysis electropherograms. Cytosine and thymine at the mutation site, correspond to wild-type and mutated X-chromosomes, respectively. (**A**) Paternal DNA as hemizygous carrier of the mutation. Presence of thymine at the mutation site. (**B**) Grandmother's DNA as heterozygous carrier of the mutation. Presence of both cytosine and thymine at the mutation site. (**C**) Chorion villus DNA. Presence of both cytosine and thymine at the mutation site. (**D**) Plasma at 10th week of gestation (template of 20 μL). Presence of cytosine at the mutation site. (**E**) Plasma at 10th week of gestation (template of 30 μL) Presence of cytosine at the mutation site. (**F**) Plasma at 19th week of gestation (template of 20 μL). Presence of both cytosine and thymine at the mutation site. (**G**) Plasma at 19th week of gestation (template of 30 μL) Presence of both cytosine and thymine at the mutation site. (**H**) Maternal DNA. Presence of cytosine at the mutation site (homozygous wild type).

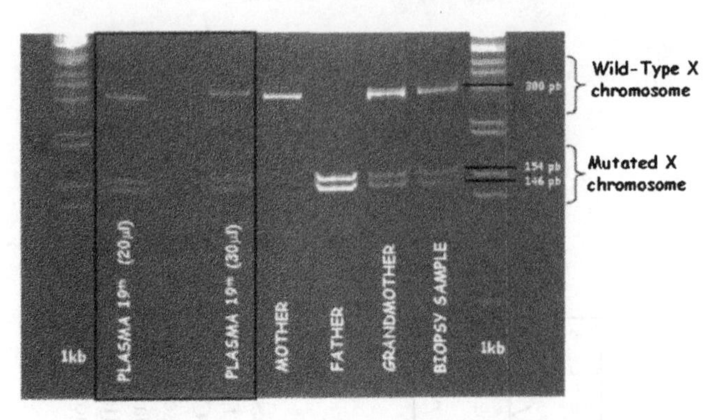

FIGURE 3. Results obtained from the restriction analysis. The 300-bp fragment corresponds to the wild-type X-chromosome and 154-bp and 146-bp fragments correspond to the mutated X-chromosome. *Lane 1*: 1 kb DNA Ladder (Lab Clinics) *Lane 2*: Plasma at 19th week of gestation (template of 20 μL). Detection of 300-, 154-, and 146-bp fragments. *Lane 3*: Blank (no amplification) *Lane 4*: Plasma at 19th week of gestation (template of 30 μL). Detection of 300-, 154-, and 146-bp fragments. *Lane 5*: Maternal DNA. Detection of 300-bp fragment. *Lane 6*: Paternal DNA as hemizygous carrier of the mutation. Detection of 154- and 146-bp fragments. *Lane 7*: Grandmother's DNA as heterozygous carrier of the mutation. Detection of 300-, 154-, and 146-bp fragments. *Lane 8*: Chorion villus DNA. Detection of 300-, 154-, and 146-bp fragments. *Lane 9*: 1 kb DNA Ladder (Lab Clinics).

The detection of point mutations by the use of restriction analysis has been widely described.[14,15] In fact, it has been used in the present study in order to confirm the results obtained by sequencing analysis. However, these types of studies required a restriction site to be created or destroyed by the mutation. The advantage of sequencing analysis versus restriction analysis is that any kind of mutation at a nucleotide level can be studied by the use of this technique. The incorporation of sequencing analysis in diagnostic laboratories has represented a great advance in genetic testing because of its sensitivity and availability. The possible application of this technique in the analysis of fetal DNA present in maternal plasma samples would represent a step forward in noninvasive prenatal diagnosis and a perfect complement to the techniques already in use. These have been described as good tools for the prenatal diagnosis in maternal plasma.[14,28–30]

In the 19th week plasma sample the paternally inherited mutation was undoubtly detected. The proportion of fetal DNA present in the total amount was sufficient for a reliable diagnosis and the results are apparently independent of the quantity of template plasma DNA used in the PCR assay (20 and 30 μL). However in the 10th week of gestation the mutation could not be detected. Previous studies have shown that there is an increase of the circulating fetal DNA throughout the gestation.[17,18,31] The techniques used in this study have been shown to be sensitive enough, so the absence of results in the 10th week plasma sample would be due to a scarcity of circulating fetal DNA in maternal blood. Nowadays, the tendency in conventional prenatal diagnosis is to have the results as soon as possible, which implies the performance of the analysis in the first trimester of gestation. This applies to noninvasive diagnosis as well. Li *et al.*[32] have reported the detection of a paternally inherited fetal mutation in the first trimester of gestation using real-time PCR. So, further studies should be done in order to know whether the quantity of fetal DNA present in the sample taken in first trimester and beginning of second trimester is adequate to be used with all the techniques available including automated sequencing.

Being a prenatal diagnosis unit, we consider that reliability is very important. Previous works have detected the presence of fetal DNA from the 6th week of gestation[18] and our group has reported the detection of fetal DNA from the 10th week of gestation.[30] However, detecting does not necessarily mean diagnostic capability in terms of reliability. The improvement of the DNA extraction, in order to decrease the maternal contamination, as well as the development of more sensitive techniques will mean the combination of both strategic points reliability and early detection.

This work opens up the possibility of incorporating automated DNA sequencing for the study of the fetal DNA present in maternal blood. To date, diverse techniques have been used for the detection of specific sequences on the circulating fetal DNA.[9,14,28–30,33,34] However, the great advantage of automated DNA sequencing is that we can not only detect this DNA, but also enable its analysis at a nucleotide level.

ACKNOWLEDGMENTS

We wish to thank to Diego Cantalapiedra for his assistance. This work is supported by the Ministry of Health (PI040218). Ana Bustamante is supported by "Fundacion Conchita Rabago de Jimenez Diaz."

REFERENCES

1. Lo, Y.M. *et al.* 1997. Presence of fetal DNA in maternal plasma and serum. Lancet **350:** 485–487.
2. Chiu, R.W. *et al.* 2001. Effects of blood-processing protocols on fetal and total DNA quantification in maternal plasma. Clin. Chem. **47:** 1607–1613.
3. Angert, R.M. *et al.* 2003. Fetal cell-free plasma DNA concentrations in maternal blood are stable 24 hours after collection: analysis of first- and third-trimester samples. Clin. Chem. **49:** 195–198.
4. Dhallan, R. *et al.* 2004. Methods to increase the percentage of free fetal DNA recovered from the maternal circulation. JAMA **291:** 1114–1119.
5. Li, Y. *et al.* 2004. Size separation of circulatory DNA in maternal plasma permits ready detection of fetal DNA polymorphisms. Clin. Chem. **50:** 1002–1011.

6. Lo, Y.M. *et al.* 1998. Quantitative analysis of fetal DNA in maternal plasma and serum: implications for noninvasive prenatal diagnosis. Am. J. Hum. Genet. **62:** 768–775.
7. Pertl, B. *et al.* 2000. Detection of male and female fetal DNA in maternal plasma by multiplex fluorescent polymerase chain reaction amplification of short tandem repeats. Hum. Genet. **106:** 45–49.
8. Guibert, J. *et al.* 2003. Kinetics of SRY gene appearance in maternal serum: detection by real time PCR in early pregnancy after assisted reproductive technique. Hum. Reprod. **18:** 1733–1736.
9. Johnson, K.L. *et al.* 2004. Interlaboratory comparison of fetal male DNA detection from common maternal plasma samples by real-time PCR. Clin. Chem. **50:** 516–521.
10. Harper, T.C. *et al.* 2004. Use of maternal plasma for noninvasive determination of fetal RhD status. Am. J. Obstet. Gynecol. **191:** 1730–1732.
11. Gautier, E. *et al.* 2005. Fetal RhD genotyping by maternal serum analysis: a two-year experience. Am. J. Obstet. Gynecol. **192:** 666–669.
12. Moise, K.J. 2005. Fetal RhD typing with free DNA in maternal plasma. Am. J. Obstet. Gynecol. **192:** 663–665.
13. Saito, H. 2000. Prenatal DNA diagnosis of a single-gene disorder from maternal plasma. Lancet **356:** 1170.
14. González-González, M.C. *et al.* 2002. Prenatal detection of a cystic fibrosis mutation in fetal DNA from maternal plasma. Prenat. Diagn. **22:** 946–948.
15. Li, Y. *et al.* 2004a. Improved prenatal detection of a fetal point mutation for achondroplasia by the use of size-fractionated circulatory DNA in maternal plasma-case report. Prenat. Diagn. **24:** 896–898.
16. Wong, B.C. & Y.M. Lo, 2003. Cell-free DNA and RNA in plasma as new tools for molecular diagnostics. Expert Rev. Mol. Diagn. **3:** 785–797.
17. Bischoff, F.Z. *et al.* 2005. Cell-free fetal DNA in maternal blood: kinetics, source and structure. Hum. Reprod. Update **11:** 59–67.
18. Galbiati, S. *et al.* 2005. Fetal DNA detection in maternal plasma throughout gestation. Hum. Genet. **117:** 243–248.
19. Lo, Y.M. *et al.* 1999. Increased fetal DNA concentrations in the plasma of pregnant women carrying fetuses with trisomy 21. Clin. Chem. **45:** 1747–1751.
20. Spencer, K. *et al.* 2003. Increased total cell-free DNA in the serum of pregnant women carrying a fetus affected by trisomy 21. Prenat. Diagn. **23:** 580–583.
21. Wataganara, T. 2003. Maternal serum cell-free fetal DNA levels are increased in cases of trisomy 13 but not trisomy 18. Hum. Genet. **112:** 204–208.
22. Lo, Y.M. *et al.* 1999b. Quantitative abnormalities of fetal DNA in maternal serum in preeclampsia. Clin. Chem. **45:** 184–188.
23. Zhong, X.Y. *et al.* 2001. Elevation of both maternal and fetal extracellular circulating deoxyribonucleic acid concentrations in the plasma of pregnant women with preeclampsia. Am. J. Obstet. Gynecol. **184:** 414–419.
24. Poon, L.L. *et al.* 2002. Differential DNA methylation between fetus and mother as a strategy for detecting fetal DNA in maternal plasma. Clin. Chem. **48:** 35–41.
25. Musarella, M.A. *et al.* 1988. Localization of the gene for X-linked recesive type of retinitis pigmentosa (XLRP) to Xp21 by linkage analysis. Am. J. Hum. Genet. **43:** 484–494.
26. Bhattacharya, S.S. *et al.* 1984. Close genetic linkage between X-linked retinitis pigmentosa and a restriction fragment length polymorphism identified by recombinant DNA probe L1.28. Nature **309:** 253–255.

27. SCHWAHN, U. *et al.* 1998. Postional cloning of the gene for X-linked retinitis pigmentosa 2. Nat.Genet. **19:** 327–332.
28. GONZÁLEZ-GONZÁLEZ, M.C. *et al.* 2003. Huntington disease-unaffected fetus diagnosed from maternal plasma using QF-PCR. Prenat. Diagn. **23:** 232–234.
29. GONZÁLEZ-GONZÁLEZ, M.C. 2003. Early Huntington disease prenatal diagnosis by maternal semiquantitative fluorescent PCR. Neurology **60:** 1214–1215.
30. GONZÁLEZ-GONZÁLEZ, M.C. 2005. Application of fetal DNA detection in maternal plasma: a prenatal diagnosis unit experience. J. Histochem. Cytochem. **53:** 307–314.
31. BIANCHI, D.W. 2004. Circulating fetal DNA: its origin and diagnostic potential-a review. Placenta **25**(Suppl. A)**:** S93–S101.
32. LI, Y. *et al.* 2005. Detection of paternally inherited fetal point mutations for beta-thalassemia using size-fractionated cell-free DNA in maternal plasma. JAMA 2005 **293:** 843–849.
33. HROMADNIKOVA, I. *et al.* 2003. Replicate real-time PCR testing of DNA in maternal plasma increases the sensitivity of non-invasive fetal sex determination. Prenat. Diagn. **23:** 235–238.
34. HO, S.S. *et al.* 2004. Non-invasive prenatal diagnosis of fetal gender using real-time polymerase chain reaction amplification of SRY in maternal plasma. Ann. Acad. Med. Singapore **33**(Suppl. 5)**:** S61–S62.

Occurrence of Neutrophil Extracellular DNA Traps (NETs) in Pre-Eclampsia

A Link with Elevated Levels of Cell-Free DNA?

ANURAG GUPTA,[a] PAUL HASLER,[b] STEPHAN GEBHARDT,[c] WOLFGANG HOLZGREVE,[a] AND SINUHE HAHN

[a]*University Women's Hospital/Department of Research, Basel, Switzerland*

[b]*Department of Rheumatology, Kantonsspital Aarau, Aargau, Switzerland*

[c]*Department of Obstetrics and Gynecology, University of Stellenbosch, Stellenbosch, South Africa*

ABSTRACT: Manifest pre-eclampsia is associated with activation of peripheral neutrophils as well as elevations in maternal cell-free DNA. For this reason, we were very intrigued by recent reports indicating that activated circulatory neutrophils secrete nuclear DNA to generate extracellular DNA lattices, termed NETs (neutrophil extracellular traps). Our preliminary data indicate that placental syncytiotrophoblast microparticles, which are released in elevated amounts in pre-eclampsia, can induce NETs in isolated neutrophils. Furthermore, we found evidence for the increased presence of NETs directly in the intervillous space of pre-eclamptic placentae. Therefore, these newly discovered entities may be implicated in the underlying etiology of this disorder.

KEYWORDS: neutrophil extracellular traps; cell-free DNA; pre-eclampsia

INTRODUCTION

Pre-eclampsia, a life-threatening disorder peculiar to human pregnancy, is a considerable health care problem worldwide.[1,2] The disorder, which usually occurs in the latter half of pregnancy, is characterized by hypertension and edema in previously normotensive pregnant women. Further complications include HELLP (hemolysis, elevated liver enzyme levels, and a low platelet count) syndrome, affecting the liver and eclampsia, resulting in seizures and possible cerebral hemorrhage and subsequent maternal death. Currently, the only effective treatment remains delivery of the fetus and removal of the placenta. Since pre-eclampsia, especially the more severe forms, usually occurs

Address for correspondence: Dr. Sinuhe Hahn, Laboratory for Prenatal Medicine, University Women's Hospital, Department of Research, Spitalstrasse 21, CH 4031, Basel, Switzerland. Voice: ++41-61-265-9224; fax: ++41-61-265-9399.
e-mail: shahn@uhbs.ch

Ann. N.Y. Acad. Sci. 1075: 118–122 (2006).
doi: 10.1196/annals.1368.015

fairly early in pregnancy (prior to 34 weeks of gestation), this results in the delivery of a very premature fetus, frequently affected by growth retardation, leading to a high rate of fetal mortality.[1,2]

The underlying etiology of the disorder is thought to center on the placenta, as most true pre-eclampsias usually resolve once the placenta is removed. Placental aberrancies in pre-eclampsia include the inadequate modification of maternal spiral arteries by extravillous cytotrophoblast, the elevated release of inflammatory syncytiotrophoblast debris, elevated transplacental fetal cell traffic and trophoblast deportation, as well as a gross imbalance in placentally derived angiogenic factors and inflammatory cytokines.[1,2] Maternal effects include damage of the endovasculature and an overt activation of the innate immune response.[1,2]

Previous studies have indicated that pre-eclampsia is associated with an elevated release of fetal cf-DNA, and that these elevations can occur prior to the onset of clinical symptoms.[3–6] These studies have also shown that the manifest disorder is also associated with an increased presence of maternal cf-DNA, and that these elevations corresponded to the severity of the disorder.[4] While the elevations in fetal cf-DNA have been interpreted as further evidence of placental dysfunction in pre-eclampsia, the source of the increased amounts of maternal cf-DNA is not that clear.[7]

NETs: A New Aspect of Neutrophil Physiology

In a startling recent publication in *Science*, Brinkmann and colleagues observed that when circulatory neutrophils are activated by bacterial endotoxin, inflammatory cytokines (IL-8), or pharmacological agents (PMA; phorbol 12-myristate 13-acetate), these cells weave a complex web which largely consists of extracellular DNA.[8] Furthermore, these investigators were able to show that these networks, which they termed NETs (neutrophil extracellular traps), contain bacteriocidal substances. In this manner, these NETs not only served to ensnare bacteria, but also possessed the ability to kill them. By being able to perform this in a discretely localized manner, it was also argued that this would aid in preventing damage to the surrounding tissue, a feature that would occur by the simple release of toxic agents into the circulation or adjacent tissue.

As the innate immune system, particularly that of neutrophils, is known to be overtly activated in pre-eclampsia,[9] we were keen to determine whether a possible link existed between these two phenomena: NETs and the elevated presence of maternal cf-DNA in pre-eclampsia.

NETs Are Induced by Placental Microparticles and Occur in Pre-eclamptic Placentae

In our preliminary investigation, we examined whether syncytiotrophoblast microparticles (frequently termed STBM) can activate isolated circulatory polymorphonuclear neutrophils (PMNs).[2] This analysis indicated that the treatment of PMNs with STBMs leads to the upregulated expression of the activation marker CD11b on these cells, to a comparable level we observed using known activators of PMNs (rIL-8 and PMA) (FIG. 1).

We also observed that the treatment of PMNs with STBMs also triggered the generation of NETs, which were very comparable to those described by Brinkmann and colleagues, in that they largely consisted of extracellular DNA and had similar dimensions. These analyses also indicated that the STBMs were being physically trapped in the NETs, a feature which could be verified by Sytox green staining, which stained the extracellular DNA in the NETs structures, as well as the fetal cf-DNA in the STBM. This finding therefore suggests that NETs not only serve to trap foreign objects such as bacteria,

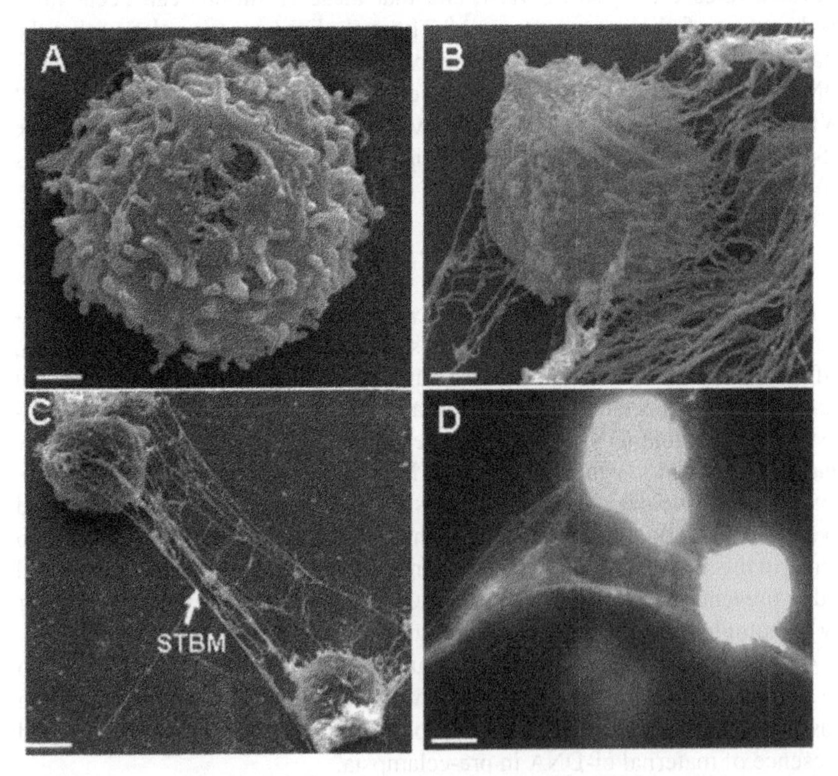

FIGURE 1. NET formation by the activated neutrophils. High-resolution scanning electron microscopy revealed that isolated circulatory untreated neutrophils were devoid of any NETs (**A**). Upon treatment with 25 nM PMA (**B**), and 150 μg/mL STBM (**C**), neutrophils readily generated NETs within 30 min of incubation. Fluorescent microscopy revealed that NETs are rich in DNA as it is evident by the staining of DNA by Sytox green, a DNA-binding dye (**D**). *Bars* in the figures **A-B, C**, and **D** are 1 μm, 5 μm, 10 μm, respectively.

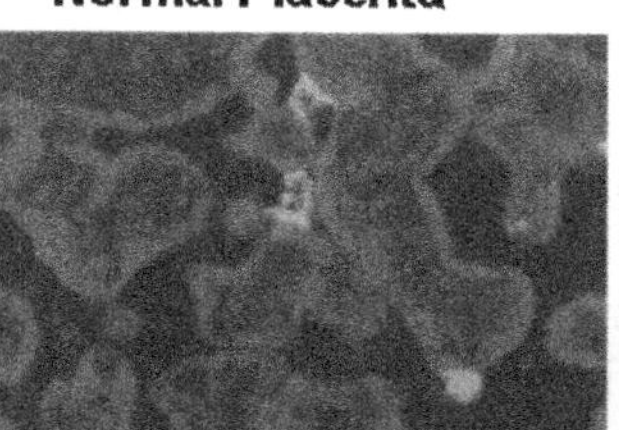

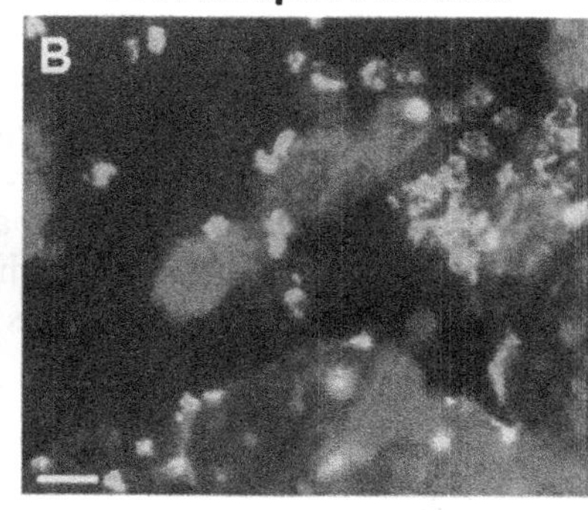

FIGURE 2. Presence of NETs in intervillous spaces of normal and pre-eclamptic placentae. *Normal placentae:* Immunofluorescence staining of neutrophil elastase and staining for extracellular DNA with DAPI reveals localization of NETs within the villi or adjacent to the syncytiotrophoblast in the normal placenta (**A**). *Pre-eclamptic placentae:* Staining for neutrophil elastase and extracellular DNA (**B**) reveals a haze over large portions of the intervillous space. This overlay indicates that these extracellular neutrophil lattices are largely localized within the intervillous space. *Bar* in the figures is 50 μm.

but may also be instrumental in the capture of cellular micro-debris, such as STBMs.

Since these experiments were performed on PMNs isolated from the periphery of healthy donors, we were keen to see whether we could find any evidence of NET formation in pre-eclampsia, especially at the site responsible for this disorder, the placenta. For this purpose, we examined frozen placental tissue sections taken from normal deliveries and pregnancies affected by pre-eclampsia. Our examination, using stains specific for DNA (DAPI; 4′6-diamidino-2-phenylindol) and neutrophil elastase, indicated that NETs were present in normal placentae, occurring in close proximity to the syncytiotrophoblast layer (FIG. 2). This is to be expected, as STBMs are deported in normal pregnancies, especially close to term. On the other hand, our analysis of pre-eclamptic placental sections indicated that the intervillous space was frequently infiltrated by numerous NETS, and that the presence of these was dramatically increased under these conditions.

DISCUSSION

Our novel findings demonstrate that NETs are not only triggered by foreign matter such as bacteria, but may also be elicited by physiological signals, such as placental syncytiotrophoblast micro-debris (STBMs). Furthermore, we find the first evidence that NETs may be implicated in a human disorder, pre-eclampsia, in that huge numbers of NETs were present directly in the intervillous space of affected placentae. As it has been shown that NETs possess sufficient strength to ensnare large bacteria (*S. aureus*),[8] it is possible that the presence of large numbers of NETs in the intervillous space may hinder the flow of blood through this critical feto-maternal interface, leading to a condition of hypoxia. Consequently, pre-eclampsia may not be only associated with a more distal condition of hypoxia at the site of the inadequately modified spiral arteries,[1,2] but may also occur directly at the site of feto-maternal exchange, the intervillous space. This hypothesis, which may have fundamental implications concerning our understanding of this enigmatic disorder, is currently being examined in more detailed studies.

REFERENCES

1. SIBAI, B. *et al.* 2005. Pre-eclampsia. Lancet **365:** 785–799.
2. REDMAN, C.W. & I.L. SARGENT. 2005. Latest advances in understanding pre-eclampsia. Science **308:** 1592–1594.
3. LO, Y.M. *et al.* 1999. Quantitative abnormalities of fetal DNA in maternal serum in pre-eclampsia. Clin. Chem. **45:** 184–188.
4. ZHONG, X.Y. *et al.* 2001. Elevation of both maternal and fetal extracellular circulating deoxyribonucleic acid concentrations in the plasma of pregnant women with pre-eclampsia. Am. J. Obstet. Gynecol. **184:** 414–419.
5. LEUNG, T.N. *et al.* 2001. Increased maternal plasma fetal DNA concentrations in women who eventually develop pre-eclampsia. Clin. Chem. **47:** 137–139.
6. ZHONG, X.Y. *et al.* 2001. Circulatory fetal and maternal DNA in pregnancies at risk and those affected by pre-eclampsia. Ann. N. Y. Acad. Sci. **945:** 138–140.
7. HAHN, S. & W. HOLZGREVE. 2002. Fetal cells and cell-free fetal DNA in maternal blood: new insights into pre-eclampsia. Hum. Reprod. Update **8:** 501–508.
8. BRINKMANN, V. *et al.* 2004. Neutrophil extracellular traps kill bacteria. Science **303:** 1532–1535.
9. SACKS, G. *et al.* 1999. An innate view of human pregnancy. Immunol. Today **20:** 114–118.

Use of Bi-Allelic Insertion/Deletion Polymorphisms as a Positive Control for Fetal Genotyping in Maternal Blood

First Clinical Experience

GODELIEVE C.M.L. PAGE-CHRISTIAENS,[a] BERNADETTE BOSSERS,[b] C. ELLEN VAN DER SCHOOT,[b] AND MASJA DE HAAS[b]

[a]*Division of Perinatology and Gynecology, University Medical Center, Utrecht, the Netherlands*

[b]*Sanquin Research and Diagnostic Services, Amsterdam, the Netherlands*

ABSTRACT: Amplification of fetal DNA in maternal plasma is a new way for non-invasive fetal genotyping in pregnancies at risk for disorders where the presence of a paternal DNA sequence contributes to the risk status of the fetus. We describe the use of a panel of 10 bi-allelic highly polymorphic markers to ascertain the presence and amplification of fetal DNA in case the fetus is negative for the targeted paternal "disease" sequence.

KEYWORDS: fetal DNA; maternal plasma; bi-allelic polymorphisms

INTRODUCTION

Iatrogenic pregnancy loss as a result of invasive diagnostic procedures, such as chorionic villus sampling or amniocentesis, is still one of the major reasons for women to decline prenatal diagnosis. During the last 10 years, many well-informed patients have shifted away from *invasive diagnostic* tests for Down's syndrome testing in favor of *non-invasive risk assessment* tests.[1] The discovery by Lo *et al.*[2] of the presence of free fetal DNA originating mainly from the placenta has opened a new window toward *non-invasive diagnostic* tests. Amplification of fetal DNA in maternal plasma theoretically allows the demonstration of any paternal sequence present in the fetus and absent in its mother. Fetal sexing and fetal *RHD* typing are hitherto the most frequent applications of this new technology. A review of the results from different published

Address for correspondence: Godelieve C.M.L. Page-Christiaens, Division of Perinatology and Gynecology, University Medical Centre, P.B. 85090, 3508 AB Utrecht, the Netherlands. Voice: +31-30-2506426; fax: +31-30-2505320.
e-mail: l.christiaens@umcutrecht.nl

Ann. N.Y. Acad. Sci. 1075: 123–129 (2006).
doi: 10.1196/annals.1368.016

series shows that the overall sensitivity of these *RHD* and *SRY* (sex-determining region Y gene) PCRs (polymerase chain reaction) is 97%. In the first trimester, the fetal DNA concentration is around the detection limit of the assays. When the targeted sequence is not amplified, it needs to be confirmed that indeed there was fetal DNA and that the assay was sensitive enough to amplify it. To this end, we have applied a set of real-time quantitative PCR assays as a positive control for the presence of fetal DNA in maternal plasma. The assay has been developed and validated to assess hematopoietic chimerism after bone marrow transplantation.[3] By dilution series, we have demonstrated that the sensitivity of these PCRs is similar to the sensitivity of the *RHD*- and *SRY* PCRs. We describe the first use of this positive control in clinical practice.

PATIENTS AND METHODS

Three groups of pregnant patients were studied. The first group (n = 23) was of women with anti-RhD alloantibodies and a presumably heterozygous partner as determined by serologic examinations. The second group (n = 23) was of carriers of a sex-linked disease, and the third group (n = 9) was of women at risk for carrying a baby with adrenogenital syndrome. All women were informed of the innovative aspect of the test. As a safeguard, fetal gender was ascertained by ultrasonography around 16 weeks in all cases with fetal sexing as an indication. Furthermore, in all cases, prenatal diagnosis was possible in chorionic villi as well as in amniotic fluid, allowing "rescue" invasive diagnosis in case the ultrasonographic gender differed from the one predicted by the plasma DNA.

For *RHD* typing, two regions of the *RHD* gene were amplified with TaqMan-technology-based RQ-PCR using the ABI-PRISM 7000/7700 Sequence Detection Systems (Applied Biosystems, Foster City, CA) as described in detail in Rijnders *et al.*[4] To check for the presence of fetal DNA, a Y-chromosome-specific region was amplified with the *SRY*- PCR.[5] If *RHD* and *SRY*- PCRs were negative, the presence of fetal DNA was ascertained through testing for sequences unique for the father. To this end, leukocyte-derived DNA from both parents was typed for 10 bi-allelic insertion-deletion polymorphisms (so-called S-alleles) to identify suitable S-allele paternal markers (*SPM*). Primer and probe sequences to amplify the *SPM* have been described elsewhere.[3] We implemented the panel composed of S01, S03, S04, S05, S06, S07, S08, S09, S10, and S11 as SPMs in our routine diagnostics. S02 is located on the Y chromosome and was therefore not used. The efficiency of the amplification of the *SPM* sequences was equal to that of the *RHD* and *SRY* RQ-PCRs. For all *SPM* PCRs, as for the *RHD* and *SRY* PCRs, the maximal theoretical sensitivity of 1 genome equivalent (6.6 pg of DNA) was reached (manuscript in preparation). If an input of 10 ng of *SPM*-allele negative DNA was used, nonspecific amplification was only obtained for three *SPM*: 1a, 8a, and 10a,

TABLE 1. Decision tree for fetal *RHD* typing in maternal plasma

RHD PCR	*SRY* PCR	*SPM*	Conclusion
Positive	Positive	Not done	RhD-positive
Positive	Negative	Not done	RhD-positive
Negative	Positive	Not done	RhD-negative
Negative	Negative	Positive	RhD-negative
Negative	Negative	Non-informative, negative, or background amplification	Inconclusive

TABLE 2. Decision tree for fetal sexing in maternal plasma

SRY PCR	*SPM*	Conclusion
Positive	Not done	♂
Negative	Positive	♀
Negative	Non-informative, negative, or background amplification	Inconclusive

SPM = S-allele paternal markers.

illustrating the high level of specificity of the *SPM* PCR assays. This background amplification was only seen after 39–40 cycles, but in view of the clinical importance, whenever background amplification was seen, the results were reported is inconclusive. For fetal sexing, the *SRY* PCR was used, again controlled by *SPM* in case of negative result. The *SPM* assay was done in maternal plasma in all cases with negative plasma results in the *RHD* and/or *SRY* PCR provided that informative markers (mother negative, father positive) were available. If all PCR assays were negative, an invasive test was advised (TABLES 1 and 2).

RESULTS

We tested 23 patients for fetal *RHD* status and 32 patients for fetal gender (TABLE 3). In 32 cases, *RHD* and/or *SRY* were positive. In 23 cases, *RHD* and/or *SRY* PCRs were negative and *SPM* was tested (FIGS. 1–3). Five of these were inconclusive either on account of the lack of suitable paternal markers (n = 1) or because of negative plasma results when the father was heterozygous for the informative markers (n = 2) or because of background amplification (n = 2),

TABLE 3. Indications for fetal genotyping in maternal plasma ($n = 55$)

Anti-RhD antibodies	23
Risk adrenogenital syndrome	9
Carrier of severe hemophilia A	8
Carrier of Duchenne muscular dystrophy	7
Carrier of other severe X-linked diseases	8

respectively, for S8a and S10a (TABLE 4). In all five cases, girls were born. In the two inconclusive cases, where the mother was RhD-negative, the newborns were also RhD-negative. The gender and/or *RHD* status of the other 50 cases were as predicted.

The clinical advantage obtained by the plasma test can be summarized as follows (TABLE 5): The PCR results made invasive tests unnecessary in 36 of the 55 cases. In 14 of 14 RhD-positive fetuses of sensitized mothers in this way boostering was avoided. In 7 of 23 sensitized mothers, the fetus was predicted RhD-negative and the pregnancy could be managed as a low-risk pregnancy. In 5 of 9 cases at risk for adrenogenital syndrome, the fetus was male and maternal dexamethasone use to prevent virilization could be tapered and stopped. In 19 of the 55 cases, fetal genotyping in maternal plasma did not result in any clinical advantage and invasive tests had to be performed. In one case, the clinician chose to perform an invasive test in a patient at risk for a fetus with X-linked disease, while the plasma test had predicted a female fetus. The fetal karyotype was indeed female as predicted.

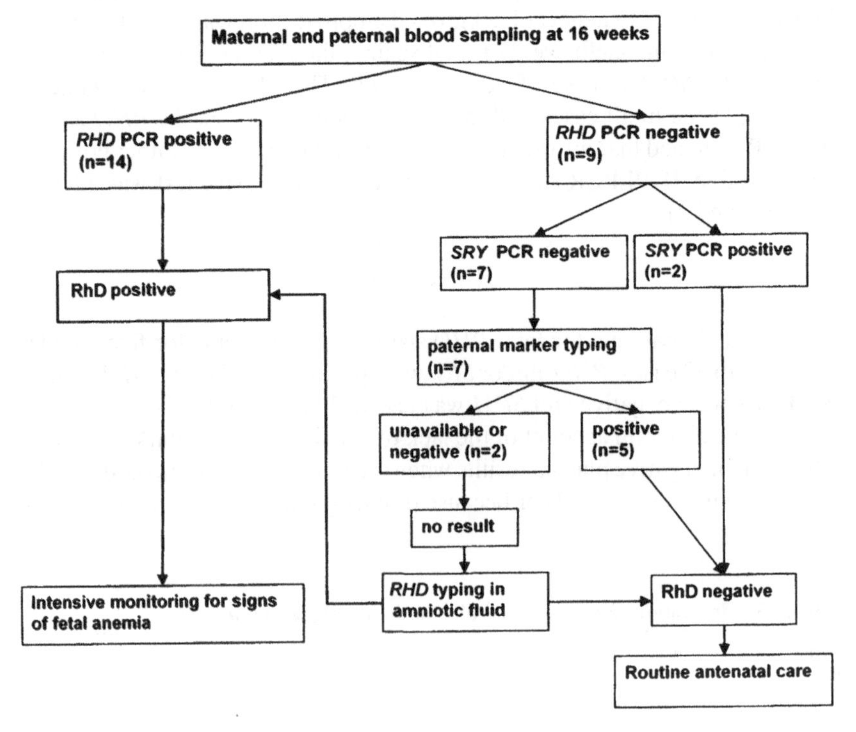

FIGURE 1. Clinical decision tree in case of anti-RhD alloantibodies in the pregnant mother.

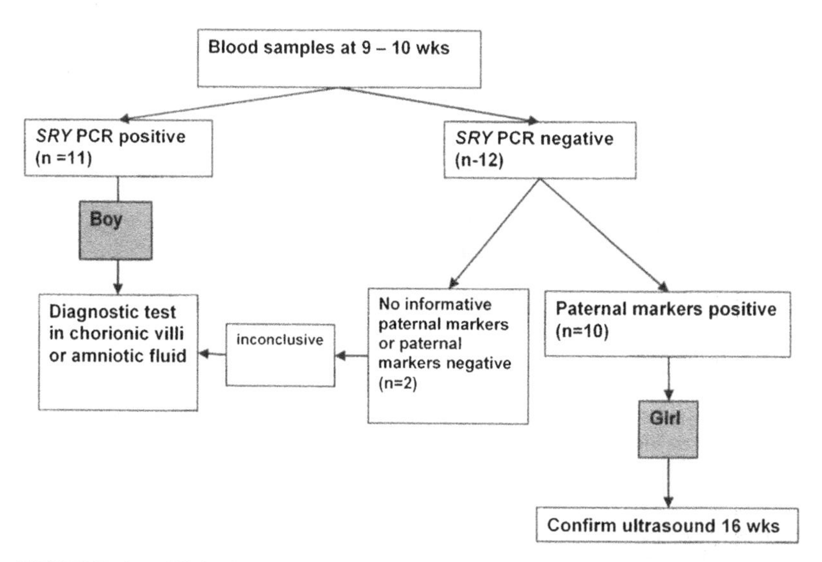

FIGURE 2. Clinical decision tree in case of an enhanced risk of fetus with X-linked recessive disease.

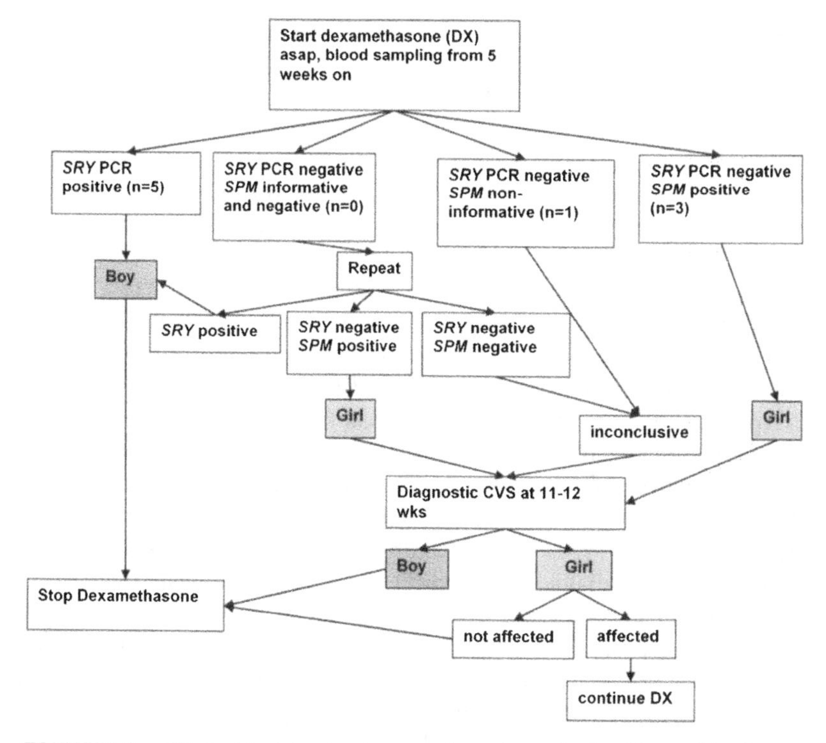

FIGURE 3. Clinical decision tree in case of an enhanced risk of fetus with adrenogenital syndrome.

TABLE 4. Information about the bi-allelic deletion/insertion polymorphisms (SPMs) in 23 families with negative *RHD* and/or *SRY* PCR in maternal plasma

	Number of cases	Paternal marker detectable in plasma	Number of failures to detect fetal DNA
No markers available	1	–	1
Father homozygous positive for ≥1 SPM	11	10	1
Father heterozygous positive for 1 *SPM* and not homozygous positive	1	0	1
Father heterozygous positive for 2 *SPM* and not homozygous positive	4	2	2
Father heterozygous positive for 3 *SPM* and not homozygous positive	2	2	0
Father heterozygous positive for 4 *SPM* and not homozygous positive	3	3	0
No paternal blood available	1	1	0

TABLE 5. Clinical gain of fetal genotyping in 55 cases

No invasive test needed	36/55
Shorter period of antenatal dexamethasone	5/9
Low-risk management	7/23
No gain	19/55

DISCUSSION

In 18 of 23 pregnancies, the use of a set of PCRs on bi-allelic insertion/deletion polymorphisms (*SPM*) were able to prove the presence and amplification of fetal DNA in sufficient amounts to serve as a positive control in case of negative *RHD* and *SRY* PCRs. In three pregnancies, the set of *SPM* was too limited to provide informative markers, and in another two pregnancies nonspecific background amplification with maternal leukocyte-derived DNA (Cts of 39-40) prompted us to be prudent. As a result of incorporation of these PCRs in the decision tree for fetal *RHD* and/or *SRY* genotyping, invasive procedures were avoided in more than 60% of cases overall and in all cases of RhD sensitization. Nevertheless, when fetal sexing was the indication, we chose to confirm fetal gender by ultrasound. Reasons for this policy are: (a) the utmost importance of accurate results, and (b) the effort required to confirm fetal gender ultrasonographically is small. The *SPM* assay can easily be implemented in a diagnostic setting. The relevant allele-specific PCRs can be performed in the same run as the diagnostic fetal genotyping PCR. For all PCRs, the maximal sensitivity of 1 genome equivalent was reached. Some precautions are necessary, however. First, if the allele-specific PCRs will be used as a control for our assays fetal genotyping that are less sensitive than the *SRY* or the *RHD* PCRs, the fetal DNA concentration as measured by the allele-specific PCR should be taken into account. Only when this concentration exceeds the detection level of the fetal genotyping assay, a negative genotyping assay is really negative. Second, it is desirable to develop an additional set of PCRs for those cases in which none of the described systems is informative, especially for non-Caucasian patients. Third, preferably more than one system should be used to prevent false-negative results due to an unknown null allele or third allele in incorrectly presumed bi-allelic systems.

CONCLUSION

We were not able to prove the presence of fetal DNA in 5 of 23 cases. However, the number of genetic marker systems tested can easily be increased to ensure the presence of an informative marker in a higher percentage of cases. In all cases, we correctly identified the fetal gender and/or *RHD* status. The use of bi-allelic insertion/deletion-based polymorphisms allowed an increase in the predictive value of a negative result of assays aimed at detecting fetal DNA sequences in maternal plasma.

REFERENCES

1. Benn, P.A. *et al.* 2005. Trends in the use of second trimester maternal serum screening from 1991 to 2003. Genet. Med. **7:** 328-331.
2. Lo, Y.M. *et al.* 1997. Presence of fetal DNA in maternal plasma and serum. Lancet **350:** 485–487.
3. Alizadeh, M. *et al.* 2002. Quantitative assessment of hematopoietic chimerism after bone marrow transplantation by real-time quantitative polymerase chain reaction. Blood **99:** 4618–4625.
4. Rijnders, R.J. *et al.* 2004. Clinical applications of cell-free fetal DNA from maternal plasma. Obstet. Gynecol. **103:** 157–164.
5. Lo, Y.M. *et al.* 1998. Quantitative analysis of fetal DNA in maternal plasma and serum: implications for non-invasive prenatal diagnosis. Am. J. Hum. Genet. **62:** 768–775.

PLAC1 mRNA in Maternal Blood Correlates with Doppler Waveform in Uterine Arteries in Normal Pregnancies at the Second and Third Trimester

ANTONIO FARINA, MANUELA CONCU, IRINA BANZOLA, ANNALISA TEMPESTA, SONIA VAGNONI, SANDRO GABRIELLI, MARA MATTIOLI, PAOLO CARINCI, GIANLUIGI PILU, DANILA MORANO, AND NICOLA RIZZO

Department of Histology and Medical Embryology, Division of Prenatal Medicine, University of Bologna, Bologna, Italy

ABSTRACT: **Doppler analysis of the uterine arteries is currently used for pre-eclampsia (PE) screening. *PLAC1* is a trophoblast-specific gene, and it is known that in normal pregnancies, trophoblastic cells are released into the maternal circulation, where specific trophoblastic mRNA can be detected. In PE, as in women who eventually develop PE, an abnormal passage of fetal and placental cells is also present. In this study, we aimed to verify whether, in normal pregnancies, Doppler waveform of the uterine arteries correlates with *PLAC1 mRNA* concentrations. Thirteen cases of normal pregnancies at 37 weeks' gestation (23–41) were enrolled in the study. *PLAC1 mRNA* was extracted from 2 mL of blood by ABI PRISM 6100 nucleic acid Prep Station (Applied Biosystems, Foster City, CA) and quantitative reverse transcriptase-polymerase chain reaction (RT-PCR) analysis was performed by a PE 5700 Sequence detection system. Bulk RNA from normal placental tissue was used as the reference curve, and the amount of *PLAC1 mRNA* in the study samples was then expressed as the "relative amount" of weight of placental tissue (ng/mL). The uterine arterial mean resistance index (RI) and presence/absence of a dicrote waveform were calculated by using a 5 MHz transabdominal probe (Tecnos, ESAOTE) at the uterine cervico-corporal junction. Doppler measurement was performed on the same day as blood collection. The median of the means of uterine arterial RI was 0.52 (0.39–0.68). RI of uterine arteries and *PLAC1 mRNA* were significantly correlated in a log-linear regression ($R^2 = 0.483$, $P = 0.024$). Our data support that in normal pregnancy, the passage of trophoblast material into the maternal circulation is correlated with the quantitative measurement of uterine hemodynamics.**

Address for correspondence: Antonio Farina, M.D., Via Belmeloro 8, 40126 Bologna, Italy. Voice: ++39051-2094097; fax: ++39051-2094110.
e-mail: antonio.farina@unibo.it

Ann. N.Y. Acad. Sci. 1075: 130–136 (2006).
doi: 10.1196/annals.1368.017

KEYWORDS: PLAC1 gene; circulating mRNA; maternal fetal trafficking; Doppler; uterine arteries

INTRODUCTION

Normal placentation relies on a proper trophoblastic invasion of the maternal decidua, myometrium, and blood vessels. Trophoblastic cells invade the uterine spiral arteries, modifying their endothelial lining and media, and convert them from small-diameter, high-resistance vessels into low-resistance nonresponsive channels.[1] The trophoblastic invasion of the spiral arteries and their conversion into low-resistance vessels is associated with a decrease in the vascular resistance in the uterine arteries as gestation progresses. In the presence of insufficient trophoblast invasion, the utero-placental circulation remains in a state of high resistance, which causes generalized endothelial cell injury, resulting in local ischemia and necrosis in the placenta.[2] Such a phenomenon is also associated with a higher degree of apoptosis and a higher passage of circulating nucleic acids.[3,4] It must be stressed that apoptosis is a normal feature of a subset of differentiating trophoblast cells in early pregnancy,[5,6] and it also appears to play a role in the normal development, remodeling, and ageing of the placenta.[7] Recent studies suggest that the vast majority of the fetal DNA and possibly also RNA present in the maternal plasma originates from apoptotic trophoblast cells.[8,9] Thus, poor trophoblast invasion, apoptosis, and circulating nucleic acids should be correlated. Since the degree of the utero-placental resistance can be assessed by uterine artery Doppler analysis, we hypothesized that utero-placental perfusion could be quantitatively associated with placental apoptosis, and could therefore correlate with trophoblast mRNA levels in the maternal circulation. In this study, we used mRNA for the *PLAC1* gene for evaluating the amount of the nucleic acid present in the maternal circulation.

MATERIALS AND METHODS

Study Subjects

All study subjects provided informed written consent for the use of biological specimens for research purposes. Peripheral venous blood samples (5 mL) were placed into ethylenediaminetetraacetic acid (EDTA) tubes from 11 low-risk women undergoing routine ultrasound. Gestational age was established from the fetal crown-rump length measurement, and expressed as days for statistical purposes. This protocol was approved by the Institutional Review Boards of the University of Bologna, Policlinico Sant'Orsola, Malpighi, Bologna, Italy.

Amplification and Measurement of Placental mRNA from Maternal Blood

The method is described in detail elsewhere.[10] Peripheral blood samples (500 μL) were collected into tubes containing EDTA and processed within 1 h. An equal volume of PBS 1X (phosphate-buffered saline solution) and a double volume of lysis reagent (Applied Biosystems, Foster City, CA) were added (final volume 2 mL). The lysis solution is part of the starter kit protocol and is needed to lyse the sample and to protect the free nucleic acids from degradation. The samples were stored at –20°C until further processing.

Total RNA was extracted with reagents and disposable starter kits for ABI Prism 6100 nucleic acid Prep Station (Applied Biosystems), which yields high-quality RNA. From 2 mL peripheral blood, 100 μL of RNA was collected. During RNA extraction, we included an additional step with Absolute RNA Wash Solution, which is designed to remove contaminating DNA and polymerase chain reaction (PCR)-inhibitory substances from purified RNA. This solution is able to directly remove background DNA *in situ* from immobilized RNA on the purification tray during the standard protocol developed by Applied Biosystems. After a brief incubation, the subsequent wash step removes the reagent.

Reverse transcription was performed using high-capacity cDNA archive kits (Applied Biosystems) according to the manufacturer's protocol. The whole sample of total RNA from peripheral blood was reverse-transcribed in a final volume of 100 μL, using RT buffer 10×, dNTPs 25×, and RT random primers 10× 50 U/μL MultiScribe RT. Incubation in a GenAmp PCR System 9600 thermal cycler was performed in two steps: 10 min at 25°C, followed by 120 min at 37°C.

Quantitative real-time PCR analysis was performed using a PE Applied Biosystems 5700 Sequence Detection System (Applied Biosystems). Primers and a dual-labeled probe were designed using the Primer Express version 2.0 software (Applied Biosystems). Sequence data were obtained from the GenBank Sequence Database. To determine the amount of cDNA, the *PLAC1* locus was used.

The primers and probe combination was:
PLAC1 forward: 5′-ATT GGC TGC AGG GAT GAA AG –3′;
PLAC1 reverse: 5′-TTT GGG GTC TCC TGA AGA TG –3′;
PLAC1 probe: 5′-(Fam) CTA CGA GGT GTT CAG CTT GTC
ACA GTC CA (Tamra)-3′.

The size of the fragment analyzed was 680 base pairs.

For PCR analysis, 50 μL of reaction volume contained 10 μL of cDNA (corresponding to 0.2 mL of the initial blood sample). The reaction mix contained the amplification primer (900 nM), the dual-labeled probe (100 nM), and the necessary components provided in the TaqMan Universal PCR Master Mix (Applied Biosystems). This corresponded to 1.25 U of AmpliTaq Gold DNA Polymerase, 0.5 U of AmpEase uracil N-glycosylase (UNG), each dATP, dCTP, and dGTP (200 μM), 400 μM dUTP, 10X TaqMan Buffer A, 25 mM MgCl2. The AmpErase UNG activity, in combination with dUTP, was used to prevent contamination by carry-over of PCR products. Buffer A contains Passive Reference I for signal normalization in all TaqMan reactions. Each sample was analyzed in triplicate and the means were used for statistical purposes. The thermal profiles were obtained by a 2-min incubation at 50°C, followed by an initial 10-min denaturation step at 95°C, and by 40 cycles of 1 min each at 60°C plus 15 sec at 95°C. RNA prepared from normal placental tissue was reverse-transcribed by the same method utilized for the blood samples and used for calibration curves. Serial dilutions of the cDNA preparation were run in triplicate with each analysis. The amount of cDNA in the samples was expressed "relative amount." The relative amount of *PLAC1 mRNA* in the samples was expressed as the weight of placental tissue. Since each sample was derived from 0.2 mL of maternal blood, data were expressed as "relative concentrations (ng/mL of maternal blood)."

Doppler Analysis

The uterine arteries' mean resistance index (RI) and presence/absence of a dicrote waveform were measured using a 5 MHz transabdominal probe (Tecnos, ESAOTE) at the uterine cervico-corporal junction. The Doppler measurement was performed on the same day as blood collection.

Statistics

A log-linear regression was used to assess the relationship between the RI of uterine arteries, and *PLAC1 mRNA*. *PLAC1 mRNA* data were adjusted for maternal body mass index.

RESULTS

Gestational age at the time of enrollment was 37 (range, 23–41) weeks or 262 (161–288) days. All the pregnant women here enrolled ended with a normal delivery and puerperium. The median RI was 0.52 (0.39–0.68). Three patients had an abnormal RI value (>0.60) but did not develop any symptom of pre-eclampsia (PE) throughout pregnancy. Median log *PLAC1 mRNA* value was 3.11 (2.63–3.18) pg/mL. RI of uterine arteries and *PLAC1 mRNA* were significantly correlated in a log-linear regression ($R^2 = 0.483$, $P = 0.024$) as shown in FIGURE 1.

DISCUSSION

It has been previously reported that at the second trimester, circulating fetal DNA is higher in subjects positive on uterine artery Doppler velocimetry

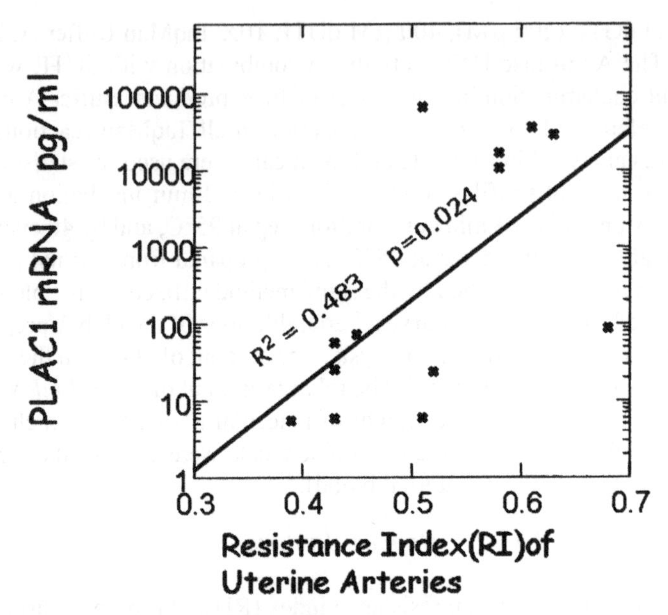

FIGURE 1. Scatterplot of Doppler image of uterine arteries versus PLAC1 mRNA. $R^2 = 0.483$, $P = 0.024$.

analysis and at high risk of developing intrauterine growth restriction.[11] In this paper, we used a specific trophoblastic mRNA to evaluate whether a quantitative correlation would also be present between Doppler findings and circulating mRNA in normal pregnancies. The fact that apoptotic trophoblastic cells from the placenta can be isolated in maternal circulation along with their specific mRNAs has been widely reported.[12,13] Previous studies suggest that enhanced apoptosis may interfere with the process of placentation and placental ischemia;[5] and that uterine artery Doppler indices and histological features of placentation are correlated. Thus, it is logical to suppose that quantitative Doppler studies, apoptosis, and mRNA are all correlated in both normal and affected pregnancies. However, several factors can affect the degree of correlation: for example, the mRNA concentration in the blood may vary with the intracellular expression and with the number of trophoblastic cells expressing that specific mRNA per unit of blood. Since we did not find any significant correlation between gestational age and *PLAC1 mRNA* during 23–41 weeks, however, we can speculate that the correlation between the Doppler measurement (that actually decreases as the gestation progresses) and *PLAC1* is due to a different degree of cellular passage into the maternal blood. However, concentration of other mRNAs in the maternal plasma, such as *CRH,* were higher than expected in pregnancies affected by PE.[14] It is possible that *PLAC1* could have a different distribution in PE patients. We previously noted that the concentration of *PLAC1* gene mRNA in the maternal blood is lower in patients suffering from early vaginal bleeding, but only before 10 weeks of gestation, because of possible underlying delayed or abnormal placental growth.[15] In conclusion, the levels of circulating mRNA in normal pregnancies at the second and third trimester is also correlated with the uterine artery Doppler values. Circulating mRNAs could be useful for the prediction of PE at the second–third trimester, either alone or in association with Doppler measurements.

ACKNOWLEDGMENTS

This work was supported by Fondazione CARISBO Progetto triennale—Molecular Genetics of Fetal DNA, Progetto E.F. 2002 and ex-60% A.F.

REFERENCES

1. Brosens, I. *et al.* 1967. The physiological response of the vessels of the placental bed to normal pregnancy. J. Pathol. Bacteriol. **93:** 569–579.
2. Khong, T.Y. & C. Mott. 1993. Immunohistologic demonstration of endothelial disruption in acute atherosis in preeclampsia. Eur. J. Obstet. Gynecol. Reprod. Biol. **51:** 193–197.
3. Zhong, X.Y. *et al.* 2002. Cell-free fetal DNA in the maternal circulation does not stem from the transplacental passage of fetal erythroblasts. Mol. Hum. Reprod. **8:** 864–870.
4. Farina, A. *et al.* 2004. Cell-free fetal DNA (SRY locus) concentration in maternal plasma is directly correlated to the time elapsed from the onset of preeclampsia to the collection of blood. Prenat. Diagn. **24:** 293–297.
5. Levy, R. & D.M. Nelson. 2000. To be, or not to be, that is the question: apoptosis in human trophoblast. Placenta **21:** 1–13.
6. Huppertz, B. & P. Kaufmann. 1999. The apoptosis cascade in human villous trophoblast: a review. Trophoblast Res. **13:** 215–242.
7. Smith, S.C. *et al.* 1997. Placental apoptosis in normal human pregnancy. Am. J. Obstet. Gynecol. **177:** 57–65.
8. Sekizawa, A. *et al.* 2004. Proteinuria and hypertension are independent factors affecting fetal DNA values: a retrospective analysis of affected and unaffected patients. Clin. Chem. **50:** 221–224.
9. Austgulen, R. *et al.* 2004. Pre-eclampsia: associated with increased syncytial apoptosis when the infant is small-for-gestational-age. J. Reprod. Immunol. **61:** 39–50.
10. Concu, M. *et al.* 2005. Rapid clearance of mRNA for PLAC1 gene in maternal blood after delivery. Fetal. Diagn. Ther. **20:** 27–30.
11. Caramelli, E. *et al.* 2003. Cell-free fetal DNA concentration in plasma of patients with abnormal uterine artery Doppler waveform and intrauterine growth restriction—a pilot study. Prenat. Diagn. **23:** 367–371.
12. Taniguchi, R. *et al.* 2000. Trophoblastic cells expressing human chorionic gonadotropin genes in peripheral blood of patients with trophoblastic disease. Life Sci. **66:** 1593–1601.

13. VAN WIJK, I.J. *et al.* 1998. Identification of HASH2-positive extravillous trophoblast cells in the peripheral blood of pregnant women. Trophoblast Res. **11:** 23–33.
14. FARINA, A. *et al.* 2004. Circulating corticotrophin-releasing hormone mRNA in maternal plasma: relationship with gestational age and severity of preeclampsia. Clin. Chem. **50:** 1851–1854.
15. FARINA, A. *et al.* 2005. Lower maternal PLAC1 mRNA in pregnancies complicated with vaginal bleeding (threatened abortion <20 weeks) and a surviving fetus. Clin. Chem. **51:** 224–227.

Different Approaches for Noninvasive Prenatal Diagnosis of Genetic Diseases Based on PNA-Mediated Enriched PCR

SILVIA GALBIATI,[a] GABRIELLA RESTAGNO,[b] BARBARA FOGLIENI,[a] SARA BONALUMI,[a] MAURIZIO TRAVI,[c] ANTONIO PIGA,[d] LUCA SBAIZ,[b] MARCELLA CHIARI,[e] FRANCESCO DAMIN,[e] MADDALENA SMID,[f] LUCA VALSECCHI,[f] FEDERICA PASI,[f] AUGUSTO FERRARI,[f,g] MAURIZIO FERRARI,[a,g,h] AND LAURA CREMONESI[a]

[a]*Genomic Unit for the Diagnosis of Human Pathologies, H. San Raffaele, Milan, Italy*

[b]*Dipartimento di Patologia Clinica, AOOIRM-S Anna, Torino, Italy*

[c]*Ospedale Maggiore Policlinico Mangiagalli Regina Elena, Milan, Italy*

[d]*Department of Pediatrics, University of Torino, Torino, Italy*

[e]*Istituto di Chimica del Riconoscimento Molecolare CNR, Milan, Italy*

[f]*Department of Obstetrics and Gynecology, San Raffaele Scientific Institute, Milan, Italy*

[g]*Università Vita-Salute San Raffaele, Milan, Italy*

[h]*Diagnostica e Ricerca San Raffaele SpA, Milan, Italy*

ABSTRACT: The aim of this work was to develop advanced and accessible protocols for noninvasive prenatal diagnosis of genetic diseases. We are evaluating different technologies for mutation detection, based on fluorescent probe hybridization of the amplified product and pyrosequencing, a technique that relies on the incorporation of nucleotides in a primer-directed polymerase extension reaction. In a previous investigation, we have already proven that these approaches are sufficiently sensitive to detect a few copies of a minority-mutated allele in the presence of an excess of wild-type DNA, In this work, in order to further enhance the sensitivity, we have employed a mutant enrichment amplification strategy based on the use of peptide nucleic acids (PNAs). These DNA analogues bind wild-type DNA, thus interfering with its amplification while still allowing the mutant DNA to become detectable. We have synthesized different PNAs, which are highly effective in clamping wild-type DNA in the beta-globin gene region, where four beta-thalassemia

Address for correspondence: Dr. Laura Cremonesi, Genomic Unit for the Diagnosis of Human Pathologies, San Raffaele Scientific Institute, Via Olgettina 58, 20132 Milan, Italy. Voice: 39-0226434779; fax: 39-022643351.
e-mail:cremonesi.laura@hsr.it

**Ann. N.Y. Acad. Sci. 1075: 137–143 (2006). © 2006 New York Academy of Sciences.
doi: 10.1196/annals.1368.018**

mutations are located (IVSI.110, CD39, IVSI.1, IVSI.6) plus HbS. The fluorescence microchip readout allows us to monitor the extent of wild-type allele inhibition, thus facilitating the assessment of the optimal PNA concentration.

KEYWORDS: noninvasive prenatal diagnosis; fetal DNA in maternal plasma; PNA-mediated enriched PCR; beta-thalassemia

INTRODUCTION

Medicine is slowly advancing towards noninvasive diagnostic and therapeutic approaches. Release of fetal DNA into maternal plasma is a well-known phenomenon which is being extensively studied in order to develop safe and reliable procedures for monitoring pregnancy complications and noninvasive prenatal diagnosis of genetic diseases. Despite the scarcity of fetal DNA in maternal plasma, the detection of unique fetal sequences, such as the Y-chromosome in male fetuses or the RhD factor in RhD-negative mothers, can be performed by real-time quantitative polymerase chain reaction (PCR) methods with an accuracy close to 100%.[1,2] In contrast, the identification of paternally inherited mutations in maternal plasma is more challenging, mainly because of the excess of maternal DNA, which may differ by as little as one base pair from the fetal one.

To date, direct detection of fetal mutations in maternal plasma has been performed only by mass spectrometry, which has been applied to the identification of Southeast Asian beta-globin gene mutations.[3] However, this approach is based on the use of a highly sophisticated and expensive apparatus which is not easily available. Therefore, alternative strategies suitable for large-scale application are still of great interest.

In an attempt to selectively amplify fetal DNA sequences, our group has explored the feasibility of binding maternal wild-type DNA via peptide nucleic acid (PNA)-mediated PCR.[4,5] PNAs are DNA homologues in which the phosphate sugar backbone is replaced by a peptide polymer. This confers upon PNAs the capacity to hybridize with high affinity to complementary DNA sequences and prevent their amplification.[6] On the basis of this approach, detection of fetal mutations was reported by allele-specific PNA-assisted amplification following enrichment of plasma DNA for fetal sequences by size separation in an agarose gel.[7] Our main goal is to develop methods easily applicable to the direct detection of fetal sequences in maternal plasma without need of size enrichment. To this end, we intend to couple different detection systems based either on microchips or on pyrosequencing with the selective amplification of the mutated fetal DNA obtained in the presence of PNAs. We have already proven that both approaches are sensitive enough to detect a few copies of a minority-mutated allele in the presence of an excess of wild-type DNA (see Ref. 5 and unpublished results). We have used the detection of mutations causing beta-thalassemia as a model system with these approaches and applied these to the analysis of maternal plasma in couples at risk of carrying a fetus with beta-thalassemia.

PNA-Mediated Enriched PCR

PNAs complementary to the wild-type sequence of each mutation of interest in the beta-globin gene (IVSI.110, CD39, IVSI.1, IVSI.6, and HbS) have been designed. Several PNA-binding formats have been explored, including those where the PNA site overlaps or is located adjacent to or at a certain distance from the PCR primer site. Additionally, various PNAs differing in base composition and length have been designed and tested for each mutation of interest. PNAs have been included in the PCR reaction mixture to obtain the maximum binding of the wild-type alleles.

DNA Microchips

For the identification of paternally inherited mutations or single nucleotide polymorphisms (SNPs), we evaluated the NMW 1000 NanoChip Molecular Biology Workstation (Nanogen, San Diego, CA).[8] In this system, negatively charged biotinylated amplicon is electronically addressed to selective pads on the array where they remain embedded through interaction with streptavidin in the permeation layer. The DNA at each pad is then hybridized with the wild-type and the mutant reporters, labeled with the fluorescent tags Cy3 and Cy5. After a step of thermal stringency to detach mismatched probes, the array is imaged and the fluorescence quantified.[9,10]

As an alternative, standard microarray technology was also evaluated. Amino-modified amplicon was printed in quadruplicate on custom-made silicon slides coated with a functional polymer by use of a Genetix QArray2 apparatus.[11] The printing solution was prepared in 1× printing buffer (150 mM sodium phosphate [pH 8.5]). After printing, the amplicon was coupled to the slide surface by overnight incubation in a NaCl humidity chamber and then washed with a blocking solution (50 mM ethanolamine, 1 M Tris, pH 9.0) for 15 min at 50°C to inactivate residual functional groups. The slides were stored at room temperature under desiccating conditions until further use. Immediately before use, printed slides were immersed for 2 min in boiling water and hybridized for 1 h at room temperature with wild-type and mutant reporters identical to those used with the Nanogen platform. The slides were scanned with a Scan Array Express scanner from Packard Bioscience (Billerica, MA, USA) for fluorescence quantification.

Pyrosequencing

Pyrosequencing (Pyrosequencing AB, Uppsala, Sweden) is a real-time sequencing method that relies on the sequential addition and incorporation of nucleotides in a primer-directed polymerase extension reaction.[12] The pyrophosphate released on incorporation of a base-paired nucleotide is coupled to the enzymes ATP sulfurylase and luciferase to generate detectable light, which is proportional to the number of nucleotides incorporated. The technology is based on immobilization of a biotinylated PCR template onto streptavidin-coated beads, followed by generation of a single-stranded template and annealing of a sequencing primer. In the pyrosequencing reaction, nucleotides are added sequentially to an enzyme buffer system containing Klenow DNA polymerase and the prepared DNA template. The incorporation of nucleotides during DNA synthesis generates pyrophosphate (Ppi), which triggers a cascade of enzymatic reactions that produce visible light in proportion to the number of nucleotides incorporated, followed by the degradation of nonincorporated nucleotides by the action of apyrase.

Plasma from a carrier of the IVS1.110 mutation

A

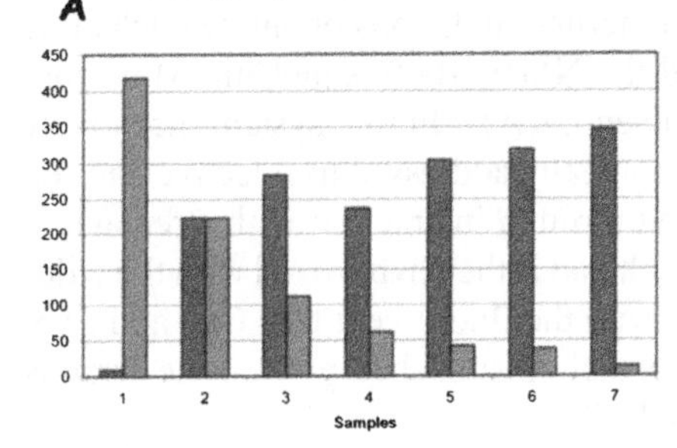

Sample	Genotype	PNA (μM)	R:G ratio
1	wt/wt	0	1.46
2	wt/mt	0	1:1
3	wt/mt	0.10	2.5:1
4	wt/mt	0.25	3.8:1
5	wt/mt	0.5	7.1:1
6	wt/mt	1	8.4:1
7	mt/mt	0	24:1

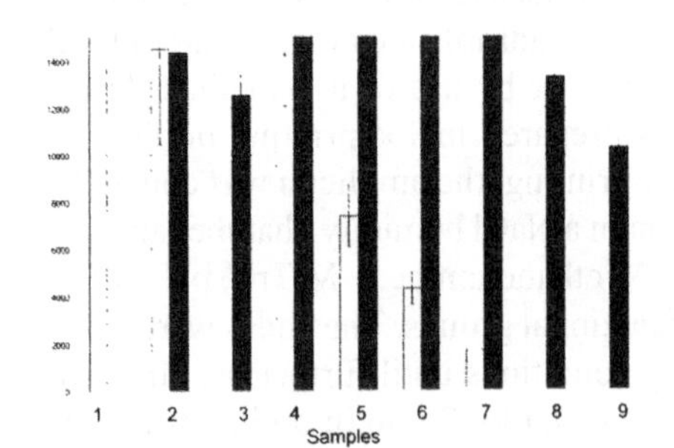

Sample	Genotype	PNA (μM)	R:G ratio
1	wt/wt	0	>1:100
2	wt/mt	0	1:1
3	mt/mt	0	>100:1
4	wt/mt	0	2:1
5	wt/mt	0.05	4.2:1
6	wt/mt	0.10	4:1
7	wt/mt	0.25	12:1
8	wt/mt	0.5	>100:1
9	wt/mt	1	>100:1

■ Mutated probe
■ Wild-type probe

FIGURE 1. Inhibition of the wild-type allele through PNA-clamping-mediated enriched PCR in the presence of increasing amounts of PNA. (**A**) Analysis performed by the Nanogen microchip system (*upper panel*) and the home-made microchip system (*lower panel*) of a subject carrying the IVSI.110 mutation in the beta-globin gene in the presence of increasing amounts of PNA. (**B**) Analysis performed by the pyrosequencing system of subjects carrying the IVSI.110 (B1) and CD39 mutations (B2) at the higher PNA concentration.

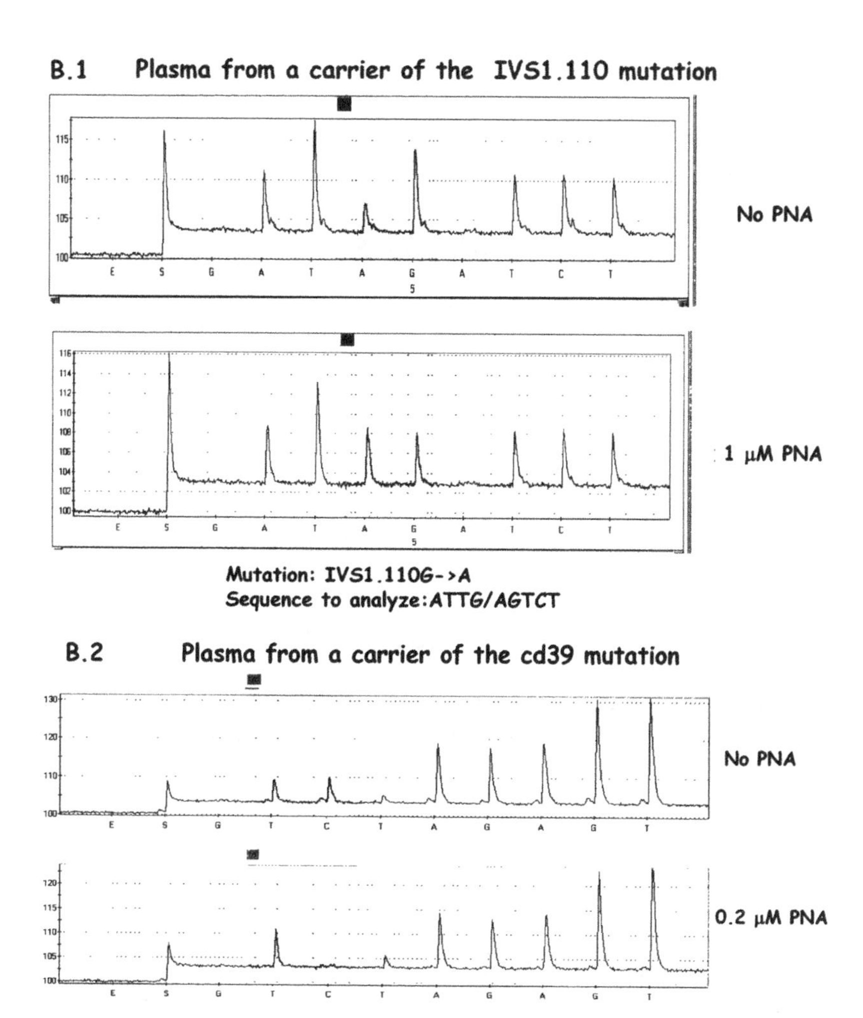

FIGURE 1. *Continued.*

RESULTS

In order to evaluate the applicability of our approach to noninvasive prenatal diagnosis of beta-thalassemia mutations, we assessed the sensitivity of the microchip and pyrosequencing systems in detecting different levels of PNA-mediated inhibition of the wild-type allele in DNA samples extracted from plasma of subjects carrying mutations in the beta-globin gene (FIG. 1). Both approaches were able to monitor the variation of wild-type-to-mutated allele fluorescence ratio in the presence of different amounts of wild-type PNAs. This may be very useful for evaluating the optimal PNA concentration in the PCR mixture to obtain the highest interference with the amplification of the wild-type allele.

CONCLUSIONS

Our preliminary data obtained by PNA-mediated mutant enrichment amplification look very promising, indicating that this approach, when optimized, could be used in combination with different microchip technologies and with the pyrosequencing system.

On the basis of these preliminary results, we are now, blindly, analyzing maternal plasma in couples undergoing invasive prenatal diagnosis for beta-thalassemia, where mother and father carry different mutations, in parallel with analysis on DNA extracted from chorionic villi.

ACKNOWLEDGMENTS

Funded by Telethon GG904016 (L.C.), MIUR Cofin 2002 (A.F.), CIPE 2003 Regione Piemonte (G.R.), and European Union FP6 Network of Excellence "SAFE" (L.C.).

REFERENCES

1. Galbiati, S. *et al.* 2005. Fetal DNA detection in maternal plasma throughout gestation. Hum. Genet. **117:** 243–248.
2. Lo, Y.M. 2005. Recent advances in fetal nucleic acids in maternal plasma. J. Histochem. Cytochem. **53:** 293–296.
3. Ding, C. *et al.* 2004. MS analysis of single-nucleotide differences in circulating nucleic acids: application to noninvasive prenatal diagnosis. Proc. Natl. Acad. Sci. USA **101:** 10762–10767.
4. Orum, H. *et al.* 1993. Single base pair mutation analysis by PNA directed PCR clamping. Nucleic Acids Res. **21:** 5332–5336.
5. Cremonesi, L. *et al.* 2004. Feasibility study for a microchip-based approach for noninvasive prenatal diagnosis of genetic diseases. Ann. N. Y. Acad. Sci. **1022:** 1–8.
6. Nielsen, P.E., M. Egholm & O. Buchardt. 1994. Peptide nucleic acid (PNA): a DNA mimic with a peptide backbone. Bioconjug. Chem. **5:** 3–7.
7. Li, Y. *et al.* 2005. Detection of paternally inherited fetal point mutations for beta-thalassemia using size-fractionated cell-free DNA in maternal plasma. JAMA **293:** 843–849.
8. Sosnowski, R.G. *et al.* 1997. Rapid determination of single base mismatch mutations in DNA hybrids by direct electric field control. Proc. Natl. Acad. Sci. USA **94:** 1119–1123.

9. Santacroce, R. *et al.* 2002. Methods for analysis of clinically relevant single nucleotide polymorphisms using microelectronic array technology. Clin. Chem. **48:** 2124–2130.
10. Foglieni, B. *et al.* 2004. Beta-thalassemia microelectronic chip: a fast and accurate method for mutation detection. Clin. Chem. **50:** 73–79.
11. Pirri, G. *et al.* 2004. Characterization of a polymeric adsorbed coating for DNA microarray glass slides. Anal. Chem. **76:** 1352–1358.
12. Fakhrai-Rad, H., N. Pourmand & M. Ronaghi. 2002. Pyrosequencing: an accurate detection platform for single nucleotide polymorphisms. Hum. Mutat. **19:** 479–485.

Detection of SNPs in the Plasma of Pregnant Women and in the Urine of Kidney Transplant Recipients by Mass Spectrometry

YING LI,[a] DEIRDRÉ HAHN,[b] FRIEDEL WENZEL,[c]
WOLFGANG HOLZGREVE,[a] AND SINUHE HAHN[a]

[a]*University Women's Hospital/Department of Research, University of Basel, Basel, Switzerland*

[b]*Division of Paediatric Nephrology, University of the Witwatersrand and Johannesburg Hospital, Johannesburg, South Africa*

[c]*Medical Genetics, Department of Research, University of Basel, Switzerland*

ABSTRACT: Recently, it has been discovered that cell-free fetal DNA is smaller than corresponding maternal DNA. Therefore, circulating fetal DNA can be enriched by size-fractionation. Such a selection improves the non-invasive prenatal diagnosis of paternally inherited single gene mutations. Recent studies showed that MALDI-TOF mass spectrometry (MS) can be used to reliably detect fetal-specific single-nucleotide polymorphisms (SNPs) in maternal plasma. In this study, we looked at whether the size-fractionation approach could improve the detection of paternally inherited SNPs by MS assay. Our results indicated that the size-fractionation approach improved the analysis of paternally inherited SNP alleles. Our previous studies showed that donor-derived STR sequences could be detected in the urine of kidney transplant recipients. Here, we also examined whether donor-specific SNPs could be detected in recipient's urine by MS.

KEYWORDS: SNPs; cell-free DNA; size-fractionation; urinary DNA

INTRODUCTION

Although circulating fetal DNA has been known to be present in maternal plasma for many years,[1] the clinical applications of its measurement have mainly focused on the detection of fetal DNA sequences, which are absent

Address for correspondence: Dr. Sinuhe Hahn, Laboratory for Prenatal Medicine, University Women's Hospital, Department of Research, Spitalstrasse 21, CH 4031 Basel, Switzerland. Voice: ++41-61-265-9224; fax: ++41-61-265-9399.
e-mail: shahn@unbs.ch

Ann. N.Y. Acad. Sci. 1075: 144–147 (2006). © 2006 New York Academy of Sciences.
doi: 10.1196/annals.1368.019

from maternal genome, such as Y chromosome–specific gene or Rhesus D gene in RhD–negative women. Since only about 3–6% of the total circulating DNA in maternal plasma is of fetal origin, the analysis of feto-maternal single gene differences is much more difficult.[2]

It has recently been shown that the majority of cell-free fetal DNA in maternal plasma is of smaller size than maternal DNA, and that on this basis circulating fetal DNA could be enriched by selecting fragments with a size of less than 300 bp.[3] The enriched circulating fetal DNA can be used to improve the detection of masked fetal polymorphic loci, such as short tandem-repeat (STR) sequences and single gene point mutations.[3–5]

Matrix-assisted laser desorption/ionization time-of-flight (MALDI-TOF) mass spectrometry (MS) is a highly sensitive technology for single-nucleotide polymorphism (SNP) genotyping. Recently, it has been shown that it can reliably detect single nucleic differences in maternal plasma, including fetal gene point mutations.[6] Such detections cannot be reliably made by more conventional polymerase chain reaction (PCR)–based methods. We were thus interested to see whether the size-fractionation approach could improve the detection of paternally inherited SNP alleles in the maternal plasma by MS analysis.

Several studies have showed that cell-free donor-derived DNA can be detected in the urine of renal transplant recipients.[7] Quantitative analysis indicated that the urinary donor-derived cell-free DNA may be useful for the monitoring of kidney transplant engraftment because increased cell-free DNA is associated with acute graft rejection.[7,8] However, these studies frequently relied on the detection of Y chromosome-specific sequences in sex-disparate donor–recipient pairs, in which the female recipient received male kidneys. We previously reported that the PCR analysis of STR loci can be used for the detection of donor-specific urinary cell-free DNA.[9] Now we examined whether the MS assay could permit the detection of donor-derived SNP alleles in urinary cell-free DNA.

Detection of Paternally Inherited SNPs in Maternal Plasma

To further confirm that size-fractionation could improve the detection of more subtle feto-maternal genetic differences, we used the highly sensitive MALDI-TOF MS to detect paternally inherited fetal SNP alleles in maternal plasma. Plasma (5–7 mL) from pregnant women was used for DNA isolation, and the purified plasma DNA was size-fractionated by agarose gel electrophoresis.[3] Maternal and fetal genomic DNA was obtained from maternal blood cells and fetal cord blood or amniotic fluid cells, respectively. The analysis of SNP loci was performed by MALDI-TOF MS with the homogenous MassEXTEND assay (Sequenom, San Diego, CA). Maternal SNP alleles and fetal SNP alleles were genotyped. The informative markers, which are present in fetal genome and absent from maternal genome, were used to examine circulating fetal DNA. Our results clearly showed that, in the analysis of total circulating plasma DNA, only maternal allele was detected, or the paternally inherited fetal allele was only slightly detectable. However, in the analysis of size-fractionated circulating DNA of less than 300 bp, the paternally inherited fetal allele was readily detected in all cases (FIG. 1).

Detection of Donor-Derived SNPs in the Urine of Kidney Transplant Recipients

We have examined whether MALDI-TOF MS can be used for the detection of donor-specific SNPs in cell-free urinary DNA. In total, we examined 20 SNP markers. These were tested on four cases involving related living-donor kidney transplants, which we had previously examined for the presence of STR loci.[9] Genomic and urinary DNA was extracted as described previously.[9] First, the donor-recipient pairs were genotyped using genomic DNA to identify informative SNP alleles. The results indicated that in 16 instances the donor-derived SNP allele was absent from the recipient genome. The examination of recipients' urinary DNA has shown that in each of those 16 instances, the pertinent donor-specific SNP alleles could be reliably detected in the corresponding urinary cell-free DNA (FIG. 2).

CONCLUSION

Our data have further confirmed our previous finding that cell-free fetal DNA exists in maternal plasma as small-size fragments. Thus, cell-free fetal DNA can be enriched by size-fractionation, which permits the improvement of the detection of paternally inherited fetal loci, which are similar to maternal background DNA, such as SNPs. The highly sensitive MALDI-TOF MS

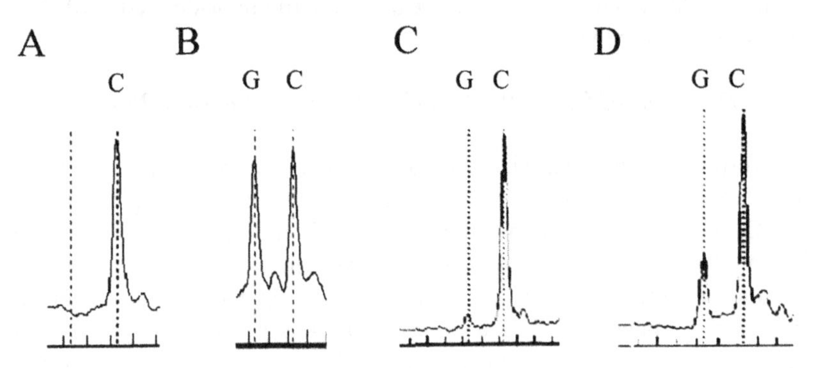

FIGURE 1. Detection of both paternally and maternally inherited SNPs in size-fractionated circulating DNA (SNP_ID:13234721). (**A**) Maternal SNP genotype with single C allele. (**B**) Fetal SNP genotype with both GC alleles. (**C**) Analysis of total circulating DNA. Paternally inherited fetal SNP G allele was only slightly detectable. (**D**) Analysis of size-fractionated circulating DNA with a size <300bp. Paternally inherited fetal SNP G allele was clearly detected.

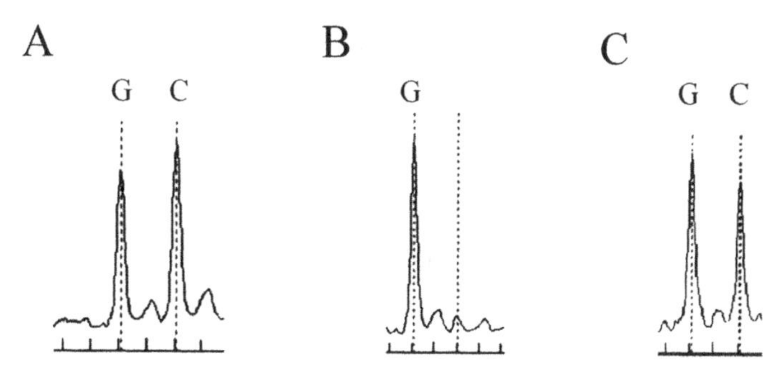

FIGURE 2. Detection of donor-derived SNP in the urine of a kidney transplant recipient (SNP_ID:222775). (**A**) Donor SNP genotype with both GC alleles. (**B**) Recipient SNP genotype with single G alleles. (**C**) Analysis of recipient's urine. The donor-derived SNP C allele was clearly detectable.

technology may provide a new alternative for non-invasive prenatal diagnosis of fetal single gene point mutations in maternal plasma.

Our study also indicated that donor-derived SNP alleles can be detected in the urine of kidney transplant recipients by the MS assay. Provided that this approach is truly amenable to quantification, this strategy could provide a new method for the precise assessment of donor-derived cell-free DNA levels.

REFERENCES

1. Lo, Y.M. *et al.* 1997. Presence of fetal DNA in maternal plasma and serum. Lancet **350:** 485–487.
2. Hahn, S. & W. Holzgreve. 2002. Prenatal diagnosis using fetal cells and cell-free fetal DNA in maternal blood: what is currently feasible? Clin. Obstet. Gynecol. **45:** 649–656.
3. Li, Y. *et al.* 2004. Size separation of circulatory DNA in maternal plasma permits ready detection of fetal DNA polymorphisms. Clin. Chem. **50:** 1002–1011.
4. Li, Y. *et al.* 2004. Improved prenatal detection of a fetal point mutation for achondroplasia by the use of size-fractionated circulatory DNA in maternal plasma: case report. Prenat. Diagn. **24:** 896–898.
5. Li, Y. *et al.* 2005. Detection of paternally inherited fetal point mutations for beta-thalassemia using size-fractionated cell-free DNA in maternal plasma. JAMA **293:** 843–849.
6. Ding, C. *et al.* 2004. MS analysis of single-nucleotide differences in circulating nucleic acids: application to noninvasive prenatal diagnosis. Proc. Natl. Acad. Sci. USA **101:** 10762–10767.
7. Zhang, J. *et al.* 1999. Presence of donor- and recipient-derived DNA in cell-free urine samples of renal transplantation recipients: urinary DNA chimerism. Clin. Chem. **45:** 1741–1746.
8. Zhong, X.Y. *et al.* 2001. Cell-free DNA in urine: a marker for kidney graft rejection, but not for prenatal diagnosis? Ann. N. Y. Acad. Sci. **945:** 250–257.
9. Li, Y. *et al.* 2003. Detection of donor-specific DNA polymorphisms in the urine of renal transplant recipients. Clin. Chem. **49:** 655–658.

Noninvasive Prenatal Diagnostic Assay for the Detection of β-Thalassemia

THESSALIA PAPASAVVA,[a] GABRIEL KALAKOUTIS,[b] IOANNIS KALIKAS,[b] ELECTRA NEOKLI,[b] SOTEROULA PAPACHARALAMBOUS,[b] ANDREANNI KYRRI,[c] AND MARINA KLEANTHOUS[a]

[a]*The Cyprus Institute of Neurology and Genetics, Nicosia, Cyprus*

[b]*Archibishop Makarios III Hospital, Nicosia, Cyprus*

[c]*Thalassaemia Center, Nicosia, Cyprus*

ABSTRACT: The development of a noninvasive method for detection of β-thalassemia in the population of Cyprus is based on the detection of paternally inherited single nucleotide polymorphisms (SNPs) as well as β-thalassemia (β-thal) mutations. We selected 11 informative SNPs for the Cypriot population linked to the β-globin locus. Two different approaches were used: allele-specific polymerase chain reaction (AS-PCR) and the arrayed primer extension (APEX) method. The AS-PCR approach is being standardized, and the method was applied in two families. The paternally inherited allele was noninvasively detected with the AS-PCR approach on maternal plasma. Some preliminary tests were performed with the APEX method on genomic DNA of parents carrying the β-thal mutation.

KEYWORDS: noninvasive prenatal diagnosis (NIPD); thalassemia; microarrays

INTRODUCTION

The discovery that during pregnancy fetal DNA is circulating in maternal plasma and constitutes 3–6% of the total plasma DNA has opened up new possibilities in the noninvasive prenatal diagnosis.[1–3] The development of a reliable and sensitive assay has to be based on its ability to discriminate fetal DNA from the coexisting background of maternal DNA by detecting differences between the two. Due to the fact that one single mutation accounts for the majority of the cases in the Cypriot population, our assay has to be based on the detection of paternally inherited polymorphisms such as single nucleotide

Address for correspondence: Thessalia Papasavva, Molecular Genetics Thalassaemia Department, The Cyprus Institute of Neurology and Genetics, 6, International Airport, 1683 Nicosia, Cyprus. Voice: +357-22-392664; fax: +357-22-392615.
e-mail: thesalia@cing.ac.cy

**Ann. N.Y. Acad. Sci. 1075: 148–153 (2006). © 2006 New York Academy of Sciences.
doi: 10.1196/annals.1368.020**

polymorphisms (SNPs), (CA)n, VNTRs linked to β-globin gene cluster, as well as mutations.

MATERIALS AND METHODS

Blood samples were collected after informed consent from 18 families at risk for β-thalassemia (β-thal) in their newborns. Approximately 10 mL of peripheral maternal blood was collected into EDTA-containing tubes before chorionic villus sampling (CVS) around 11 weeks of gestation.

Isolation of DNA from Plasma

Plasma was separated from whole blood after centrifugation at 1,600 × *g* for 10 min following a second high-speed centrifugation step at 16,000 × *g* for 10 min. DNA was isolated from plasma (400 μL) by using the QIAamp DNA blood mini kit (QIAGEN). Genomic DNA was extracted through desalting procedures.

Selection of SNPs

We selected 11 informative polymorphisms (SNPs) for our population linked to the β-globin locus. Some of the SNPs that we have selected are routinely used in our lab for prenatal diagnosis. The rest of the SNPs were selected on the basis of their degree of heterozygosity in the Cypriot population, which was determined after sequencing of random samples. Therefore, more searching and testing needs to be performed to find more SNPs that can be used for all families.

Allele-Specific Polymerase Chain Reaction (AS-PCR) and Design of Primers

AS-PCR is a PCR-based method that uses allele-specific priming. Primers were designed to be specific for each SNP. Each reaction was carried out with genomic DNA or ~20 ng/μL of maternal plasma DNA in a total reaction volume of 25 μL that contained 1× PCR buffer with 1.5 mμ μgcl_2, 200μM d NTPs, and 1 unit of AmpliTaq DNA polymerase. AmpliTaq DNA polymerase is manufactured for Applied Biosystems by Roche Molecule Systems dnc., (Branchburg, NJ). Annealing temperature was set at 57°C for 1 min for 40 cycles.

Arrayed Primer Extension (APEX) Reaction

The APEX approach that allows for the parallel one-shot detection of multiple mutations/SNPs was also used for the development of an NIPD for β-thal. Through collaboration with ASPER Biotech (Tartu, Estonia), we have developed a microarray (DNA chip) system called "thalassochip." It contains all the common and frequent mutations and polymorphisms of the Mediterranean region. A fragment of β-globin is amplified using the primers BG1F/R and BG2F/R and then fragmented, followed by the APEX reaction using fluorescently labeled dideoxynucleoside triphosphate.[4] The processed slides harboring an APEX array are imaged in a microarray detector called Genorama™ QuattroImager (ASPER Biotech, Tartu, Estonia). The procedure for the APEX reaction was carried out according to the instructions of ASPER Biotech.

RESULTS

Sensitivity of AS-PCR

To determine the sensitivity of AS-PCR, we made serial dilutions of genomic DNA from 1:10 to $1{:}10^4$. The diluted DNAs were amplified using the primers for HinfI site for both alleles. The detection limit of the assay was $1{:}10^4$, which corresponds to 210 pg/reaction (30 genome equivalents). The extracted maternal plasma DNA is about 3–4 ng/μL, whereas fetal DNA in maternal plasma is about 90 pg/μL. Therefore, AS-PCR proves to be a sensitive assay for the detection of fetal DNA in maternal plasma.

Specificity of AS-PCR

Genomic DNA having the c/c genotype for the AvaII polymorphism was spiked with 3, 2, and 0% genomic DNA having the g/g genotype and vice versa. The DNA used had similar concentration to that of plasma DNA (~4 ng/μL). Spiking with 3% corresponds to fetal DNA in maternal plasma DNA. After a number of trials for optimizing the procedure, we were able to detect the allele even at the lowest concentration. Moreover, we did not have amplification of the allele that was missing from the reaction (0%) proving, therefore, the specificity of the approach.

AS-PCR for SNP Analysis of the Plasma Samples

We applied the procedure on maternal plasma DNA samples. The genotypes of the parents and the CVS were determined by direct RFLP analysis.

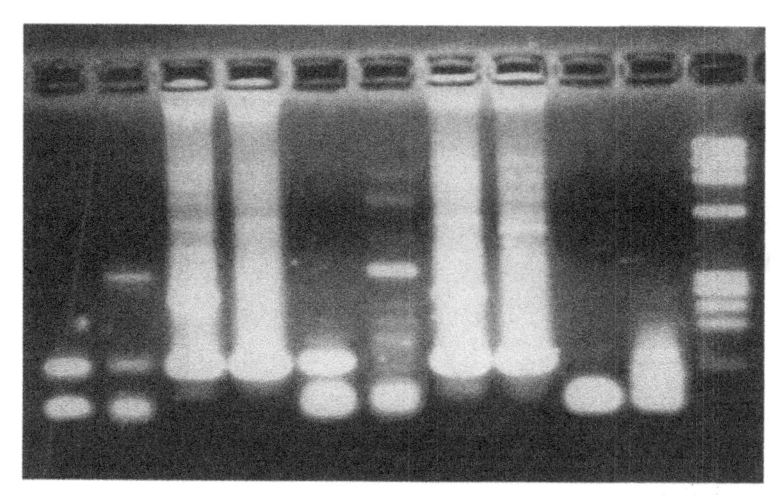

FIGURE 1. AS-PCR for HinfI site on maternal plasma. Lanes 1, 3, 5, 7 are the "a" allele-specific reactions and lanes 2, 4, 6, 8 are the "t" allele-specific reactions. Family 9; lanes 1,2: maternal plasma, 3, 4: paternal genomic DNA. Family 13; lanes 5,6: maternal plasma DNA, 7, 8: paternal genomic DNA. Lanes 9, 10: Blanks.

FIGURE 1 shows the results of analysis of samples from two families for the HinfI site. We can see that for the Family 9, the fetus has inherited the "t" allele from the father, while for Family 13, the fetus did not inherit the "t" allele. These findings are in agreement with the CVS analysis. For our SNP analysis, we selected the SNPs for which the mother is homozygous. We have examined two families, and we have managed to determine the paternally inherited allele. Maternal plasma DNA from Family 13 analyzed by AS-PCR for SNPs HinfI and HindIIIG/γ is shown in FIGURE 2. It can be seen that the fetus has inherited the β-thal allele of the father; therefore, the fetus is β-thal trait or β-thal major.

Sensitivity of the APEX Reaction

Serial dilutions of genomic DNA having the IVSI-110 mutation from 1:10 to 1:10,000 were carried out for the APEX reaction. The initial concentration of DNA was 478 ng/μL. The thalassochip was visualized on the Genorama™ QuattroImager (ASPER Biotech). The results were analyzed using the Genorama Basecaler Software. We had a strong APEX signal at the $1{:}10^3$ dilution (FIGURE 3). The dilution $1{:}10^4$ gave a very weak signal. In other words, the detection limit of the APEX reaction was $1{:}10^3$, which corresponds to 2.4 ng/reaction and is approximately the same as that for maternal plasma DNA. The results are preliminary and the reaction needs further optimization to be able to detect fetal DNA in maternal plasma.

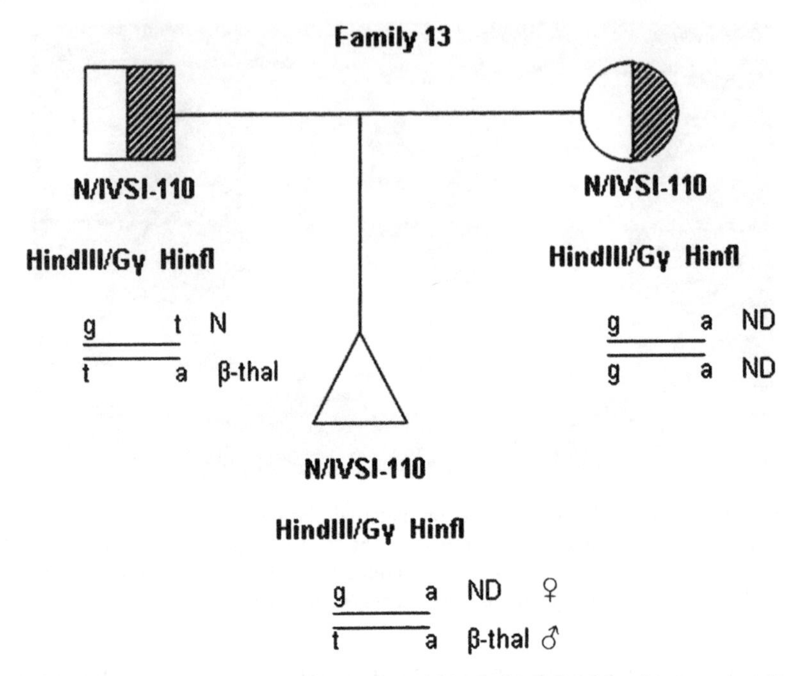

FIGURE 2. Haplotype analysis using AS-PCR method. Fetal haplotypes were determined by analysis on maternal plasma DNA. The genotypes were determined using AS-PCR analysis on genomic DNA from parents and CVS.

DISCUSSION

In this study, we took advantage of the allele specificity and sensitivity of the AS-PCR. We managed to develop the conditions required for the AS-PCR to detect and distinguish the paternally inherited allele in the maternal plasma for

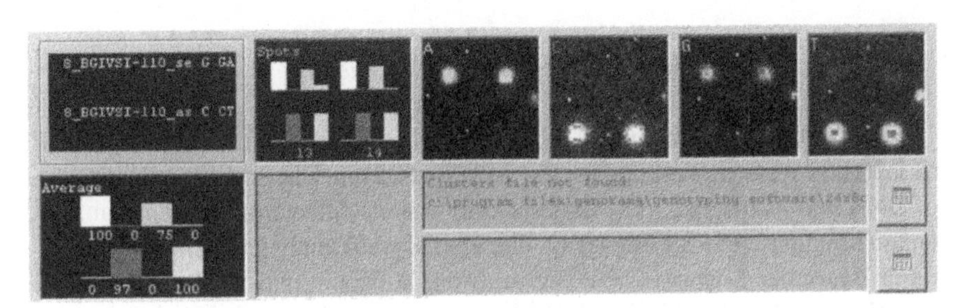

FIGURE 3. Sensitivity of APEX reaction. Mutated position IVSI-110. Genomic DNA heterozygous for IVSI-110, 1:10^3 dilution (2.4 ng). Wild-type allele: G (sense strand) and C (antisense strand). Mutated allele: A (sense strand) and T (antisense strand). The signals in A and G square are for sense strand, while the signals in C and T square are from antisense strand.

three SNPs. Moreover, the APEX assay seems to be very promising according to our very preliminary results which showed that we were able to detect very low DNA copy number. The aim of this work is to develop and optimize assays for the detection of even lower DNA copy number as well as differentiating between two alleles. This can be performed by using different approaches, such as PNAs[5], which suppress maternal DNA amplification, thus enhancing fetal DNA amplification. Other possible approaches will be based on differences between maternal DNA and fetal DNA from plasma. Nevertheless, more work needs to be performed for developing and validating these two assays into efficient, precise, and reliable assays for the NIPD of β-thal because the study is at an early stage and the results are preliminary.

ACKNOWLEDGMENTS

The study is supported by the European Union FP6 Network of Excellence "SAFE" as well as funds from the Cyprus Institute of Neurology and Genetics, Nicosia-Cyprus.

REFERENCES

1. Lo, Y.M.D. *et al.* 1997. Presence of fetal DNA in maternal plasma and serum. Lancet **350**: 485–487.
2. Chan, K.C.A. *et al.* 2004. Size distribution of maternal and fetal DNA in maternal plasma. Clin. Chem. **50**: 88–92.
3. Lo, Y.M.D. *et al.* 1998. Quantitative analysis of fetal DNA in maternal plasma and serum: implications for non-invasive prenatal diagnosis. Am. J. Hum. Genet. **62**: 768–775.
4. Gemignani, F. *et al.* 2002. Reliable detection of β-thalassaemia and G6PD mutations by DNA microarray. Clin. Chem. **48**: 2051–2054.
5. Li, Y. *et al.* 2005. Detection of paternally inherited fetal point mutations for β-thalassaemia using size-fractionated cell-free DNA in maternal plasma. JAMA **293**: 843–849.

Circulating DNA and Lung Cancer

XIAOYAN XUE, YONG M. ZHU, AND PENELLA J. WOLL

Department of Clinical Oncology, University of Sheffield, Sheffield, United Kingdom

ABSTRACT: Lung cancer is the leading cause of cancer death worldwide. The majority of patients is diagnosed too late for curative treatment. There is an urgent need for a noninvasive test to identify early lung cancer. Although levels of circulating cell-free DNA in plasma or serum are higher in patients with lung cancer than in healthy controls, it is not yet clear whether this will be of diagnostic or prognostic significance. The finding that circulating DNA in lung cancer patients exhibits genetic and epigenetic changes typical of the tumor (including chromosome loss, oncogene activation, and tumor-suppressor gene inactivation by methylation) has led to intense efforts to determine whether these are sensitive and specific enough to be used clinically. Here we review the evidence on circulating DNA in lung cancer and consider possible future applications in patient management.

KEYWORDS: plasma DNA; serum DNA; lung cancer; gene mutation; gene methylation; loss of heterozygosity; microsatellite instability

INTRODUCTION

Lung cancer is the most common cause of cancer death worldwide. In the developed world, it accounts for 1 in 15 of all deaths and 31% of cancer deaths. Overall, the prognosis is poor. In the United Kingdom, only 20% of patients survive for 1 year after the diagnosis of lung cancer. However, earlier tumor detection could increase the cure rate, as surgical resection of early-stage tumors can achieve 60–80% 5-year survival. Unfortunately, most patients have advanced disease at presentation. Current diagnostic procedures are not good at detecting early disease and require invasive investigations. Thus, there is an urgent need for new methods to facilitate the early and rapid detection of lung cancer and preneoplasia. Recent interest in the study of circulating DNA in a variety of diseases has led to interest in it as a prognostic factor, a diagnostic or screening tool for lung cancer. Here we will review the evidence on circulating DNA and its potential applications in the management of lung cancer.

Address for correspondence: Penella J. Woll, Department of Clinical Oncology, Weston Park Hospital, Whitham Road, Sheffield S10 2SJ, United Kingdom. Voice: +44 114 226 3235; fax: +44 114 226 5678.
e-mail: p.j.woll@shef.ac.uk

Ann. N.Y. Acad. Sci. 1075: 154–164 (2006).
doi: 10.1196/annals.1368.021

Quantitation of Circulating DNA in Lung Cancer Patients

Raised levels of circulating cell-free DNA in cancer patients were first reported in 1987 and have now been reported in many cancer types. Although the precise mechanism of DNA release into the blood remains obscure, it seems clear that much of it is derived from apoptotic and necrotic tumor cells.[1] The DNA concentration in plasma or serum is measured either directly or indirectly following DNA extraction. In early studies, DNA was quantitated directly by DNA dipstick, PicoGreen, dot-hybridization, or nick translation. Because plasma DNA circulates as nucleosomes (complexes of histones and DNA), the detection of nucleosomes by enzyme-linked immunosorbent assay (ELISA) is an alternative method to assess circulating DNA concentrations, which has also shown elevated levels in lung cancer patients.[2,3] More recently, quantitative real-time polymerase chain reaction (qPCR) has been widely used to measure DNA concentrations.

The results of studies reporting quantitative estimates of circulating DNA in lung cancer patients are summarized in TABLE 1. There is considerable variation in the reported results, with estimates of mean circulating DNA levels ranging from 3.7 to 594 ng/mL in lung cancer patients and 1.8 to 78 ng/mL in controls, reflecting a number of problems in interpreting these data. First, the populations of lung cancer patients are heterogenous, including different pathologies (small cell lung cancer [SCLC] and nonsmall cell lung cancer [NSCLC]) and stages of disease. For example, the studies of Sozzi *et al.*[4,5] were in patients with early-stage NSCLC awaiting surgery, whereas those of Sirera *et al.*[6] were in advanced NSCLC and those of Chen *et al.*[7] were in SCLC. Second, most of the studies included some controls, but these were often "healthy controls" and were rarely matched for age, sex, smoking history, and comorbidity. Therefore, even where significant differences were found between the patient and control groups, it cannot be taken to indicate that the test would be of diagnostic use in a clinical population. Third, a variety of different methods were used to estimate DNA concentrations, as described above. In general, more recent studies using qPCR have found the concentration of circulating DNA to be lower than that of earlier studies using other methods. For example, two studies by Sozzi *et al.*[4,5] using dip-stick and qPCR methods in similar patient populations obtained estimates of plasma DNA of 318 and 24 ng/mL, respectively, in lung cancer patients, and 18 and 3 ng/mL in controls. Further inconsistency may be explained by the use of different genes for qPCR, such as hTERT, GAPDH, or β-actin. Fourth, none of the studies report the efficiency of the methods used to extract DNA from blood. In addition, a recent report from one of the most experienced groups working in this field indicates that freeze-thawing and storage at −20°C or −80°C can lead to reduction in plasma DNA concentrations of up to 70%.[8] Future studies should use standardized methodology and large, clearly defined patient and control populations in order for the results to be interpretable.

TABLE 1. Quantitative estimates of circulating DNA in lung cancer patients and controls

Author	Material	Method	Subjects, number	DNA, ng/mL (range)	Significance
Maebo, 1990[9]	Plasma	Dot-hybridization (Alu-DNA)	Lung cancer, 45	71% > 80 ng/mL	
			Controls, 59	None < 80 ng/mL	
Fournie, 1995[10]	Plasma	Nick translation labeling	Lung cancer, 68	30 (23–38)	$P < 0.05$
			Controls, 26	10 (7–14)	
Sozzi, 2001[4]	Plasma	DNA Dipstick KM Kit	NSCLC, 84	318	Yes
			Healthy controls, 43	18	
Bearzatto, 2002[11]	Plasma	DNA Dipstick KM Kit	NSCLC, 35	66% > 125 ng/mL	
			Controls, 15	None > 125 ng/mL	
Beau-Faller, 2003[12]	Plasma	PicoGreen dsDNA Kit	Lung cancer, 34	157 (14–1,054)	
			SCLC, 11	594 (150–1,054)	
			SQC, 11	128 (14–332)	
			ADC, 12	87 (44–120)	
			Controls, 20	78 (23–128)	
Sozzi, 2003[5]	Plasma	qPCR (hTERT)	NSCLC, 100	24	Yes
			Matched controls, 100	3.1	
Gormally, 2004[13]	Plasma	PicoGreen dsDNA Kit	Lung cancer, 82	25 (2–1,128)	No
			Controls, 776	26 (0–10)	
Gautschi, 2004[14]	Plasma	qPCR (GAPDH)	NSCLC, 185	3.7 (0.1–94)	$P < 0.01$
			Healthy controls, 46	1.8	
Gautschi, 2004[14]	Serum	qPCR (GAPDH)	NSCLC, 185	40 (0.3–641)	
			Healthy controls, 46	13	
Herrera, 2005[15]	Plasma	qPCR (ß-actin)	Lung cancer, 25	14.6 (3–30)	No
			Healthy controls, 11	10.6 (7–14)	
Sirera, 2005[6]	Serum	qPCR (hTERT)	NSCLC stage IIIB, 29	21 (4.9–152)	No
			NSCLC stage IV, 71	18 (2.9–849)	
Chen, 2005[7]	Plasma	qPCR	SCLC, 10	8.4 (0.5–59)	

NSCLC: nonsmall-cell lung cancer; SCLC: small-cell lung cancer; hTERT: human telomerase reverse transcriptase; GAPDH: glyceraldehyde-3-phosphate dehydrogenase; SQC: squamous cell carcinoma; ADC: adenocarcinoma.

Despite these limitations, some inferences can be drawn from these data. As in other tumor types, higher concentrations of circulating DNA are obtained in serum than in plasma. Gautschi *et al.*[14] demonstrated that serum DNA in lung cancer patients was tenfold higher than plasma DNA (39.6 vs. 3.7 ng/mL). However, this is attributable to leukocyte lysis' releasing genomic DNA during clotting, so plasma DNA is regarded as a better reflection of the underlying tumor. In general, higher concentrations of circulating DNA are found in lung cancer patients than controls, although not all the studies achieved statistical significance. The results shown here are consistent with a large prospective European multicenter collaborative study of plasma DNA from healthy controls, patients with chronic obstructive pulmonary disease, and patients with various types of cancer; this study found that plasma DNA concentration was not sufficiently sensitive or specific for cancer screening or diagnosis.[13]

Some of the studies have addressed the question of whether the circulating DNA concentration is of prognostic significance in lung cancer, but the evidence is incomplete. In patients with early-stage NSCLC, no correlation was found between plasma DNA levels and tumor size, stage, cell type, whether the patient was relapse-free, or overall survival.[4,5] A further study in advanced NSCLC found no difference in serum DNA concentration between patients with stage IIIB and stage IV disease.[6] The levels of plasma DNA fall dramatically after successful surgical resection, but evidence for elevation at relapse is scanty.[4,5] However, in patients with advanced disease receiving chemotherapy, increasing levels of plasma DNA were associated with disease progression, and higher plasma DNA levels (>10 ng/mL) were associated with reduced survival.[14]

Genetic and Epigenetic Changes in Circulating DNA in Lung Cancer

Lung cancer arises through the accumulation of multiple genetic changes, including chromosome deletions, oncogene mutations, and methylation of tumor-suppressor gene promoters. Studies of normal controls, smokers, and patients with bronchial dysplasia and invasive lung cancer have demonstrated the frequency and sequence of acquisition of these genetic changes.[16] The reports by Chen *et al.*[17] and Sozzi *et al.*[18] that circulating DNA in lung cancer patients demonstrated genetic alterations typical of the tumor led to intense interest in the use of circulating DNA in the management of lung cancer.

A number of studies have compared genetic alterations in circulating DNA and tumor DNA. In lung cancer patients with specific genetic changes in their tumor, 40–90% exhibit the same changes in circulating DNA. For example, An *et al.*[19] found that 88% of 73 lung cancer patients with methylated P16 in tumor DNA had the same abnormality in plasma DNA. Usadel *et al.*[20] found that 47% of 89 patients with methylated APC in tumor DNA had the same abnormality in plasma DNA. Andriani *et al.*[21] found that 73% of 64 lung cancer patients with mutated P53 in tumor DNA, 58% with inactivated FHIT, and 48% with 3p deletions had the same abnormality in plasma DNA. Thus, it is clear that circulating DNA in lung cancer patients is derived from tumor DNA, but not all the genetic alterations found in the tumor can be detected in circulating DNA.

Microsatellite Instability and Loss of Heterozygosity

Loss of genetic material from chromosome 3p is one of the earliest and most frequent genetic changes occurring during bronchial carcinogenesis. Microsatellite instability (MSI) and loss of heterozygosity (LOH) can be assessed by PCR to visualize band shifts and allele losses. Detection of MSI and LOH in circulating DNA is attractive because direct comparison can be made with genomic leukocyte DNA from the same blood sample, but it can be difficult to achieve accurate and satisfactory results because the circulating DNAs are not exclusively derived from tumor cells. Therefore, false-negative results are possible, particularly when the proportion of circulating DNA derived from cancer cells is low.

TABLE 2 summarizes the results of studies of MSI and LOH in lung cancer patients. It shows the proportion of circulating DNA samples positive for specific genetic alterations (positivity) and the proportion of positives among patients in whom the tumor was shown to be positive (sensitivity). Positivity rates for individual markers are variable and often low, so the majority of authors have used a panel of markers and described the overall positivity rate. In general, genetic alterations (LOH or MSI) were found in circulating DNA of 27–88% of lung cancer patients, and in 48–94% of patients whose tumors were known to be positive for the markers. In our own study, we assessed the performance of 16 markers in plasma DNA from 32 lung cancer patients.[22] The markers were chosen to represent a wide range of genetic alterations commonly described in lung cancer on chromosomes 3p, 8p, 9p, 13q, and 17p. The positivity rate for individual markers ranged from 0% to 59%. Combining all 16 markers, the overall positivity rate was 69%, but when analysis was restricted to the three markers with the highest frequency of genetic alteration, 60% of samples were positive. These three markers (D3S1300, D3S1560, and D8S201) were therefore used in a larger study of 86 lung cancer patients and 120 patients with other respiratory disorders attending a respiratory clinic.[22] Genetic alterations were found in 69% of patients with lung cancer and 42% of patients with other respiratory disorders ($P < .001$). The positivity rate of 69% in lung cancer patients is consistent with the other studies shown here. The unexpectedly high positivity rate in patients with other respiratory disorders suggests that this marker panel is unlikely to be useful in distinguishing lung cancer patients from controls. However, it raises the intriguing possibility of identifying individuals at high risk of lung cancer who might be suitable

TABLE 2. Microsatellite alterations detected in circulating DNA of lung cancer patients

Author	Tumor	Material	Method	Genes analyzed	Positive (%)	Sensitivity (%)
Chen, 1996[17]	SCLC	Plasma	PCR	MSA (3 markers)	15/21 (71%)	15/16 (94%)
Sanchez-Cespedes, 1998[23]	NSCLC	Serum	PCR	D3S1038, D3S1611, D3S1067, D3S1284	6/22 (28%)	/
Sozzi, 1999[18]	NSCLC	Plasma	PCR	D3S1234, D3S1300, D3S4103, D21S1245	35/87 (40%)	30/49 (61%)
Gonzalez, 2000[24]	SCLC	Plasma	PCR	ACTBP2, UT762, AR[a]	25/35 (71%)	25/35 (71%)
Cuda, 2000[25]	SCLC	Serum	PCR	ACTBP2, UT762, AR[a]	7/11 (64%)	7/8 (88%)
	NSCLC	Serum	PCR	UT762, D3S4103, D3S1300	7/17 (41%)	7/10 (70%)
Sozzi, 2001[4]	NSCLC	Plasma	PCR	D3S1300, D3S1289, D3S1266, D3S2338, D3S1304	9/33 (27%)	9/20 (45%)
Bearzatto, 2002[11]	NSCLC	Plasma	PCR	D3S1300, D3S1289, D3S1266, D3S2338, D3S1304	10/35 (29%)	10/20 (50%)
Beau-Faller, 2003[12]	NSC/ SCLC	Plasma	Allelotyping	D3S1283, D3S1293, D3S3700, D5S346, D9S171, D9S179, *RB*, TP53, D17S1818, D17S800, D20S108, D20S170.	30/34 (88%)	17/20 (85%)
Khan, 2004[22]	NSC/ SCLC	Plasma	PCR	D3S1300, D3S1560, D8S201	59/86 (69%)	/
Andriani, 2004[21]	NSCLC	Plasma	F-PCR	D3S1300 (FHIT)	9/56 (16%)	7/22 (32%)
	NSCLC	Plasma	F-PCR	D3S1300, D3S1289, D3S1266, D3S2338, D3S1304	23/64 (36%)	19/40 (48%)

[a]The three polymorphic markers are located on chromosomes 5 (ACTBP2), 21 (UT762), and X (AR).
MSA: microsatellite alteration; F-PCR: fluorescence-PCR; *RB*: retinoblastoma (13q14).

TABLE 3. Gene mutations detected in circulating DNA of lung cancer patients

Author	Tumor	Material	Method	Genes analyzed	Positivity (%)	Sensitivity (%)
Bearzatto, 2002[11]	NSCLC	Plasma	Mutant-enriched PCR	K-ras mutation	0/35	0/11 (0%)
Ramirez, 2003[27]	NSCLC	Serum	Enriched PCR		12/50 (24%)	
Kimura, 2004[28]	NSCLC	Plasma	RFLP-PCR		19/163 (12%)	
Silva, 1999[29]	SCLC	Plasma	PCR-SSCP/ sequencing	P53 mutation	1/10 (10%)	1/2 (50%)
Andriani, 2004[21]	NSCLC	Plasma	Sequencing/ REA/MASA		19/64 (30%)	19/26 (73.1%)

SCLC: small-cell lung cancer; REA: restriction endonuclease analysis; MASA: mutation allele-specific amplification.

for screening or chemoprevention studies. We are now undertaking a more detailed, population-based study that will permit longitudinal follow-up of controls to determine whether they are indeed at increased risk of developing lung cancer.

K-ras and p53 Mutations

Activating mutations in oncogenes and inactivating mutations in tumor-suppressor genes are key genetic changes in the pathway to cancer. These can be detected in circulating DNA by mutant-enriched PCR and direct sequencing. Those most commonly studied are K-ras, which is mutated in about 30% of NSCLCs, and p53, which is mutated in over 90% of SCLCs and 50% of NSCLCs.[26] Studies of p53 mutations are more difficult because of the large number of possible mutations in several exons. TABLE 3 summarizes the results of studies of K-ras and p53 mutations in circulating DNA of lung cancer patients. In three studies of K-ras, positivity was found in 31 of 248 (12.5%) circulating DNA samples, while in two studies of mutated p53, positivity was found in 20 of 74 (27%) of samples. Interestingly, discordant mutations were found in tumor and circulating DNA, suggesting that some circulating DNA may be derived from metastatic organs with different clonal changes to the primary tumor.

Gene Hypermethylation

In addition to changes in the genes themselves, gene expression can be modified by epigenetic changes such as alterations in promoter methylation. Hypermethylation represses transcription of the promoter regions of genes, leading to gene silencing, and is particularly important in inactivating tumor-suppressor

TABLE 4. Methylated gene alterations detected in the plasma/serum of lung cancer patients and controls

Author	Tumor	Material	Methods	Genes analyzed	Positivity (%)	Sensitivity (%)
Esteller, 1999[30]	NSCLC	Serum	MSP	p16INK4a	9/22 (41%)	3/9 (33.3%)
Bearzatto, 2002[11]	NSCLC	Plasma	F-MSP	p16INK4a	12/35 (34%)	12/22 (55%)
An, 2002[19]	NSCLC	Plasma	Semi-nested MSP	p16INK4a	77/105 (73%)	64/73 (87.7%)
Esteller, 1999[30]	NSCLC	Serum	MSP	DAPK	4/22 (18%)	4/5 (80%)
Ramirez, 2003[27]	NSCLC	Serum	MSP	DAPK	20/50 (40%)	
Esteller, 1999[30]	NSCLC	Serum	MSP	MGMT	4/22 (18%)	4/6 (66%)
Esteller, 1999[30]	NSCLC	Serum	MSP	GSTP1	1/22 (5%)	1/2 (50%)
Usadel, 2002[20]	NSCLC	Plasma/serum	MSP	APC	42/89 (47%)	
Ramirez, 2003[27]	NSCLC	Serum	MSP	TMS1	17/50 (34%)	
Ramirez, 2003[27]	NSCLC	Serum	MSP	RASSF1	17/50 (34%)	

APC: adenomatous polyposis coli; TMS1: target of methylation-induced silencing; RASSFF1: ras association domain family protein 1; DAPK: death-associated protein kinase; GSTP1: glutathione S-transferase P1; MGMT: O6 methylguanine methyl-transferase; MSP: methylation-specific PCR; F-MSP: fluorescence methylation-specific PCR.

genes in cancer. Methylation-specific PCR (MSP) exploits the use of bisulfite to convert all unmethylated, but not methylated, cytosines to uracil, and subsequent amplification with primers specific for methylated versus unmethylated DNA. The method requires relatively large amounts of DNA, and bisulfite treatment itself can cause DNA degradation, so it may be difficult to apply to routinely collect clinical samples. TABLE 4 summarizes the results of studies of methylated gene alterations in circulating DNA of lung cancer patients.

In three studies, $p16^{INK4a}$ methylation was detected in 98 of 162 (60%) circulating DNA samples from NSCLC patients. The positivity rate was higher in studies using fluorescent MSP assays[11] or seminested MSP.[19] Among the other gene studies, hypermethylation was most commonly detected in APC (47%),[20] TMS1 and RASSF1 (34%),[27] and DAPK (33%).[27,30] Esteller *et al.*[30] analyzed promoter methylation of four genes ($p16^{INK4a}$, DAPK, GSTP1, and MGMT) in 22 NSCLC patients, but found the overall positivity rate to be only 50%.

Potential Applications of Circulating DNA in the Management of Lung Cancer

The discovery that cell-free tumor DNA can be detected in the circulation of lung cancer patients held out the promise of a noninvasive "blood test" for cancer. This promise has not yet been realized, but recent results have brought us closer to this goal, and have allowed a number of new questions to be articulated. First, is the quantity of circulating DNA in lung cancer of diagnostic or prognostic value? As discussed earlier, recent reports suggest that elevated levels of circulating DNA are not likely to be high enough to distinguish lung cancer patients from other patients with chronic respiratory disorders. At present there is little consensus on how samples should be collected, stored, or processed. Most published studies provide no data on reproducibility or quality control, making comparisons between them difficult. At present there is little evidence to suggest that the quantity of circulating DNA is of any prognostic significance.

Second, are the genetic and epigenetic alterations in circulating DNA in lung cancer of diagnostic or prognostic value? Many studies have demonstrated that circulating DNA has the features of tumor DNA, although it is contaminated with genomic DNA. The proportion of tumor-derived DNA ranges from 3% to 97%,[1] leading to difficulties in the interpretation of results. We have reviewed the positivity and sensitivity rates for a number of genetic and epigenetic markers, finding wide variation between different reports. Most authors favor using a panel of markers, but even then obtain positivity rates of 40–80%. It remains to be determined whether a marker panel of sufficient specificity and sensitivity to be diagnostic for lung cancer can be devised. No combination of markers has yet been shown to be of prognostic value.

Third, can circulating DNA be used to identify a population at high risk of lung cancer? Our finding that 42% of patients with chronic respiratory diseases exhibited genetic alterations in circulating DNA[22] suggests that individuals with early genetic changes of bronchial carcinogenesis might be identified in this way. This would allow them to be offered smoking cessation programs, and screening and chemoprevention studies. Further longitudinal studies will help answer this question.

The goal of developing a noninvasive test for lung cancer that will increase the diagnosis of the disease at a curable stage remains elusive. The main conclusion from the present review is that standardized methodologies and larger studies in defined patient populations are urgently required.

REFERENCES

1. JAHR, S. *et al.* 2001. DNA fragments in the blood plasma of cancer patients: quantitations and evidence for their origin from apoptotic and necrotic cells. Cancer Res. **61**: 1659–1665.
2. HOLDENRIEDER, S. *et al.* 2001. Nucleosomes in serum of patients with benign and malignant diseases. Int. J. Cancer **95**: 114–120.
3. HOLDENRIEDER, S. *et al.* 2004. Circulating nucleosomes predict the response to chemotherapy in patients with advanced non-small cell lung cancer. Clin. Cancer Res. **10**: 5981–5987.
4. SOZZI, G. *et al.* 2001. Analysis of circulating tumour DNA in plasma at diagnosis and during follow-up of lung cancer patients. Cancer Res. **61**: 4675–4678.
5. SOZZI, G. *et al.* 2003. Quantification of free circulating DNA as a diagnostic marker in lung cancer. J. Clin. Oncol. **21**: 3902–3908.
6. SIRERA, R. *et al.* 2005. The analysis of serum DNA concentration by means of hTERT quantification: a useful prognostic factor in advanced non-small cell lung cancer (NSCLC). Lung Cancer **49** (Suppl. 2): S74.
7. CHEN, L. *et al.* 2005. Biomarkers for small cell lung cancer (SCLC): DNA methylation and GD2 synthase transcript levels. Lung Cancer **49** (Suppl. 2): S316.
8. ANDRIANI, F. *et al.* 2005. Storage of plasma or isolated plasma DNA affects the results of circulating DNA quantification assays. Lung Cancer **49** (Suppl. 2): S17.
9. MAEBO, A. 1990. [Plasma DNA level as a tumor marker in primary lung cancer]. Nihon Kyobu Shikkan Gakkai Zasshi **28**: 1085–1091.
10. FOURNIE, G. J. *et al.* 1995. Plasma DNA as a marker of cancerous cell death: investigations in patients suffering from lung cancer and in nude mice bearing human tumours. Cancer Lett. **91**: 221–227.
11. BEARZATTO, A. *et al.* 2002. p16 (INK4A) Hypermethylation detected by fluorescent methylation-specific PCR in plasmas from non-small cell lung cancer. Clin. Cancer Res. **8**: 3782–3787.
12. BEAU-FALLER, M. *et al.* 2003. Plasma DNA microsatellite panel as sensitive and tumour-specific marker in lung cancer patients. Int. J. Cancer **105**: 361–370.
13. GORMALLY, E. *et al.* 2004. Amount of DNA in plasma and cancer risk: a prospective study. Int. J. Cancer **111**: 746–749.

14. GAUTSCHI, O. *et al.* 2004. Circulating deoxyribonucleic acid as prognostic marker in non-small-cell lung cancer patients undergoing chemotherapy. J. Clin. Oncol. **22**: 4157–4164.
15. HERRERA, L. J. *et al.* 2005. Quantitative analysis of circulating plasma DNA as a tumour marker in thoracic malignancies. Clin. Chem. **51**: 113–118.
16. FONG, K. M. *et al.* 2003. Molecular biology of lung cancer: clinical implications. Thorax **58**: 892–900.
17. CHEN, X. Q. *et al.* 1996. Microsatellite alterations in plasma DNA of small cell lung cancer patients. Nat. Med. **2**: 1033–1035.
18. SOZZI, G. *et al.* 1999. Detection of microsatellite alterations in plasma DNA of non-small cell lung cancer patients: a prospect for early diagnosis. Clin. Cancer Res. **5**: 2689–2692.
19. AN, Q. *et al.* 2002. Detection of p16 hypermethylation in circulating plasma DNA of non-small cell lung cancer patients. Cancer Lett. **188**: 109–114.
20. USADEL, H. *et al.* 2002. Quantitative adenomatous polyposis coli promoter methylation analysis in tumour tissue, serum, and plasma DNA of patients with lung cancer. Cancer Res. **62**: 371–375.
21. ANDRIANI, F. *et al.* 2004. Detecting lung cancer in plasma with the use of multiple genetic markers. Int. J. Cancer **108**: 91–96.
22. KHAN, S. *et al.* 2004. Genetic abnormalities in plasma DNA of patients with lung cancer and other respiratory diseases. Int. J. Cancer **110**: 891–895.
23. SANCHEZ-CESPEDES, M. *et al.* 1998. Detection of chromosome 3p alterations in serum DNA of non-small-cell lung cancer patients. Ann. Oncol. **9**: 113–116.
24. GONZALEZ, R. *et al.* 2000. Microsatellite alterations and TP53 mutations in plasma DNA of small-cell lung cancer patients: follow-up study and prognostic significance. Ann. Oncol. **11**: 1097–1104.
25. CUDA, G. *et al.* 2000. Detection of microsatellite instability and loss of heterozygosity in serum DNA of small and non-small cell lung cancer patients: a tool for early diagnosis? Lung Cancer **30**: 211–214.
26. GIRARD, L. *et al.* 2000. Genome-wide allelotyping of lung cancer identifies new regions of allelic loss, differences between small cell lung cancer and non-small cell lung cancer, and loci clustering. Cancer Res. **60**: 4894–4906.
27. RAMIREZ, J. L. *et al.* 2003. Methylation patterns and K-ras mutations in tumour and paired serum of resected non-small-cell lung cancer patients. Cancer Lett. **193**: 207–216.
28. KIMURA, T. *et al.* 2004. Mutant DNA in plasma of lung cancer patients: potential for monitoring response to therapy. Ann. N. Y. Acad. Sci. **1022**: 55–60.
29. SILVA, J. M. *et al.* 1999. TP53 gene mutations in plasma DNA of cancer patients. Genes Chromosomes Cancer **24**: 160–161.
30. ESTELLER, M. *et al.* 1999. Detection of aberrant promoter hypermethylation of tumour suppressor genes in serum DNA from non-small cell lung cancer patients. Cancer Res. **59**: 67–70.

Circulating Nucleic Acids in Plasma/Serum and Tumor Progression

Are Apoptotic Bodies Involved? An Experimental Study in a Rat Cancer Model

JULIA SAMOS,[a] DOLORES C. GARCÍA-OLMO,[a] MARÍA G. PICAZO,[a] ANTONIO RUBIO-VITALLER,[b] AND DAMIÁN GARCÍA-OLMO[c]

[a]*Experimental Research Unit, General University Hospital of Albacete, Albacete, Spain*

[b]*Department of Haematology, General University Hospital of Albacete, Albacete, Spain*

[c]*Department of Surgery, Universidad Autónoma de Madrid and La Paz University Hospital, Madrid, Spain*

ABSTRACT: The "genometastasis hypothesis" proposes that cell-free tumor nucleic acids might be able to transform host stem cells, and that this might be a pathway for the development of metastases. This theory is supported by previous experimental findings and is consistent with observations of other authors. It has been suggested that tumor DNA might be horizontally transferred by the uptake of apoptotic bodies and initiate the genetic changes that are necessary for tumor formation. In addition, apoptotic bodies have been proposed as possible vehicles that protect the nucleic acids circulating in the plasma from enzymatic degradation. In the present study, we analyzed the presence of apoptotic bodies in serum and its relationship with tumor progression in a heterotopic model of colon cancer in the rat. We injected DHD/K12-PROb cancer cells subcutaneously into BD-IX rats and divided the animals into three groups according to the time between the injection of tumor cells and euthanasia. A control group of healthy animals was included ($n = 6$). After euthanasia, macroscopic metastases were assessed and samples of blood were collected. To detect apoptotic bodies in the sera, each sample was mixed with FITC-conjugated annexin V antibody in combination with propidium iodide and then analyzed by flow cytometry. Detection of apoptotic bodies was only significantly increased in the sera of a few tumor-bearing animals in late stages of tumor development. Thus, such particles appear not to be the vehicle of the cell-free tumor nucleic acids that are detected at early stages of cancer.

Address for correspondence: Prof. Damián García-Olmo, Servicio de Cirugía General—C, Hospital Universitario "La Paz," Paseo Castellana 261, 28046 Madrid, Spain. Voice: +34-917-27-70-00; fax: +34-967-24-39-52.

e-mail: damian.garcia@uam.es

**Ann. N.Y. Acad. Sci. 1075: 165–173 (2006). © 2006 New York Academy of Sciences.
doi: 10.1196/annals.1368.022**

KEYWORDS: apoptotic bodies; colon cancer; circulating nucleic acids; genometastasis

INTRODUCTION

The "Genometastasis" Hypothesis: An Overview

The presence of cell-free DNA in the blood of cancer patients has frequently been assumed to be a simple consequence of tumor growth rather than a phenomenon that might be involved in tumor progression. However, it has been suggested that cell-free tumor nucleic acids might have a role in the progression of tumors and in the development of metastases by horizontal gene transfer or related mechanisms.[1–3]

In fact, in 1999, our group proposed a new hypothesis that suggests an alternative pathway to explain the development of metastases in distant targets. Specifically, this theory proposed: "Metastasis might occur via transfection of susceptible cells, located in distant target organs, with dominant oncogenes that are derived from the primary tumor and are circulating in the plasma."[1–3]

This hypothesis has been considered by other authors as one of the main models that might explain the discrepancies in the experimental findings concerning the putative selective nature of the metastatic phenotype.[4]

The idea that nucleic acids circulating in plasma are not inert molecules, but might have a biological activity, is supported by previous experimental findings and, moreover, is consistent with observations of other authors. In fact, the presence of high amounts of tumor DNA in plasma or serum of patients with diverse kinds of cancer has been repeatedly confirmed by a variety of techniques[5–7] and the integrity of such DNA, as well as that of RNA in the plasma, has been verified.[8,9] The persistence of this phenomenon has led to suggesting the possible utility of the analysis of cell-free nucleic acids as a valuable diagnostic[10,11] and prognostic tool.[12,13]

In an animal model, we found that cell-free tumor DNA was detected sooner and more frequently than circulating tumor cells.[14] In this study, we observed that dissemination of tumor DNA in the plasma seemed to be much more common than detectable hematogenic tumor cells during the spread of colorectal cancer.[14] These results are in agreement with the above-mentioned hypothesis and with others who have questioned the hematogenic dissemination of tumor cells as the only pathway to the development of metastasis. For example, we could verify that the site of injection of tumor cells in rats did not influence the subsequent distribution of metastases and, moreover, our results suggested the absence of any influence of the capillary bed on tumor invasion.[15]

In the same animal model of colon cancer, we also showed that plasma from tumor-bearing rats was able to stably transfect and transform cultured cells.[1–3] This transformation was demonstrated in the genotype and in the phenotype of those cultured cells. Moreover, we could detect the tumor marker in plasma and parenchymas of healthy rats that had been injected with plasma from tumor-bearing rats.[1,3]

These results were consistent with previous *in vitro* studies that showed the potential of cell-free nucleic acids. Many years ago, Pulciani *et al.* showed that extracts from human tumor could transfect cultured cells and transform it.[16] Anker *et al.* reported that NIH/3T3 mouse cells were transformed when the supernatant from a culture of colon cancer cells (SW480 cells) was added to the medium.[17]

Summarizing our experimental observations, together with the fact that it is biologically possible that cell-free nucleic acids can be stably transferred to cells, led to us to propose the "genometastases hypothesis." We suggested that genomic DNA in apoptotic bodies might be a possible mechanism for the genometastatic pathway. This is supported by the observations of Holmgren *et al.*,[18] who demonstrated that genomic DNA from apoptotic bodies could be transferred to the nuclear compartment of phagocytosing cells and that this transferred DNA was stable over time. Subsequently, evidence has been provided that tumor DNA might be horizontally transferred by the uptake of apoptotic bodies, suggesting that lateral transfer of DNA between eukaryotic cells might result in aneuploidy and the accumulation of genetic changes that are necessary for tumor formation.[19] These exciting observations led us to a new question: Might apoptotic bodies be a key for the tumor progression and development of metastasis?

Apoptotic Bodies in Serum and Tumor Progression: An Experimental Study

Many years ago, it was proposed that tumor-derived circulating plasma RNA of cancer patients was protected within vesicles[20] or complexes.[21–23] These possibilities might explain the fact that such RNA has sufficient integrity to permit reverse transcription-PCR,[8] despite the high concentration of RNase found in cancer patients.[24] More recently, the resistance of cell-free plasma RNA of cancer patients to degradation has been corroborated.[25]

On the basis of an *in vitro* study, Hasselmann *et al.* proposed that particles that protect RNA from enzymatic degradation might be apoptotic bodies.[26] This interesting suggestion is consistent with other evidence, such as the presence of filterable DNA and mRNA in serum and plasma of healthy subjects and cancer patients.[27,28] In addition, it has been reported that, in the course of apoptosis, RNA and DNA are packaged into separate apoptotic bodies.[29] Although the origin of nucleic acids in plasma remains unclear, apoptosis has been proposed as one of the major sources for plasma DNA in cancer patients.[7] From this and other evidence, many authors are in agreement with the suggestion that cell-free nucleic acids might circulate within apoptotic bodies.[25,27]

These observations, together with the above-mentioned ideas about the possible implication of apoptotic bodies in tumor progression, led us to design a new study to try and detect apoptotic bodies in serum of tumor-bearing rats and to attempt to correlate this detection with tumor progression. To our knowledge, no previous study has addressed this issue.

MATERIALS AND METHODS

We used DHD/k12-PROb cells (also called DHD/k12-TRb cells) and both male and female BD-IX rats. The animals were taken from a colony established at the authors' animal facility from founders purchased from a commercial breeder (Charles River Laboratories, Barcelona, Spain). Breeding was performed in compliance with European Community Directive 86/609/CEE for the use of laboratory animals. As recommended by the Federation of European Laboratory Animal Science Associations (FELASA), rats in the animal facility are tested periodically to ensure that the colony is free of pathogens, such as *Mycoplasma pulmonis*, *Salmonella* sp., Sendai virus, Hantaan virus, and Toolan H1 virus.

From birth to the end of the experiments, all rats had unlimited access to water and standard rat chow (Panlab s. l., Barcelona, Spain). At the beginning of the experiments, rats were 6–8 weeks old and weighed 90–225 g.

Tumors were generated in the thoracic region of each rat by unilateral subcutaneous injection of a million DHD/k12-PROb cells, following a previously established procedure.[14] The growth of subcutaneous tumors was monitored in all animals and recorded weekly.

After injection, animals were sacrificed randomly after 5 (group 5; $n = 7$), 7 (group 7; $n = 6$) or 11 (group 11; $n = 10$) weeks. To avoid possible confusion with an increase of apoptotic bodies in serum due to other pathologic processes, a control group ($n = 6$) of healthy animals of the same age as the experimental animals was sacrificed at the same time.

At the time of sacrifice, rats were anesthetized with an intraperitoneal injection of a mixture of ketamine (75 mg/kg) and xylazin (10mg/kg). Then blood (approximately 3.5–4.0 mL per rat) was withdrawn by cardiac puncture and a lethal dose of intracardiac injection of sodium thiopental was administered immediately afterwards. All samples of blood were centrifuged and serum was separated. Presence or absence of lung metastases was recorded after visual inspection.

Immediately after separation, each serum sample was mixed with fluorescein isothiocyanate (FITC)-conjugated annexin V antibody in combination with propidium iodide (PI; Annexin-V-FLUOS Staining Kit, Roche, Mannhein, Germany). For this, 100 μL of serum was incubated in the dark with 100 μL of the staining solution for 15 min. The analysis of this mixture was immediately carried out by flow cytometry using a FACScalibur flow cytometer (Beckton Dickinson, San Diego, CA). Apoptotic bodies were identified as the particles that were stained positive with both FITC and PI.[30] Quantification of these particles was recorded as the percentage of events that were doubly stained, which, in a dual-color dot plot of FITC and PI, were in the upper right square (FIG. 1). For each lot of serum samples, a sample of phosphate buffered saline (PBS) was analyzed by flow cytometry to avoid false-positive results.

RESULTS AND DISCUSSION

Tumors were detectable from the 1st week after inoculation and grew at an apparently continuous rate for the duration of the experiment. The tumor diameter was 0.1 ± 0.0 cm (mean ± SD) after the 1st week, 0.4 ± 0.2 cm after the 2nd week, 1.0 ± 0.2 cm after the 3rd, 1.4 ± 0.4 cm after the 4th, 1.9 ± 0.5 cm after the 5th, 2.3 ± 0.7 cm after the 6th, 2.6 ± 0.7 cm after the 7th, 2.9 ± 0.9 cm after the 8th, 3.1 ± 0.9 cm after the 9th, 3.9 ± 1.0 cm after 10th, and 3.7 ± 1.0 after the 11th week.

No metastasis was observed in abdominal tissues; they were only found in the lungs. Lung macrometastases were detected in no animal in group 5, in two animals in group 7 (33%), and in four animals in group 11 (40%).

The rates of apoptotic bodies detected in the serum samples from animals in groups 5 and 7 were similar to those in control group (FIGS. 1 and 2). However, in some animals in group 11 (6/10), the mean of rates of apoptotic bodies detected in serum was similar to control group (1.80 ± 1.06%; mean ± SD), while in the remaining, this was significantly higher (9.82 ± 0.94%; $P > 0.001$). We therefore divided the group 11 into two subgroups, named as 11a and 11b, respectively (FIG. 2).

No statistically significant differences were found between the rates of serum apoptotic bodies detected in group 5, group 7, subgroup 11a, and the control

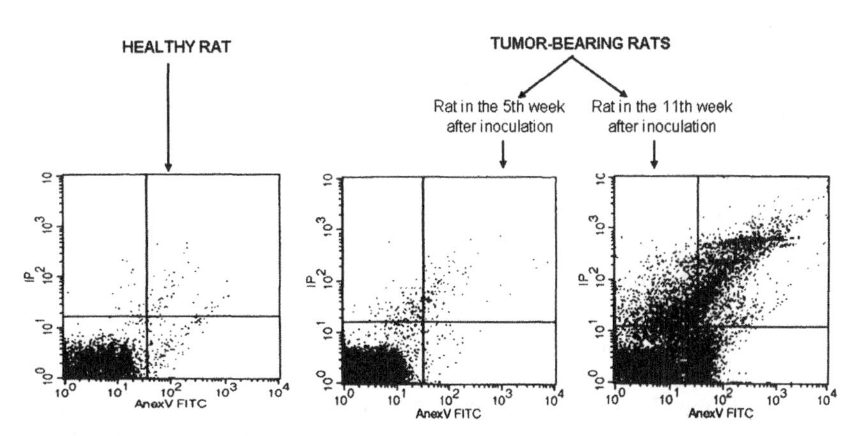

FIGURE 1. Flow cytometry of rat sera: Annexin V (AnexV-FITC)/propidium iodide (PI) assay to detect apoptotic bodies. The *left panel* shows the result of the analysis of a serum from a healthy rat; the *middle* and the *right panels* show the result obtained with the samples of two tumor-bearing rats, at the 5th and the 11th week after inoculation, respectively.

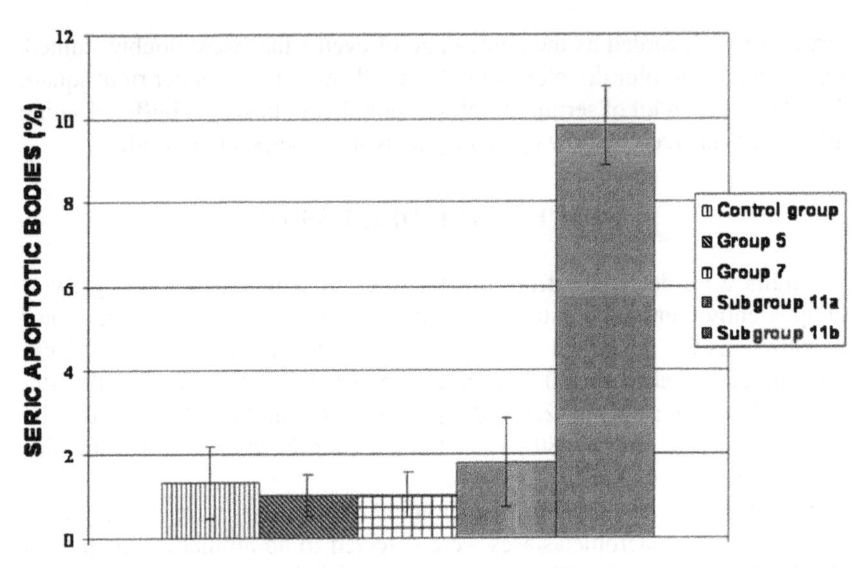

FIGURE 2. Histogram showing differences in the rates of seric apoptotic bodies detected in healthy and tumor-bearing rats.

group. However, subgroup 11b was significantly different from all other groups ($P < 0.001$ in all cases). Thus, detection of apoptotic bodies in serum was only increased in a few tumor-bearing animals at 11 weeks. Also, this was not related to the presence of lung macrometastases.

The use of plasma or serum to detect particles, such as apoptotic bodies, should be discussed. Serum may not be appropriate for this purpose, since particles might be fractioned or captured during clotting. However, previous studies have shown that filtration of plasma, as well as of serum samples, caused a decrease in the concentrations of DNA and RNA in the filtrates, due to the retention of particles containing nucleic acids by the filters.[25,28] Thus, such particles are present in both serum and plasma.

In a previous study with the same animal model, the detection rates of tumor DNA in plasma of tumor-bearing rats was highest from the end of the 1st week to the end of the 5th after inoculation.[14] These results were in agreement with those that showed the early presence of tumor DNA in the plasma of cancer patients and led to the proposal that detection of such DNA might be a complementary marker for a cancer screening.[10,11] Also, in our previous experimental study, tumor DNA in plasma of tumor-bearing rats persisted from the 5th to the 11th week after inoculation.[14] By contrast, in the present study, apoptotic bodies were only found in a few animals at the 11th week after inoculation. Therefore, it appears that there was no relation between the time of the highest detection of cell-free tumor DNA in plasma and that of apoptotic bodies in serum. In the light of our results, it seems unlikely that the tumor DNA, which is frequently detected in plasma and serum at early stages of tumor progression, circulates within apoptotic bodies.

We cannot discard the possibility that apoptosis might be a source of cell-free nucleic acids in plasma. In this study, the increase of apoptotic bodies in serum appeared to be related to cancer, since in all of the healthy animals the rates of apoptotic bodies in serum were low. However, we have not demonstrated that apoptotic bodies are the vehicles of cell-free nucleic acids during tumor progression.

The Current Challenges

Many questions need to be answered to clear up the puzzle of the implication of the cell-free tumor nucleic acids in tumor progression and in the development of metastases. Further studies are needed to identify the particles in which cell-free nucleic acids circulate protected from enzymatic degradation. The elucidation of this issue will give more clues about the origin of such nucleic acids. It has been suggested that the presence of cell-free nucleic acids in the blood stream might be the result, in variable proportions, of the different mechanism which produces leakage or excretion of DNA.[5]

Also, the potential role of apoptotic bodies in tumor progression by gene transfer should not be discarded, since their ability to transfer genes has been demonstrated.[18] These particles might be involved in the local tumor growth as Bergsmedh *et al.* proposed,[19] and, moreover, perhaps they circulate in the blood stream within phagocytosing cells.

Another question arises: if we assume that cell-free nucleic acids can be transferred to cells and transform them, what kind of cell is susceptible to being transformed? Since cancer cells and stem cells display striking similarities,[31] might stem cells be involved in the cancer progression? If yes, then what kind of stem cells are involved?

ACKNOWLEDGMENTS

This work was supported by a grant (03/1540) from "Fondo Europeo de Desarrollo Regional" (FEDER) and "Fondo de Investigaciones Sanitarias" (FIS) and a grant (03-047) from the Government of Castilla-La Mancha (Junta de Comunidades de Castilla-La Mancha), Spain. J. Samos and M.G. Picazo were recipients of fellowships from the Government of Castilla-La Mancha.

REFERENCES

1. García-Olmo, D. *et al.* 1999. Tumour DNA circulating in the plasma might play a role in metastasis. The hypothesis of the genometastasis. Histol. Histopathol. **14:** 1159–1164.
2. García-Olmo, D. *et al.* 2000. Horizontal transfer of DNA and the "genometastasis hypothesis." Blood **95:** 724–725.

3. García-Olmo, D. & D.C. García-Olmo. 2001. Functionality of circulating DNA: the hypothesis of genometastasis. Ann. N. Y. Acad. Sci. **945:** 265–275.
4. Weigelt, B. *et al.* 2005. Breast cancer metastasis: markers and models. Nat. Rev. Cancer **5:** 591–602.
5. Anker, P. *et al.* 1999. Detection of circulating tumour DNA in the blood (plasma/serum) of cancer patients. Cancer Metastasis Rev. **18:** 65–73.
6. Kopreski, M.S. *et al.* 2000. Somatic mutation screening: identification of individuals harbouring *K-ras* mutations with the use of plasma DNA. J. Nat. Cancer Inst. **92:** 918–923.
7. Jahr, S. *et al.* 2001. DNA fragments in the blood plasma of cancer patients: quantitations and evidence for their origin from apoptotic and necrotic cells. Cancer Res. **61:** 1659–1665.
8. Kopreski, M.S. *et al.* 1999. Detection of tumour messenger RNA in the serum of patients with malignant melanoma. Clin. Cancer Res. **5:** 1961–1965.
9. Wang, B.G. *et al.* 2003. Increased plasma DNA integrity in cancer patients. Cancer Res. **63:** 3966–3968.
10. Sozzi, G. *et al.* 2001. Analysis of circulating tumour DNA in plasma at diagnosis and during follow-up of lung cancer patients. Cancer Res. **61:** 4675–4678.
11. Kopreski, M.S. *et al.* 2000. Somatic mutation screening: identification of individuals harbouring K-ras mutations with the use of plasma DNA. J. Natl. Cancer Inst. **92:** 918–923.
12. Lecomte, T. *et al.* 2002. Detection of free-circulating tumour-associated DNA in plasma of colorectal cancer patients and its association with prognosis. Int. J. Cancer **100:** 542–548.
13. Silva, J.M., *et al.* 2002. Tumour DNA in plasma at diagnosis of breast cancer patients is a valuable predictor of disease-free survival. Clin. Cancer Res. **12:** 3761–3766.
14. García-Olmo, D.C. *et al.* 2005. Detection of circulating tumour cells and of tumour DNA in plasma during tumour progression in rats. Cancer Lett. **217:** 115–123.
15. García-Olmo, D.C. *et al.* 2003. The site of injection of tumour cells in rats does not influence the subsequent distribution of metastases. Oncol. Rep. **10:** 903–907.
16. Pulciani, S. *et al.* 1982. Oncogenes in human tumour cell lines: molecular cloning of a transforming gene from human bladder carcinoma cells. Proc. Natl. Acad. Sci. USA **79:** 2845–2849.
17. Anker, P. *et al.* 1994. Transformation of NIH/3T3 cells and SW 480 cells displaying K-ras mutation. C. R. Acad. Sci. III **317:** 869–874.
18. Holmgren, L. *et al.* 1999. Horizontal transfer of DNA by the uptake of apoptotic bodies. Blood **93:** 3956–3963.
19. Bergsmedh, A. *et al.* 2001. Horizontal transfer of oncogenes by uptake of apoptotic bodies. Proc. Natl. Acad. Sci. USA **98:** 6407–6411.
20. Ceccarini, M. *et al.* 1989. Biochemical and NMR studies on structure and release conditions of RNA-containing vesicles shed by human colon adenocarcinoma cells. Int. J. Cancer **44:** 714–721.
21. Stroun, M. *et al.* 1978. Presence of RNA in the nucleoprotein complex spontaneously released by human lymphocytes and frog auricles in culture. Cancer Res. **38:** 3546–3554.
22. Wieczorek, A.J. *et al.* 1985. Isolation and characterization of an RNA–proteolipid complex associated with the malignant state in humans. Proc. Natl. Acad. Sci. USA **82:** 3455–3459.

23. ROSI, A. *et al.* 1988. RNA-lipid complexes released from the plasma membrane of human colon carcinoma cells. Cancer Lett. **39:** 153–160.
24. REDDI, K.K. & J.F. HOLLAND. 1976. Elevated serum ribonuclease in patients with pancreatic cancer. Proc. Natl. Acad. Sci. USA **73:** 2308–2310.
25. EL-HEFNAWY, T. *et al.* 2004. Characterization of amplifiable, circulating RNA in plasma and its potential as a tool for cancer diagnostics. Clin. Chem. **50:** 564–573.
26. HASSELMANN, D.O. *et al.* 2001. Extracellular tyrosinase mRNA within apoptotic bodies is protected from degradation in human serum. Clin. Chem. **47:** 1488–1489.
27. NG, E.K. *et al.* 2002. Presence of filterable and nonfilterable mRNA in the plasma of cancer patients and healthy individuals. Clin. Chem. **48:** 1212–1217.
28. TSUI, N.B. *et al.* 2002. Stability of endogenous and added RNA in blood specimens, serum, and plasma. Clin. Chem. **48:** 1647–1653.
29. HALICKA, H.D. *et al.* 2000. Segregation of RNA and separate packaging of DNA and RNA in apoptotic bodies during apoptosis. Exp. Cell Res. **260:** 248–256.
30. VERMES, I. *et al.* 2000. Flow cytometry of apoptotic cell death. J. Immunol. Meth. **243:** 167–190.
31. GARCÍA-OLMO, D.C. *et al.* 2004. Circulating nucleic acids in plasma and serum (CNAPS) and its relation to stem cells and cancer metastasis: state of the issue. Histol. Histopathol. **19:** 575–583.

Plasma RNA Integrity Analysis

Methodology and Validation

BLENDA C.K. WONG AND Y.M. DENNIS LO

Department of Chemical Pathology, The Chinese University of Hong Kong, Prince of Wales Hospital, Shatin, New Territories, Hong Kong SAR, China

ABSTRACT: The detection of cell-free RNA in plasma and serum of human subjects has found increasing applications in the field of medical diagnostics. However, many questions regarding the biology of circulating RNA remain to be addressed. One issue concerns the molecular nature of these circulating RNA species. We have recently developed a simple and quantitative method to investigate the integrity of plasma RNA. Our results have suggested that cell-free RNA in plasma is generally present as fragmented molecules instead of intact transcripts, with a predominance of 5′ fragments. In this article, we summarize the basic principles in the experimental design for plasma RNA integrity analysis and highlight some of the important technical considerations for this type of investigation.

KEYWORDS: plasma RNA; integrity; quantitative analysis

INTRODUCTION

Over the past decade, cell-free circulating nucleic acids have represented a promising source of material for noninvasive molecular diagnosis.[1] In particular, the detection and quantification of circulating RNA species, including those of tumoral[2–5] and fetal origin,[6–8] has opened up new investigational opportunities for noninvasive analysis of gene expression. Despite the rapid development of plasma RNA as a potential diagnostic tool, many of the biological aspects of these RNA species remain to be elucidated. The recent demonstration of robust RNA detection in plasma at room temperature for up to 24 hours[9] has suggested the relative stability of such circulating molecules, contrary to the conventional belief that RNA is labile in nature. It has been postulated that plasma RNA is associated with particular matter and thus protected from degradation by nucleases.[10,11] More recently, we have developed

Address for correspondence: Prof. Y. M. Dennis Lo, Department of Chemical Pathology, The Chinese University of Hong Kong, Prince of Wales Hospital, Rm 38023, 1/F Clinical Sciences Building, 30-32 Ngan Shing Street, Shatin, New Territories, Hong Kong SAR, China. Voice: +852-2632-2563; fax: + 852-2636-5090.

e-mail: loym@cuhk.edu.hk; blend@cunk.edu.hk.

Ann. N.Y. Acad. Sci. 1075: 174–178 (2006). © 2006 New York Academy of Sciences.
doi: 10.1196/annals.1368.023

a quantitative approach to determine the integrity of cell-free RNA in plasma and found that plasma RNA is associated with a preponderance of 5′ end fragments.[12] The strategies used for measurement of plasma RNA integrity and verification of the observed data are reviewed in this article. It is hoped that this information would facilitate the future study of plasma RNA integrity as a new field of investigation for circulating nucleic acids.

Quantitative Analysis of Plasma RNA Integrity

The aim of plasma RNA integrity analysis is to determine whether cell-free RNA is present as intact molecules in the circulation. Because intact transcripts should theoretically consist of both the 5′ and 3′ ends, we have hypothesized that intact transcripts would show a one-to-one ratio when the two ends are compared. Hence, we have developed one-step real-time quantitative reverse transcription polymerase chain reaction (RT-PCR) assays to amplify the 5′ and 3′ ends of a target transcript (e.g., the housekeeping gene, glyceraldehyde-3-phosphate dehydrogenase, or *GAPDH*).[12] Whether the transcript is intact or fragmented would depend on the observed abundances of the two amplicons. Basically, a discrepancy in the measured quantities or detection rates between the two assays would indicate the presence of incomplete mRNA fragments. For assay design, the 5′ and 3′ amplicons should be comparable in size. This would minimize the possibility of amplicon size variation in causing a difference in RT-PCR efficiencies, thus leading to quantitative bias in the observations.

Regarding data analysis, the absolute mRNA concentrations measured by the 5′- and 3′-specific assays can be directly compared using standard statistical methods. Alternatively, the expression of fractional mRNA concentrations relative to the absolute concentration of the 5′ amplicon may provide further information on the integrity of circulating RNA, indicating the fold difference in abundance between the two observed regions. In the study of the *GAPDH* transcript in the plasma of healthy human subjects, we have obtained a fractional concentration of 0.051, thereby suggesting that 5′ mRNA fragments are predominant in the circulation, and that the 5′ molecules are some 20-fold higher in concentration when compared to the 3′ molecules.[12]

Calibration and Optimization

Quantification and comparison of the 5′ and 3′ ends of a transcript should ideally be based on the same calibrator. In recent years, the use of a single-stranded synthetic oligodeoxynucleotide that could be commercially synthesized has simplified the process of obtaining a calibration curve for amplicons of up to 100 nucleotides.[13] However, for the purpose of studying multiple regions along a transcript and comparing different quantitative RT-PCR systems, the use of multiple individual calibration curves of such kind may potentially introduce quantitative bias in the results. To avoid this problem, it is ideal to develop a single calibrator that could be used for measuring the concentrations of all the targeted regions on the transcript of interest.

One way to prepare a "multiassay" calibrator involves subcloning into a plasmid vector the target cDNA construct, which should contain the sequences of both the 5′ and 3′ amplicons.[12] If the size of a transcript is a factor that inhibits the process of TA cloning, the targeted regions of that transcript may be prepared separately by PCR and subcloned at different sites of the vector.[12] Quantification of different regions within a transcript based on this type of calibrator would reflect more accurately the molecular characteristics of cell-free RNA in plasma.

Real-time RT-PCR assays are optimized with the synthesized calibrator over a range of known concentrations. The optimized amplification conditions are generally those that result in the lowest detection limit. For valid comparison of the 5′ and 3′ measurements, however, it is also important that we are able to achieve similar detection limits for the two systems using the calibration curve.[12]

Precautions

Other aspects of assay design are also critical for accurate data interpretation during the analysis of plasma RNA integrity. First, it should be noted that primer design across the intron–exon junction may not be achievable for certain RT-PCR assays as amplicons must target a particular region of the transcript under study (e.g., the 5′ or the 3′ end). It is therefore necessary to ensure the absence of genomic DNA contamination in the tested samples. This can be done by performing DNase I treatment and including negative controls with the omission of the reverse transcription step. Second, the presence of homologous sequences could increase the chance of nonspecific amplification. In this regard, assay design must take into consideration that the primers are uniquely specific for the sequence of interest. Technical precautions such as those described above would minimize false-positive amplification during RT-PCR and therefore reduce the possibility of introducing quantitative bias for plasma RNA integrity analysis.

Validation of Plasma RNA Integrity Analysis

Through RNA integrity analysis of multiple transcripts, we have demonstrated that the majority of plasma mRNA population is represented by 5′ end transcripts.[12] In view of importance of this observation, we have performed additional experiments to validate this finding. One experiment involves a two-step RT-PCR analysis of the 5′ and 3′ amplicons in RNA samples using random and oligo-dT primers during reverse transcription, followed by gene-specific real-time PCR.[12] The rationale for this experiment is based on the fact that oligo-dT priming could potentially introduce 3′ end bias in the amplified products, as 3′ end degraded transcripts could not be reverse-transcribed successfully in this system. Demonstration of the reverse phenomenon (i.e., 3′ end predominance) in the oligo-dT-primed system would suggest that the observed results are unlikely to be due to technical bias favoring the 5′ RT-PCR assay. The random-primed system, on the other hand, would theoretically identify the 5′ amplicon as the predominant species in support of the previous findings based on one-step real-time analysis. Further validation of plasma RNA integrity analysis entails the detection of 5′ and 3′ amplicons in serially diluted RNA samples using one-step real-time RT-PCR.[12] Given that both assays have been optimized for equal sensitivities with the calibrator, this experiment actually provides a physical means for demonstrating the predominance of one species over another in the circulation. As described previously, demonstration of the persistent detection of the 5′ amplicon and the rapid disappearance of the 3′ amplicon after rounds of serial dilution has confirmed our conclusion regarding the 5′ predominance of the circulating transcripts.[12]

CONCLUSIONS

Plasma RNA integrity analysis has opened up a new approach for exploring the biological characteristics of cell-free circulating RNA. The finding that plasma RNA is associated with a preponderance of 5′ end transcripts[12] has raised additional biological questions regarding such molecular species. For example, it would be of interest to investigate the functional role, if any, of the observed incomplete circulating RNA fragments. Here, we have described in detail the methodology for plasma RNA integrity analysis and addressed the need for adopting technical precautions and validation studies to achieve accurate data interpretation. This information would be valuable for the future study of plasma RNA integrity, which could further improve our understanding of the molecular nature of cell-free circulating RNA.

ACKNOWLEDGMENTS

This work was supported by the Kadoorie Charitable Foundations (under the auspices of the Michael Kadoorie Cancer Genetics Research Program). The Chinese University of Hong Kong holds patents or patent applications on aspects of circulating nucleic acids in plasma.

REFERENCES

1. Lo, Y.M.D. 2001. Circulating nucleic acids in plasma and serum: an overview. Ann. N. Y. Acad. Sci. **945**: 1–7.
2. Lo, K.W. *et al.* 1999. Analysis of cell-free Epstein-Barr virus associated RNA in the plasma of patients with nasopharyngeal carcinoma. Clin. Chem. **45**: 1292–1294.
3. Kopreski, M.S. *et al.* 1999. Detection of tumor messenger RNA in the serum of patients with malignant melanoma. Clin. Cancer Res. **5**: 1961–1965.
4. Dasi, F.S. *et al.* 2001. Real-time quantification in plasma of human telomerase reverse transcriptase (hTERT) mRNA: a simple blood test to monitor disease in cancer patients. Lab. Invest. **81**: 767–769.
5. Wong, S.C. *et al.* 2004. Quantification of plasma beta-catenin mRNA in colorectal cancer and adenoma patients. Clin. Cancer Res. **10**: 1613–1617.
6. Poon, L.L.M. *et al.* 2000. Presence of fetal RNA in maternal plasma. Clin. Chem. **46**: 1832–1834.
7. Ng, E.K.O. *et al.* 2003. mRNA of placental origin is readily detectable in maternal plasma. Proc. Natl. Acad. Sci. USA **100**: 4748–4753.
8. Ng, E.K.O. *et al.* 2004. Evaluation of human chorionic gonadotropin beta-subunit mRNA concentrations in maternal serum in aneuploid pregnancies: a feasibility study. Clin. Chem. **50**: 1055–1057.
9. Tsui, N.B.Y. *et al.* 2002. Stability of endogenous and added RNA in blood specimens, serum, and plasma. Clin. Chem. **48**: 1647–1653.
10. Halicka, H.D. *et al.* 2000. Segregation of RNA and separate packaging of DNA and RNA in apoptotic bodies during apoptosis. Exp. Cell Res. **260**: 248–256.
11. Ng, E.K.O. *et al.* 2002. Presence of filterable and nonfilterable mRNA in the plasma of cancer patients and healthy individuals. Clin. Chem. **48**: 1212–1217.
12. Wong, B.C.K. *et al.* 2005. Circulating placental RNA in maternal plasma is associated with a preponderance of 5′ mRNA fragments: implications for noninvasive prenatal diagnosis and monitoring. Clin. Chem. **51**: 1786–1795.
13. Bustin, S.A. 2000. Absolute quantification of mRNA using real-time reverse transcription polymerase chain reaction assays. J. Mol. Endocrinol. **25**: 169–193.

Molecular Diagnostic Markers for Lung Cancer in Sputum and Plasma

YI-CHING WANG,[a,b] HAN-SHUI HSU,[c] TSZ-PEI CHEN,[a] AND JUNG-TA CHEN[d]

[a]*Department of Life Sciences, National Taiwan Normal University, Taipei 11699, Taiwan, ROC*

[b]*Department of Pharmacology, College of Medicine, National Cheng Kung University, Tainan, Taiwan, ROC*

[c]*Division of Thoracic Surgery, Taipei Veterans General Hospital; and National Yang-Ming University School of Medicine, Taipei, Taiwan, ROC*

[d]*Department of Pathology, Taichung Veterans General Hospital, Taichung, Taiwan, ROC*

ABSTRACT: Lung cancer is the leading cause of cancer deaths worldwide. This study was designed to select multiple DNA markers, which have high sensitivity and specificity to serve as biomarkers for diagnosis of lung cancer. We examined the promoter hypermethylation of three tumor suppressor genes by methylation-specific PCR (MSP), and the instability of eight microsatellite markers by loss of heterozygosity (LOH) and microsatellite instability (MSI) analyses in lung tumor tissues and matched sputum specimens from 79 lung cancer patients. On the basis of the results of sensitivity, specificity, and concordance from each marker analyzed, we selected seven biomarkers, which are LOH of D9S286, D9S942, GATA49D12, and D13S170, MSI of D9S942, and methylation of *p16^{INK4a}* and *RARβ*, from the sputum analyses. These selected etiologically associated biomarkers can potentially be used as supplemental diagnostic biomarkers for early lung cancer detection.

KEYWORDS: LOH; MSI; promoter hypermethylation; biomarker; sputum; plasma; NSCLC

INTRODUCTION

Lung cancer is the number one cause of cancer-related death in the world. In spite of improvements in the surgical, radiotherapeutic, and chemotherapeutic modalities used to treat the various forms of lung cancer, there is no

Address for correspondence: Yi-Ching Wang, Ph.D., Department of Life Sciences, National Taiwan Normal University, No. 88, Sec. 4, Tingchou Road, Taipei 11699, Taiwan, ROC. Voice: +886-2-29336876, ext. 373; fax: +886-2-29312904.
e-mail: t43017@ntnu.edu.tw

**Ann. N.Y. Acad. Sci. 1075: 179–184 (2006). © 2006 New York Academy of Sciences.
doi: 10.1196/annals.1368.024**

validated screening method for lung cancer.[1] There is an urgent need to develop a more sensitive molecular marker panel for lung cancer diagnosis and risk assessment. For biomarker-based screening and diagnosis, access of adequate biological materials is critical. Sputum and plasma are the most commonly utilized materials since they can be collected in a non-invasive manner. By means of a PCR-based strategy, DNA alterations such as gene mutation, microsatellite alteration, and promoter hypermethylation have been demonstrated to be more sensitive and specific than sputum cytology and chest X-ray.[2]

We and others have discovered common genetic abnormalities in both pre- and invasive lung cancer lesions including loss of heterozygosity (LOH) of chromosomal regions on 3p, 5q, 9p, and 12p, genomic instability detected as microsatellite instability (MSI) using markers on chromosomes 3p, 9p, 13q, and 17p, as well as promoter hypermethylation in *BLU, CDH13, FHIT, p16,*^{INK4a} *RAR*β, and *RASSF1A* tumor suppressor genes.[3–5] To test these biomarkers for their potential use in lung cancer diagnosis, we investigated the alteration of these multiple DNA markers in cytologically negative sputum specimens and their corresponding primary tumor and normal lung tissues. In addition, plasma and matched tumor samples were examined for these multiple DNA markers. Markers that have high sensitivity and specificity have the potential to be used as biomarkers for lung cancer diagnosis.

MATERIALS AND METHODS

Study Population, Sputum Samples, and Plasma Samples

The study subjects for sputum analysis consisted of 79 patients diagnosed with primary non–small cell lung cancer (NSCLC) admitted to Taichung Veterans General Hospital. Cytologically negative sputum samples were obtained 1–18 months before surgery. The study subjects for plasma analysis consisted of 63 NSCLC patients admitted to Taipei Veteran General Hospital. Heparinized venous blood was collected from lung cancer patients before surgery. Plasma DNA was extracted using standardized procedures. In addition, 22 cancer-free individuals were recruited for both sputum and plasma analyses from Taipei Veteran General Hospital.

Eight microsatellite polymorphic markers included D3S1234, D3S1285, D5S1456, D9S286, D9S942, GATA49D12, D13S170, and D17S786. They were used for LOH and MSI analyses in sputum and corresponding microdissected lung tumor and normal lung tissue from 79 NSCLC patients. The sputum from 22 cancer-free individuals was also analyzed.

The promoter methylation of six tumor suppressor genes including *BLU, CDH13, FHIT, p16*^{INK4a}, *RAR*β, and *RASSF1A* were also examined in plasma and corresponding microdissected lung tumor and normal lung tissue from 63 NSCLC patients and 22 cancer-free individuals.

RESULTS

Biomarker Associations in Cytologically Negative Sputum Samples

Prevalence of LOH was 26~50% in tumors, 26~48% in sputum from NSCLC patients, and 0~11% in sputum from cancer-free individuals (FIG. 1A). The sensitivity, specificity, and concordance of each LOH marker were then determined. Concordance was defined as matched result of sputum samples with corresponding resected tumors. Four markers D9S286, D9S942, GATA49D12, and D13S170 showed the highest sensitivity, specificity, and concordance (FIG. 1A). The patients with LOH in D9S942 in sputum showed an increased odds ratio (OR) (4.92; 95% confidence interval, CI, 1.04–4.50; $P = 0.02$) of having lung cancer compared to the cancer-free individuals. Although the averaged sensitivity, specificity, and concordance of MSI detected in sputum could reach the range of ~64–98%, the prevalence of MSI was only ~4–31% in tumors, ~4–24% in sputum from NSCLC patients, and ~0–5% in sputum from cancer-free individuals (FIG. 1B).

The methylation status in the *FHIT, p16,*^{INK4a} and *RAR*β genes of sputum and corresponding tumor and normal lung tissue was determined by methylation-specific PCR (MSP) from 79 NSCLC patients and sputum from 22 cancer-free individuals (FIG. 1C). Prevalence of methylation in these three genes was ~34–45% in tumors, ~31–36% in sputum from NSCLC patients, and ~14–18% in sputum from cancer-free individuals. Promoter methylation in the *p16*^{INK4a} and *RAR*β genes showed the high sensitivity, specificity, and concordance. In addition, *p16*^{INK4a} methylation in sputum showed an increased OR (3.29; CI, 1.00–14.93; $P = 0.05$) of having lung cancer compared to the cancer-free individuals.

Biomarker Associations in Plasma Samples

Prevalence of methylation in six genes was ~36–54% in tumors, ~31–39% in plasma from NSCLC patients, and ~14–17% in plasma from cancer-free individuals (FIG. 2). Promoter methylation in the *p16,*^{INK4a} *RAR*β, and *RASSF1A* genes showed the high sensitivity, specificity, and concordance. In addition, methylation of the *p16*^{INK4a} and *RASSF1A* in plasma showed significantly increased ORs (5.56 for the *p16*^{INK4a} and 5.48 for the *RASSF1A*) of having lung cancer compared to the cancer-free individuals.

DISCUSSION

For the sputum analysis, we selected seven molecular markers including LOH in D9S286, D9S942, GATA49D12, and D13S170, MSI in D9S942, and

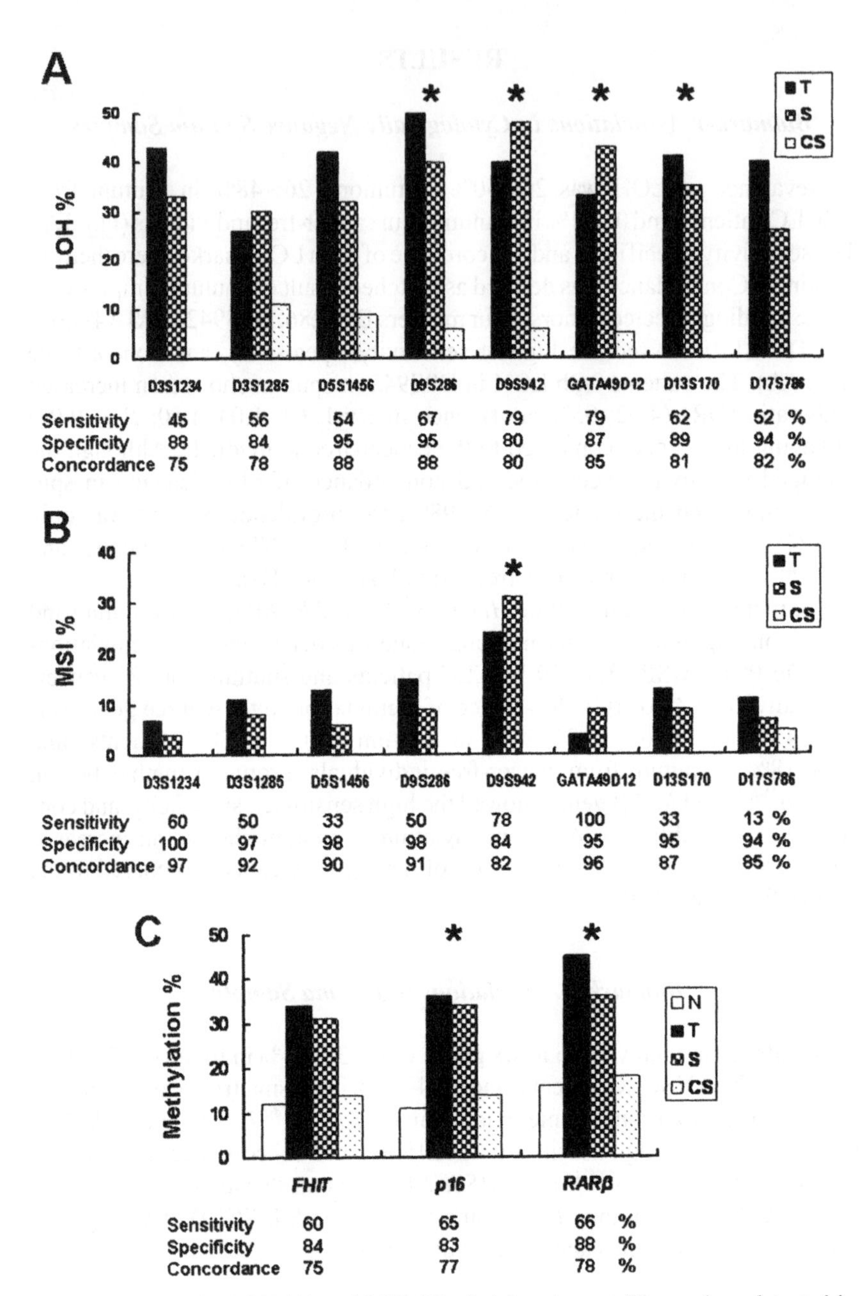

FIGURE 1. The LOH (**A**) and MSI (**B**) of eight microsatellite markers detected in lung tumor (T) and matched sputum (S), and MSP (**C**) of three tumor suppressor genes from 79 patients and control sputum (CS) from 22 cancer-free individuals. The DNA from the corresponding normal cell (N) was used as a control for all experiments. Sensitivity, specificity, and concordance are shown below the *x* axis for each marker. The markers with sensitivity, specificity, and concordance all greater than 60% are marked with symbol*.

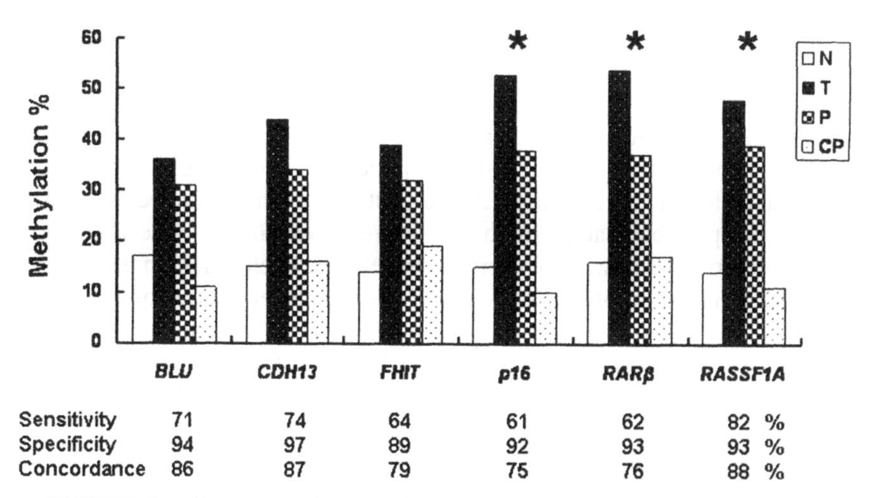

FIGURE 2. The promoter hypermethylation analysis of six genes detected in normal lung (N) cells, tumor lung (T) cells, and matched plasma (P) from 63 patients and control plasma (CP) from 22 cancer-free individuals. Sensitivity, specificity, and concordance were calculated individually for each marker. The three markers with high methylation frequency are marked with symbol*.

methylation in $p16^{INK4a}$ and *RAR*β based on their high prevalence, sensitivity, specificity, and concordance. A combined analysis, where more than two of the seven selected biomarkers showed alteration in patients with cancer risk, showed a sensitivity of 81%, a specificity of 72%, and a concordance of 77%. This combined analysis showed 91% (20/22) accuracy to distinguish the cancer-free individuals from cancer patients. The data suggested that multiple biomarkers could enhance sensitivity and specificity.

For the plasma analysis, we selected three molecular markers including methylation in $p16,^{INK4a}$ *RAR*β, and *RASSF1A* based on their high prevalence, sensitivity, specificity, and concordance. A combined analysis, where more than one of the three selected biomarkers showed alteration in patients with cancer risk, showed a sensitivity of 74%, a specificity of 78%, and a concordance of 75%. However, when applied to the cancer-free population, the combined analysis only showed 64% (14/22) accuracy. We therefore performed a logistic regression model for all six methylation biomarkers in 63 NSCLC patients and 22 cancer-free individuals. A hazard function equation was determined as follows: Y = 0.19 + 0.52 (*BLU* methyl) + 1.92 ($p16^{INK4a}$ methyl) + 1.52 (*RASSF1A* methyl), where the calculated Y reflected the hazard rate of developing cancer if number 1 is for methylation and 0 for no methylation in each selected gene. The number in front of each biomarker is a coefficient for its relative risk. This regression model gave a sensitivity of 77%, a specificity of 90%, and a concordance of 79%.

Our study using the combined analysis or logistic regression for selected panel of biomarkers in lung cancer and matched sputum or plasma samples strongly suggests that the analysis of DNA methylation patterns could become a powerful tool for accurate and early lung cancer diagnosis, with reasonable specificity and sensitivity. Note that some patients exhibited positive alteration in tumor cells, but no alteration in the sputum or plasma collected before surgery. It is possible that the tumor cells have not exfoliated to the sputum or genetic abnormalities have not occurred in cell-free nucleic acid in plasma. In addition, alteration in some DNA markers was detected in sputum or plasma of NSCLC patients whose tumors did not contain such DNA alteration. This could be explained by the presence of several distinct tumor subpopulations, one of which has methylation and microsatellite alteration, and the others having no alteration. In addition, DNA alterations in sputum or plasma might derive from clinically unidentified small second primary lung cancers.

It has been shown that cytological analysis is less accurate at detecting adenocarcinoma, peripheral cancer, or early-staged tumors.[6] In addition, the detection rate may be different between smoking and non-smoking patients. We therefore compared the sensitivity, specificity, and concordance of biomarker analysis in sputum study and plasma study according to the clinicopathological parameters including tumor stages, tumor types, and smoking habits of patients analyzed. The results suggested that the biomarker method is effective for lung cancer diagnosis for various subtypes.

Our findings suggest that DNA biomarkers, such as LOH in microsatellites and methylation in tumor suppressor genes in sputum and plasma, might track with a very high-risk status or with the actual presence of cancer, and may be used as a sensitive and reliable molecular diagnostic method for lung cancer. This should be confirmed in larger patient subsets. A population-based prospective study to determine the potential benefit of these diagnostic biomarkers is worthy of investigation.

REFERENCES

1. Deppermann, K.M. 2004. Lung cancer screening—where we are in 2004 (take home messages). Lung Cancer **45**(Suppl. 2)**:** S39–S42.
2. Mao, L. 2002. Recent advances in the molecular diagnosis of lung cancer. Oncogene **21:** 6960–6969.
3. Chen, J.-T. *et al.* 2002. Alterations of the *p16*^{INK4a} gene in resected non-small cell lung tumors and exfoliated cells within sputum. Int. J. Cancer **98:** 724–731.
4. Tzao, C. *et al.* 2004. 5′CpG island hypermethylation and aberrant transcript splicing both contribute to the inactivation of the *FHIT* gene in resected non-small cell lung cancer. Eur. J. Cancer **40:** 2175–2183.
5. Tseng, R.-C. *et al.* 2005. Genome-wide loss of heterozygosity and its clinical associations in non-small cell lung cancer. Int. J. Cancer **17:** 17–23.
6. Kennedy, T.C. & F.R. Hirsch. 2004. Using molecular markers in sputum for the early detection of lung cancer: a review. Lung Cancer **45**(Suppl. 2)**:** S21–S27.

Quantitative Analysis of Plasma DNA in Colorectal Cancer Patients

A Novel Prognostic Tool

MILO FRATTINI,[a,f] GIANFRANCESCO GALLINO,[b] STEFANO SIGNORONI,[a] DEBORA BALESTRA,[a] LUIGI BATTAGLIA,[b] GABRIELLA SOZZI,[c] ERMANNO LEO,[b] SILVANA PILOTTI,[d] AND MARCO A. PIEROTTI[c,e]

[a]*Department of Experimental Oncology and Unit of Experimental Molecular Pathology, Department of Pathology, Istituto Nazionale per lo Studio e la Cura dei Tumori, Milan, Italy*

[b]*Colorectal Surgery Unit, Istituto Nazionale per lo Studio e la Cura dei Tumori, Milan, Italy*

[c]*Department of Experimental Oncology, Istituto Nazionale per lo Studio e la Cura dei Tumori, Milan, Italy*

[d]*Unit of Experimental Molecular Pathology, Department of Pathology, Istituto Nazionale per lo Studio e la Cura dei Tumori, Milan, Italy*

[e]*FIRC Institute of Molecular Oncology (IFOM), Milan, Italy*

ABSTRACT: **Extracellular DNA in the plasma or serum of cancer patients has been recently proposed as a source of analyzable cancer-related gene sequences (qualitative approach). Furthermore, patients with different tumor types show high levels of cell-free circulating DNA both in plasma and serum (quantitative approach) at the time of surgery. Our aim was to verify whether the level of cell-free DNA in plasma might help in detecting recurrences during follow-up of colorectal cancer (CRC) patients. We studied 70 patients undergoing surgery for primary CRC. Plasma samples were obtained at the time of surgery and during follow-up. The cell-free circulating DNA in plasma was quantified by the Dipstick Kit method. At the time of surgery, in all patients, cell-free DNA levels in plasma were about 25 times higher in comparison with 20 healthy donors. In contrast, the carcinoembryonic antigen (CEA) value of this cohort of patients was altered in only about 37% of cases. During follow-up, cell-free DNA levels decreased progressively in tumor-free patients, while it increased in those developing recurrences or metastases.**

Address for correspondence: Marco A. Pierotti, Department of Experimental Oncology, Istituto Nazionale per lo Studio e la Cura dei Tumori, 20133 Milano, Italy. Voice: +39-02-2390-2236; fax: +39-02-2390-2764.

e-mail: marco.pierotti@istitutotumori.mi.it

[f] Present address: Laboratory of Diagnostic and Experimental moleculer Pathology, Clinic Pathology, Istituto Cantonale de Patologia, Locarno, Switzerland.

Ann. N.Y. Acad. Sci. 1075: 185–190 (2006). © 2006 New York Academy of Sciences. doi: 10.1196/annals.1368.025

The results were further supported by qualitative analysis of circulating tumor-specific DNA, such as K-Ras mutations and p16^{INK4a} promoter hypermethylation. These preliminary data confirm that plasma tumor DNA levels (i) are significantly higher in patients with CRC, (ii) decrease progressively in the follow-up period in tumor-free patients, and (iii) increase in patients with recurrence or metastasis. We suggest, therefore, that the quantification of plasma cell-free DNA might represent a useful tool for monitoring of CRC and, prospectively, for identifying high-risk individuals.

KEYWORDS: **colorectal cancer (CRC); carcinoembryonic antigen (CEA)**

INTRODUCTION

Colorectal cancer (CRC) is one of the most frequent causes of cancer death in Western countries. Over the last two decades, much progress has been made in the identification and characterization of the genetic changes involved in malignant colorectal transformation.[1] However, the impressive increase in knowledge of molecular markers has not prompted the development of new noninvasive diagnostic tools able to improve the detectability of CRC currently based on carcinoembryonic antigen (CEA) evaluation. Similar to other serum cancer antigens, CEA is higher in patients with CRC and decreases after curative tumor resection. Nevertheless, the use of CEA test in early diagnosis and monitoring of CRC is limited by relatively poor sensitivity (it is elevated in only about 40% of CRC patients) and specificity.[2] Therefore, a DNA marker that could be used to monitor or to predict a relapse in a presymptomatic phase of the follow-up period could have a great impact on the management of CRC patients and probably on survival.

The finding that tumors can release DNA into the circulation has opened new areas in translational cancer research. The precise mechanism of the release of circulating DNA into the bloodstream remains to be proven; however both apoptosis and necrosis of cancer cells might represent two major sources of cell-free DNA, which has been shown to display characteristics of tumor DNA.[3–4] Accordingly, it is possible to detect tumor-specific DNA in the plasma of patients carrying various cancers, such as oncogenes and tumor suppressor gene mutations, microsatellite and epigenetic alterations, and chromosomal translocations, identical to those found in primary tumor DNA.[5] However, for CRC, the choice of a unique molecular marker for plasma-based analyses is difficult. *APC* and *TP53* genes could represent two good candidates, but their mutational spectra are too wide, which makes their analysis very difficult, time-consuming, and expensive. More promising seem to be *K-Ras* mutation and *p16^{INK4a}* promoter hypermethylation, even if altered in a smaller proportion of CRC patients. Recently, a new parameter has been shown to play a relevant role in plasma (and serum) analysis: the level of cell-free circulating DNA.

Many techniques have been used to quantitate cell-free DNA,[3] but none of them has so far been evaluated in terms of reproducibility, except for the colorimetric Dipstick Kit (Invitrogen) method.[6] It is now well established that the level of cell-free DNA is increased in patients with several tumor types in comparison with healthy donors.[7] Currently, correlation between fluctuation of cell-free DNA levels in plasma and the follow-up of cancer patients has been reported only for lung cancer.[8] Sozzi *et al.* clearly demonstrated that cell-free DNA levels in lung cancer patients are significantly higher than those found in healthy donors and that circulating DNA decreased progressively in tumor-free patients during follow-up, while it increased in those in whom a recurrence occurs.[8]

The purpose of this study is to verify whether the measurement of cell-free circulating DNA correlates with the presence of primary tumor or it predicts its recurrence or occurrence of metastasis.

MATERIALS AND METHODS

Patient Cohort

Peripheral blood samples were collected in a lithium–heparin tube from 20 healthy donors and from 70 sporadic untreated CRC patients who underwent primary surgical resection at the Istituto Nazionale per lo Studio e la Cura dei Tumori of Milan. Blood samples were obtained on the day before surgery and during the follow-up period of 4 and 10 months after surgical intervention with informed consent. Plasma was immediately separated from the cellular fraction by centrifugation two times at 2,500 rpm for 10 min at 4°C and then frozen at −80°C. During the follow-up, serum CEA, colonoscopy, abdominal ecography, thoracic X ray, computed tomography, or positron emission tomography were performed.

DNA Extraction

DNA from plasma samples was extracted using the QIAamp Blood Extraction Kit (Qiagen, Chatsworth, CA) according to the manufacturer's instruction, starting from 1 mL of plasma, followed by five passages of purification on the same column and by a final elution in 50 μL of sterile double-distilled water. DNA was then stored at −20°C until further analysis.

Cell-Free DNA Quantification in Plasma

Quantification of plasma cell-free DNA was performed by using the DNA DipStick Kit (Invitrogen, Carlsbad, CA) according to the manufacturer's recommendations, as already reported.[6,8] In brief: appropriate dilutions (undiluted, 1:10, and 1:100) of control DNA and plasma DNA were prepared with sterile water and spotted onto membranes, which were subsequently incubated with coupling and developing solutions, and then dried. Color intensities of the sample spots on the membranes were compared with the control DNA. At least three independent quantification assays were performed for each plasma sample.

RESULTS AND DISCUSSION

TABLE 1 summarizes the results. Cell-free circulating DNA levels were measurable in all controls and patients. In healthy donors, the mean value of plasma cell-free circulating DNA was 10.3 ng/mL (range: 5–50 ng DNA/mL plasma), with a median value of 5 ng/mL. These data were similar to those found by Sozzi *et al.*[8] in 43 Italian controls. In CRC patients, the mean value of plasma DNA was 495.7 ng/mL (range 100–1,750 ng/mL), with a median value of 450 ng/mL. In the same cohort of patients, the CEA value was altered in only 26 cases (37%), a finding in keeping with the literature. No correlation was observed between DNA levels in plasma and CEA. Furthermore, no association was observed among plasma DNA values and various clinico-pathological features such as age and gender of patient, location and size of tumor, histologic grading, and Dukes' stage.

At the first follow-up visit, 4 months after primary resection, CRC patients displayed a significant decrease in plasma DNA levels: the mean value observed was 170.6 ng/mL (range 15–500 ng/mL), with a median value of 110 ng/mL. At 10 months after surgical intervention, mean plasma DNA value was 240.9 ng/mL with a range of 5–1,000 ng/mL, while the median value was 110 ng/mL. In patients with recurrence of tumor a dramatic increase in DNA level was observed, while in disease-free patients a continuous decrease was seen.

The rate of decrease in plasma DNA was not identical in all cases; some showed a dramatic initial decrease, while others showed a linear or constant diminution, or a slight increase followed by a dramatic decrease. These differences could be due to different surgical interventions (which might lead to postoperative complications, such as the presence of fistulae or inflammation) rather than different chemotherapic treatments, or they could be due to different DNA clearances in the various patients. The routine application of the quantitative approach complemented by a qualitative one, based on the analysis of the specific markers mainly involved in colorectal carcinogenesis, could help in clarifying this issue.

CONCLUSIONS

Our preliminary data confirm that plasma DNA levels (i) are significantly higher in patients with CRC, (ii) decrease progressively in the follow-up

TABLE 1. Cell-free circulating DNA quantification by the Dipstick method in 20 healthy donors and 70 colorectal cancer patients

	Number of subjects	t_0: day at surgery	t_1: 4 months follow-up	t_2: 10 months follow-up	t_2: 10 months follow-up disease-free patients	t_2: 10 months follow-up with recurrence patients
Healthy donors	20	10.3 ± 10.5 median: 5 range: 5–50				
Colorectal cancer patients	70	495.7 ± 357.6 median: 450 range: 100–1,750	170.6 ± 160.1 median: 110 range: 15–500	240.9 ± 276.8 median: 110 range: 5–1,000	136.2 ± 160.2 median: 62.5 range: 5–500	694.4 ± 208.3 median: 750 range: 500–1,000

Results are expressed as ng DNA/mL plasma.

period in tumor-free patients, and (iii) increase in patients with recurrence or metastasis. These observations suggest that plasma DNA may be useful (i) to confirm the presence of CRC; (ii) to identify the presence of a recurrence or metastasis during follow-up; and (iii) prospectively, to identify high-risk individuals. These findings are in contrast with CEA, which is informative only in a small number of cases.

Our data, therefore, once confirmed on a larger series of patients with CRC and with a longer follow-up period, support the usefulness of the introduction of cell-free circulating DNA quantification analysis in a clinical setting.

ACKNOWLEDGMENTS

This work was supported by grants from the Italian Association for Cancer Research (AIRC).

REFERENCES

1. ILYAS, M. *et al.* 1999. Genetic pathways in colorectal and other cancers. Eur. J. Cancer **35**: 335–351.
2. GION, M. *et al.* 1996. Dynamic use of tumour markers: rational-clinical applications and pitfalls. Anticancer Res. **16**: 2279–2284.
3. ZIEGLER, A. *et al.* 2002. Circulating DNA: a new diagnostic gold mine? Cancer Treat. Res. **28**: 255–271.
4. LINDER, S. *et al.* 2004. Determining tumour apoptosis and necrosis in patient serum using cytokeratin 18 as a biomarker. Cancer Lett. **214**: 1–9.
5. JOHNSON, P. J. *et al.* 2002. Plasma nucleic acids in the diagnosis and management of malignant disease. Clin. Chem. **48**: 1186–1193.
6. FRATTINI, M. *et al.* 2005. Reproducibility of a semiquantitative measurement of circulating DNA in plasma from neoplastic patients. J. Clin. Oncol. **430**: 3163–3164.
7. WU, T.L. *et al.* 2002. Cell-free DNA: measurement in various carcinomas and establishment of normal reference range. Clin. Chim. Acta **321**: 77–87.
8. SOZZI, G. *et al.* 2001. Analysis of circulating tumour DNA in plasma at diagnosis and during follow-up of lung cancer patients. Cancer Res. **61**: 4675–4678.

Circulating DNA and DNase Activity in Human Blood

SVETLANA N. TAMKOVICH, ANNA V. CHEREPANOVA, ELENA V. KOLESNIKOVA, ELENA Y. RYKOVA, DMITRII V. PYSHNYI, VALENTIN V. VLASSOV, AND PAVEL P. LAKTIONOV

Institute of Chemical Biology and Fundamental Medicine, Siberian Division of the Russian Academy of Sciences, 8, Lavrentiev Ave., Novosibirsk 630090, Russia

ABSTRACT: The concentration of circulating DNA (cirDNA) and deoxyribonuclease activity in blood plasma of healthy donors and patients with colon or stomach cancer were analyzed. The concentration of DNA was measured using Hoechst 33258 fluorescent assay after the isolation by the glass–milk protocol. A 1-kbp PCR product labeled with biotinylated forward and fluorescein-labeled reverse primers was used as a substrate for DNase. DNase activity was estimated from the data of immunochemical detection of the nonhydrolyzed amplicon. The average concentration of cirDNA in the plasma of healthy donors was low (34 $\pm$ 34 ng/mL), and was accompanied with high DNase activity (0.356 $\pm$ 0.410 U/mL). The increased concentrations of cirDNA in blood plasma of patients with colon and stomach cancer were accompanied by a decrease in DNase activity below the detection level of the assay. The data obtained demonstrate that low DNase activity in blood plasma of cancer patients can cause an increase in the concentration of cirDNA.

KEYWORDS: circulating DNA; deoxyribonuclease activity; amplicon-based immunoassay; blood plasma; cancer

INTRODUCTION

The concentration of circulating DNA (cirDNA) is low in blood plasma of healthy donors[1,2] but is increased in patients with autoimmune disorders,[3] trauma,[4] and tumors.[1,2] Mechanisms leading to the appearance of cirDNA in blood are not clear to date. However, processes like apoptosis and necrosis were shown to contribute to the generation of cirDNA.[5] In the bloodstream extracellular DNA is under pressure from factors influencing its circulation and clearance, including hydrolyzing enzymes. There are few enzymes capable of degrading DNA, such as deoxyribonuclease II,[6] phosphodiesterase I,[7]

Address for correspondence: Pavel P. Laktionov, Institute of Chemical Biology and Fundamental Medicine, Siberian Division of the Russian Academy of Sciences, 8, Lavrentiev Ave., Novosibirsk 630090, Russia. Voice: +7-383-3304654; fax: +7-383-3333677.
e-mail: lakt@niboch.nsc.ru

**Ann. N.Y. Acad. Sci. 1075: 191–196 (2006). © 2006 New York Academy of Sciences.
doi: 10.1196/annals.1368.026**

DNA hydrolyzing autoantibodies,[8] and neutral deoxyribonuclease I, which is responsible for no less than 90% of deoxyribonuclease activity of blood plasma.[9]

In this study we compared neutral deoxyribonuclease activity with the concentration of cirDNA in blood plasma of healthy donors and patients with malignant gastrointestinal disease.

MATERIALS AND METHODS

Blood samples from patients with previously untreated gastrointestinal cancer were obtained from the Novosibirsk Regional Oncologic Dispensary. Tumor staging was performed according to the TNM classification. Blood samples of healthy donors were obtained from the Novosibirsk Central Clinical Hospital. Blood (8 mL) was collected into tubes containing 2 mL of sterile phosphate-buffered saline solution with 50 mM EDTA. All blood samples were stored at 4°C before treatment, which occurred within 4 h after collection of the blood. Plasma was collected after pelleting of cells by two subsequent centrifugations (400 *g*, 10 min, at 4°C). Plasma samples were stored frozen at −20°C in aliquots and were thawed once before investigation.

The glass–milk-based protocol providing quantitative isolation of nucleic acids was used for isolation of DNA from plasma.[10] The concentration of DNA was measured using Hoechst 33258 dye as described.[11] The detection limit of the assay was 10 ng/mL of plasma.

A 5′-labeled 974 bp DNA fragment of 28S rRNA gene was prepared using biotinilated forward 5′-GGT$^{C6-NH-Bio}$CCAAGAATTTCACC TCTAGC-3′ and fluorescein-labeled reverse 5′-TACCTC$^{6-NH-Flu}$GGTTGAT CCTGCCAGTAG-3′ primers by 35 cycles of PCR under standard conditions. The PCR product was purified by electrophoresis in the 1% low melting agarose gel and isolated by glass–milk adsorption.[10] One hundred microliters of 4 ng/mL solution of amplicon in 150 mM NaCl, 10 mM Tris-HCl pH7.5, 0.2% gelatin, 0.05% Tween-20 were incubated with avidin-coated wells of 96-well microtiter plates (Maxisorp, Nunc Roskilde, Denmark) for 2 h at room temperature. After washing off the excess of amplicon, serial dilutions of DNase I (Fermentas, EN0531) or 5 μL of blood plasma were added into wells containing 100 μL of 150 mM NaCl, 10 mM Tris-HCl pH 7.5, supplied with 2.5 mM $MgCl_2$, 0.1 mM $CaCl_2$ and incubated at room temperature for 15 h. The unhydrolyzed amplicon was detected after incubation with rabbit anti-fluorescein antibodies[12] followed by incubation with peroxidase-conjugated anti-rabbit immunoglobulin. The measurement of peroxidase activity was performed with *o*-phenylenediamine as chromophore.[13] DNase activities of the samples were estimated from the calibration curve obtained after incubation of the substrate with serial dilutions of DNase I (FIG. 1). Concentration of EDTA in plasma samples was estimated by titration with ferric salicylate and free EDTA was neutralized by addition of equimolar amount of $MgCl_2$.

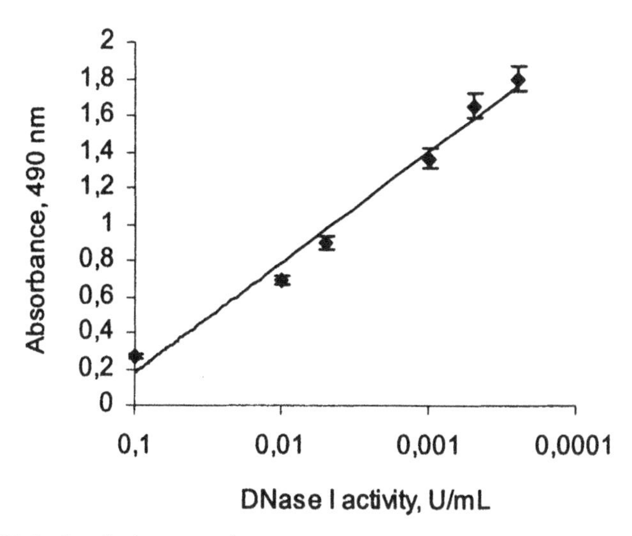

FIGURE 1. Standard curve for DNase I, obtained with the amlicone-based immunoassay.

To investigate the influence of inhibitors on DNase activity, 0.005 U of DNase I (Fermentas, EN0531) were added into the wells containing 5μL of tested blood plasma sample. DNase activity was tested as described previously.

RESULTS AND DISCUSSION

Deoxyribonuclease activity in blood plasma of healthy donors varies from 0.004 to 6 U/mL according to previous publications.[14,15] To measure DNase activity we have used hydrolysis of PCR amplification product modified by its 5′ ends with biotin and fluorescein with subsequent immunochemical detection of nondigested amplicon. The sensitivity of the assay was 0.004 U/mL (DNase I), the coefficient of variation for each point was less than 4 % ($n = 6$). Using 5 μL of human plasma the detection limit of the assay was 0.08 U/mL of blood plasma.

The data of DNase activity and concentration of cirDNA in plasma of healthy donors are summarized in TABLE 1. The average concentration of cirDNA in the plasma of healthy donors was found to be about 30 ng/mL (41 ± 30 ng/mL in healthy men and 21 ± 21 ng/mL in healthy women), in accordance with published data.[1] DNase activities in the plasma samples from male and female donors were 0.307±0.249 U/mL ($n = 10$) and 0.405±0.509 U/mL ($n = 10$), respectively. The two-tailed Student's t-test does not demonstrate statistically significant difference in DNase activity between men and women.

Cancer development is accompanied by a change of DNase activity in blood plasma. Patients with malignant lymphomas were characterized by the decrease

TABLE 1. Concentration of circulating DNA and DNase activity in plasma of healthy subjects

n	CirDNA (ng/mL)	DNase activity (U/mL)
Women		
1	24	0.276
2	10	0.49
3	38	0.498
4	10	0.098
5	0	0.902
6	24	0.59
7	0	0.336
8	0	0.15
9	0	0.1
10	0	0.614
Men		
1	41	0.256
2	50	0.216
3	41	0.249
4	39	0.104
5	43	0.425
6	66	0.425
7	39	0.366
8	0	0.472
9	0	0.384
10	11	0.168

NOTE: Concentration of cirDNA below the sensitivity of the assay is indicated as 0.

of DNase activity,[16] whereas patients with breast cancer demonstrated higher levels of DNase activity in comparison with healthy donors.[17]

CirDNA concentration and DNase activity in plasma of 9 patients with colon cancer and 9 with stomach cancer are summarized in TABLE 2. The average concentration of cirDNA in plasma of gastrointestinal cancer patients was found to be about 350 ng/mL (302 [0–594]) ng/mL in colon cancer patients and 427 [0–1,054] ng/mL in stomach cancer patients), similar to published data.[18] DNase activity in plasma samples from cancer patients was lower than the detection limit of the assay (0.08 U/mL plasma).

DNase activity in blood plasma depends not only on the concentration of DNA hydrolyzing enzymes, but also on the presence of inhibitors of DNase activity. One of the most known DNase I inhibitors is actin, which forms an inactive 1:1 stoichiometric complex with DNase with a binding constant 5×10^8 M^{-1}.[19] We have tested the presence of inhibitors in blood plasma of cancer patients by incubation of plasma samples together with 0.005 U of added DNase I. It was shown that the activity of the enzyme added to plasma samples was completely inhibited by 5 μL of the blood plasma from cancer patients. Thus, inhibitors of DNase activity can account

TABLE 2. Concentration of circulating DNA and DNase activity in the plasma of cancer patients

n	TNM	cirDNA (ng/mL)	DNase activity* (U/mL)
Colon cancer			
1	$T_3N_0M_0$	96	0
2	$T_3N_0M_0$	594	0
3	$T_3N_2M_0$	408	0
4	$T_3N_XM_0$	256	0
5	$T_3N_XM_0$	280	0
6	$T_3N_XM_X$	476	0
7	$T_4N_1M_0$	204	0
8	$T_4N_1M_0$	100	0
9	$T_4N_XM_1$	0	0
Stomach cancer			
1	$T_1N_0M_0$	556	0
2	$T_3N_0M_0$	0	0
3	$T_3N_0M_0$	388	0
4	$T_3N_0M_0$	852	0
5	$T_3N_1M_0$	64	0
6	$T_3N_2M_0$	90	0
7	$T_4N_1M_1$	60	0
8	$T_4N_2M_1$	356	0
9	$T_4N_XM_1$	1,054	0

NOTE: Concentration of cirDNA and DNase activities below the sensitivity of the assay is indicated as 0.

for the decreased level of DNase activity in the blood of gastrointestinal cancer patients.

Irrespective of the mechanisms leading to a decrease in DNase activity in plasma of patients with gastrointestinal cancer, the reduction in DNase activity increases the integrity of cirDNA and its concentration in blood plasma.

ACKNOWLEDGMENTS

The present work was supported by the Russian Academy of Sciences program "Science to Medicine," Russian Foundation for Basic Research Grant 06-04-49732, Interdisciplinary Project from the Siberian Division of the Russian Academy of Sciences No. 13, Scientific Schools Grant SS-1384, 2003, 4 Grant for young Scientists from the Siberian Division of the Russian Academy of Sciences 2006 and in part by Award REC-008 from CRDF.

REFERENCES

1. SOZZI, G. *et al.* 2001. Analysis of circulating tumor DNA in plasma at diagnosis and during follow-up of lung cancer patients. Cancer Res. **61:** 4675–4678.

2. LAKTIONOV, P. *et al.* 2004. Free and cell surface bound nucleic acids in blood of healthy donors and breast cancer patients. Ann. N. Y. Acad. Sci. **1022:** 221–227.
3. RAPTIS, L. *et al.* 1980. Quantitation and characterization of plasma DNA in normals and patients with systemic lupus erythematosus. J. Clin. Invest. **66:** 1391–1399.
4. RAINER, T. *et al.* 2001. Derivation of a prediction rule for posttraumatic organ failure using plasma DNA and other variables. Ann. N. Y. Acad. Sci. **945:** 211–220.
5. JAHR, S. *et al.* 2001. DNA fragments in the blood plasma of cancer patients: quantitations and evidence for their origin from apoptotic and necrotic cells. Cancer Res. **61:** 1659–1665.
6. YASUDA, T. *et al.* 1992. Human urine deoxyribonuclease II (DNase II) isoenzymes: a novel immunoaffinity purification, biochemical multiplicity, genetic heterogeneity and broad distribution among tissues and body fluids. Biochem. Biophys. Acta. **1119:** 185–193.
7. FRITTITTA, L. *et al.* 1999. A soluble PC-1 circulates in human plasma: relationship with insulin resistance and associated abnormalities. J. Clin. Endocrinol. Metab. **84:** 3620–3625.
8. SHUSTER, A. *et al.* 1992. DNA hydrolyzing autoantibodies. Science **256:** 665–667.
9. NADANO, D. *et al.* 1993. Measurement of deoxyribonuclease I activity in human tissues and body fluids by a single radial enzyme-diffusion method. Clin. Chem. **39:** 448–452.
10. TAMKOVICH, S. *et al.* 2004. Simple and rapid procedure suitable for quantitative isolation of low and high molecular weight extracellular nucleic acids. Nucleosides Nucleotides Nucleic Acids **23:** 873–877.
11. LABARCA, C. *et al.* 1980. A simple, rapid, and sensitive DNA assay procedure. Anal. Biochem. **102:** 344–352.
12. CHELOBANOV, B. 2003. Interaction of keratin k1 with nucleic acids on the cell surface. Biochemistry (Mosc.) **68:** 1239–1246.
13. CISAR, L. *et al.* 1985. Enzyme-linked immunosorbent assay for rat hepatic triglyceride lipase. J. Lipid Res. **26:** 380–386.
14. NADANO, D. *et al.* 1993. Measurement of deoxyribonuclease I activity in human tissues and body fluids by a single radial enzyme-diffusion method. Clin. Chem. **39:** 448–452.
15. DEWEZ, B. *et al.* 1993. Serum alkaline deoxyribonuclease activity, a sensitive marker for the therapeutic monitoring of cancer patients: methodological aspects. Eur. J. Clin. Chem. Clin. Biochem. **31:** 793–797.
16. ECONOMIDOU-KARAOGLOU, A. *et al.* 1988. Variations in serum alkaline DNase activity: a new means for therapeutic monitoring of malignant lymphomas. Cancer **61:** 1838–1843.
17. RAMANDANIS, G. *et al.* 1982. Correlation between serum and tissue deoxyribonuclease levels in breast cancer patients. Anticancer Res. **2:** 213–218.
18. SHAPIRO, B. *et al.* 1983. Determination of circulating DNA levels in patients with benign or malignant gastrointestinal disease. Cancer **51:** 2116–2120.
19. GIBSON, U. *et al.* 1992. An antibody capture assay (ACB) for DNase in human serum samples. J. Immunol. Methods **155:** 249–256.

Comparative Analysis of Mesenteric and Peripheral Blood Circulating Tumor DNA in Colorectal Cancer Patients

BRET TABACK,[a] SUKAMAL SAHA,[b] AND DAVE S. B. HOON[c]

[a]*Division of Surgical Oncology, Columbia University Medical Center, New York, New York, USA*

[b]*Department of Surgical Oncology, McLaren Regional Medical Center, Flint, Michigan, USA*

[c]*Department of Molecular Oncology, John Wayne Cancer Institute, Santa Monica, California, USA*

ABSTRACT: An increasing number of reports have demonstrated the presence of tumor-specific DNA in cancer patients' plasma/serum. These findings offer the prospective of serologic tumor markers that may aide in early disease detection, predict subclinical disease progression, and monitor treatment responses. However, for patients with colorectal cancer (CRC), there are few reports using this approach, with most revealing poor sensitivity. In contrast to tumors of other organ systems, CRCs drain predominantly via the mesenteric/portal veins (MV) to the liver. We hypothesize that because of this unique relation, tumor DNA may be less abundant in CRC patients' systemic circulation as compared to the mesenteric/portal system. At the time of surgery, paired blood was collected from both the peripheral vein (PV) and MV from 33 CRC patients. DNA was isolated from serum, quantified and assessed for loss of heterozygosity (LOH) using a panel of 11 microsatellite markers corresponding to regions on six chromosomes frequent for LOH in CRC. In addition, 16 samples were assessed for the presence of hypermethylated DNA for tumor suppressor genes: *MGMT*, *P16*, *RAR-β2*, *RASSF1A*, and *APC*. Circulating tumor DNA associated with LOH or methylation was more frequently detected in the MV of patients, 11 (33%) and 6 (38%), as compared to PV, 9 (27%) and 1 (6%), respectively. This study is the first to identify the presence of increased tumor DNA in the direct efferent venous drainage system of CRC and its variation as compared to systemic circulation. The findings provide important evidence supporting the origin of tumor-associated DNA in circulation, which merits consideration when devising blood-based nucleic acid assays for the assessment of CRC.

Address for correspondence: Dr. Dave S.B. Hoon, Department of Molecular Oncology, John Wayne Cancer Institute, 2200 Santa Monica Blvd, Santa Monica, CA 90404, USA. Voice: 310-449-5267; fax: 310-449-5282.

e-mail: Hoon@jwci.org

Ann. N.Y. Acad. Sci. 1075: 197–203 (2006).
doi: 10.1196/annals.1368.027

Keywords: colorectal cancer; circulating tumor DNA; loss of heterozygosity; methylation; serum

INTRODUCTION

Colorectal cancer (CRC) is the third leading cause of new cancer cases and death affecting men and women equally in the United States.[1] At the time of initial treatment, the majority of patients will present at a stage where all clinically evident disease can be surgically resected. Unfortunately, despite ever-improving therapies, one-third of these patients will subsequently develop recurrent disease. Therefore, additional methods are needed to monitor subclinical disease progression and distinguish patients at increased risk for the development of systemic metastasis.

A variety of genetic alterations have been associated with the initiation and progression of CRC.[2,3] More recently, epigenetic events, which result in a change in the pattern of gene expression that is mediated by mechanisms other than a variation in a gene's primary nucleotide sequence, have been characterized during CRC development.[4,5] It has been suggested that identification of genes hypermethylated in CRCs may provide novel therapeutic targets as well as molecular surrogates for tumor progression.[6–8]

A number of studies have identified circulating tumor-specific DNA in the blood of cancer patients.[9–18] Molecular-based assays offer a minimally invasive method that may assist in the diagnosis and management of cancer patients.[19] However, among patients with CRC, there exist very few reports describing this approach.[20,21] Furthermore, most demonstrate low sensitivity for detecting tumor DNA in CRC patients' serum or plasma. Unique to the gastrointestinal tract is that the majority of the vascular drainage is initially to the liver via the mesenteric and portal systems. This is in distinct contrast to the direct systemic circulation that drains tumors associated with other organs. The aim of this study is to assess the incidence and characterize the presence of circulating tumor DNA in the mesenteric/portal venous system draining patients' primary CRCs relative to that in the systemic circulation.

MATERIALS AND METHODS

At the time of surgery, paired blood was collected simultaneously from both the peripheral vein (PV) and mesenteric vein (MV) from 33 CRC patients as follows: 1 patient with carcinoma *in situ*, 7 American Joint Committee on Cancer (AJCC) stage I patients, 8 AJCC stage II patients, 8 AJCC stage III patients, 5 AJCC stage IV patients, and 4 patients with adenomas. All patients provided informed consent. Genomic DNA was isolated from serum using QIAmp DNA Mini Kit (Qiagen, Valencia, CA) and quantified with Picogreen (Molecular

Probes, Eugene, OR) as previously described.[16,22] Lymphocytes were collected from each patient, and DNA was extracted using DNAzol (Molecular Research Center, Cincinnati, OH) and quantified with Picogreen (Molecular Probes), which served as reference DNA for comparison. Polymerase chain reaction (PCR) was performed on each patient's paired samples (serum DNA and normal lymphocyte DNA) using FAM-labeled primer sets (Research Genetics, Huntsville, AL) for a panel of 11 microsatellite markers (D18S58, D18S61, TP53, D17S1832, D17S796, D15S127, D8S133, D8S264, D5S299, D4S175, and D4S1586) corresponding to regions on six chromosomes where loss of heterozygosity (LOH) is frequently observed in CRC.

Post-PCR product separation was performed using capillary array electrophoresis (CAE, CEQ 8000XL, Beckman Coulter, Fullerton, CA) and peak signal intensity and relative size were generated and interpreted by fragment analysis software (CEQ 8000XL) as previously described.[23] Briefly, LOH was scored when one allele in serum DNA demonstrated a 40% or greater reduction in peak intensity as compared with its corresponding allele identified in the control lymphocyte DNA. Patient's serum samples demonstrating homozygosity or microsatellite instability for a particular marker were considered uninformative.

In addition, paired PV–MV serum samples from 16 CRC patients contained sufficient DNA for the assessment of promoter hypermethylation for the following putative tumor suppressor genes: *MGMT*, *P16*, *RAR*-β*2*, *RASSF1A*, and *APC*. Isolated genomic DNA was subject to sodium bisulfite treatment, and methylation-specific PCR (MSP) was performed using methylated and unmethylated fluorescent labeled primers as described previously.[24,25] Methylated and unmethylated PCR products from each sample were assessed simultaneously by labeling forward primers with one of three Beckman Coulter WellRED Phosphoramidite (PA)-linked dyes (Genset Oligos, Boulder, CO). One microliter of methylated and 1 μL of unmethylated post-PCR products were mixed with 40 μL loading buffer and 0.5 μL dye-labeled size standards (Beckman Coulter) and analyzed using CAE (Beckman Coulter), which distinguishes the presence of the corresponding peaks by the respective colors.[24] *Sss*I-treated healthy donor lymphocytes and cell lines shown to contain promoter region methylation for the gene of interest served as positive controls, whereas lymphocyte DNA and sterile water constituted the negative and blank controls, respectively.

RESULTS

LOH could be identified in MV serum samples from 11 (33%) of 33 CRC patients, whereas 9 (27%) patients exhibited LOH for any one microsatellite marker in their PV serum samples. Sixteen (49%) of 33 patients demonstrated LOH in either MV and/or PV serum samples. In only four patients could LOH

TABLE 1. LOH frequency in colorectal cancer patients' serum samples

Microsatellite marker	LOH Frequency[a] MV	PV
D4S175	4/23 (17%)	4/23 (17%)
D4S1586	1/19 (5%)	0
D5S299	3/25 (12%)	3/25 (12%)
D8S133	2/19 (11%)	1/19 (5%)
D8S264	1/22 (5%)	0
D15S127	0	2/29 (7%)
TP53	1/28 (4%)	2/28 (7%)
D17S796	2/22 (9%)	1/22 (5%)
D17S1832	2/28 (7%)	0
D18S58	1/23 (4%)	0
D18S61	0	1/32 (3%)

[a] Positive LOH/total informative patients.
MV: mesenteric vein; PV: peripheral vein.

be found for the same microsatellite marker in both their MV and paired PV serum samples. The most frequently detected microsatellite marker demonstrating LOH was D4S175, occurring in both MV and PV serum samples from 4 (17%) of 23 informative patients (TABLE 1). In total, in 17 instances, a microsatellite marker for LOH was detected in the MV circulation from CRC patients' tumors, versus 14 in patients' systemic circulation.

According to disease stage, LOH was identified in the MV serum from the one patient with carcinoma *in situ*, three (43%) of seven AJCC stage I patients, three (38%) of eight AJCC stage II patients, two (25%) of eight AJCC stage III patients, two (40%) of five AJCC stage IV patients, and one (25%) of four patients with adenomas. LOH was not identified in the PV serum from any of the AJCC stage IV patients, patients with adenomas, or the patient with carcinoma *in situ*; however, LOH was detected in PV serum from four (57%) AJCC stage I patients, four (50%) AJCC stage II patients, and one (13%) AJCC stage III patient.

In 16 patients, sufficient DNA was present in paired MV–PV serum samples for the evaluation of methylated tumor DNA. Using a five-gene panel (*MGMT*, *P16*, *RAR*-β*2*, *RASSF1A*, and *APC)*, methylated DNA was identified in 6

TABLE 2. Frequency of methylated DNA in colorectal cancer patients' serum samples

Methylated DNA	Frequency MV	PV
APC	3 (19%)	1 (6%)
MGMT	2 (13%)	0
RASSFIA	1 (6%)	0
P16	0	0
RAR-β2	0	0

MV: mesenteric vein; PV: peripheral vein.

(38%) of 16 patients' MV serum samples, whereas only 1 patient demonstrated gene methylation in their PV serum sample. *APC* methylation was the most frequent marker detected circulating in serum (TABLE 2). Methylated DNA was identified in the MV from two AJCC stage I patients, three AJCC stage II patients, and one AJCC stage IV patient, who also demonstrated methylated PV DNA for a different gene.

Circulating tumor DNA associated with LOH or methylation was more frequently detected in the MV of patients, 11 (33%) and 6 (38%), as compared to PV, 9 (27%) and 1 (6%), respectively. No LOH or methylated DNA for any of the markers tested were found in PV serum from 10 healthy normal donors.

DISCUSSION

A number of genetic and epigenetic alterations have been described that occur during the evolution of CRC development. These events promote the considerable genetic heterogeneity of colorectal tumors and may in part contribute to the diverse clinical behavior and treatment responses associated with this disease. LOH and gene promoter hypermethylation can affect critical pathways that regulate cell cycling, apoptosis, DNA repair, growth, and invasion.[26,27] Characterization of these events will contribute to our understanding of the pathogenesis of CRC and may identify potential targets for therapy. Detection of these genetic and epigenetic alterations circulating in the plasma and serum of cancer patients offers the promise of a minimally invasive assay for serial assessment, which may provide additional diagnostic and prognostic information.

Inconsistently, in light of the worldwide prevalence of CRC, there remains a relative paucity of studies evaluating circulating nucleic acids in plasma and serum from patients with this disease. More so, the reported frequency of tumor DNA in CRC patients' plasma/serum appears lower as compared with solid tumors in other organ systems.[19,22,28] We speculated that the unique initial drainage pathway of CRC, via the mesenteric and portal veins to the liver, may account for these disparate findings. Using a specific panel of microsatellite markers for LOH that is found frequently in colorectal tumors, we were able to demonstrate a greater incidence of tumor DNA in patients' MV serum as compared with PV serum. There was an approximately 20% greater frequency of LOH in MV serum as compared with PV serum. Microsatellite markers that map to chromosomes 17 and 18, where LOH occurs frequently in primary CRC tumors, were found infrequently in patients' serum. This finding is consistent with other reports in the literature[20] and may demonstrate the variability by which different chromosome segments are released into serum and/or their potential half-lives in circulation.

As compared with circulating LOH, methylated DNA was much more common in MV serum than paired PV serum; however, this may be related to the fewer markers and number of patients in which the latter was assessed. Consistent with other studies,[13,20,21] we found a relatively low incidence of circulating tumor-associated DNA in CRC patients' systemic circulation, which was detected in patients with both early and late disease stages. Interestingly, these genetic alterations were identified, although infrequently, in MV circulation in one patient with a premalignant lesion and one patient with a preinvasive tumor. These findings demonstrate that the shedding of tumor DNA can occur quite early in the disease course.

The clinical significance of assessing circulating tumor-associated nucleic acids in plasma and serum of CRC will require longer-term follow-up in a larger series of patients. Nevertheless, this study contributes several original findings to this field. DNA microsatellites with LOH and methylated DNA can be detected in sera obtained from the MV and PV of CRC patients. In addition, both LOH and methylated DNA are more frequent in MV than PV serum. This is the first study to demonstrate the presence of increased tumor DNA in the direct efferent venous drainage system of CRC and its variation as compared to systemic circulation. Our results support the findings of other investigators demonstrating a relatively low frequency of tumor DNA in the systemic circulation of patients with CRC,[13,18,20,21] yet may provide a physiologic explanation to account for this based on the unique vascular anatomic pathway that drains the colon and rectum. The findings provide important evidence supporting the origin of tumor-associated DNA in circulation, which merits consideration when devising serum/plasma-based nucleic acid assays for the assessment of CRC.

REFERENCES

1. JEMAL, A. *et al.* 2005. Cancer statistics, 2005. C. A. Cancer J. Clin. **55**: 10–30.
2. VOGELSTEIN, B. *et al.* 1988. Genetic alterations during colorectal-tumor development. N. Engl. J. Med. **319**: 525–532.
3. BAKER, S.J. *et al.* 1989. Chromosome 17 deletions and p53 gene mutations in colorectal carcinomas. Science **244**: 217–221.
4. TOYOTA, M. *et al.* 1999. CpG island methylator phenotype in colorectal cancer. Proc. Natl. Acad. Sci. USA **96**: 8681–8686.
5. ESTELLER, M. *et al.* 2001. A gene hypermethylation profile of human cancer. Cancer Res. **61**: 3225–3229.
6. KOPELOVICH, L. *et al.* 2003. The epigenome as a target for cancer chemoprevention. J. Natl. Cancer Inst. **95**: 1747–1757.
7. KONDO, Y. & J.P. ISSA. 2004. Epigenetic changes in colorectal cancer. Cancer Metastasis Rev. **23**: 29–39.
8. HERMAN, J.G. 2002. Hypermethylation pathways to colorectal cancer: implications for prevention and detection. Gastroenterol. Clin. North Am. **31**: 945–958.
9. CHEN, X.Q. *et al.* 1996. Microsatellite alterations in plasma DNA of small cell lung cancer patients. Nat. Med. **2**: 1033–1035.

10. GOESSL, C. *et al.* 1998. Microsatellite analysis of plasma DNA from patients with clear cell renal carcinoma. Cancer Res. **58**: 4728–4732.
11. KAWAKAMI, K. *et al.* 2000. Hypermethylated APC DNA in plasma and prognosis of patients with esophageal adenocarcinoma. J. Natl. Cancer Inst. **92**: 1805–1811.
12. NAWROZ, H. *et al.* 1996. Microsatellite alterations in serum DNA of head and neck cancer patients. Nat. Med. **2**: 1035–1037.
13. KOLBLE, K. *et al.* 1999. Microsatellite alterations in serum DNA of patients with colorectal cancer. Lab. Invest. **79**: 1145–1150.
14. SORENSON, G.D. 2000. Detection of mutated KRAS2 sequences as tumor markers in plasma/serum of patients with gastrointestinal cancer. Clin. Cancer Res. **6**: 2129–2137.
15. TABACK, B. *et al.* 2001. Prognostic significance of circulating microsatellite markers in the plasma of melanoma patients. Cancer Res. **61**: 5723–5726.
16. TABACK, B. *et al.* 2001. Microsatellite alterations detected in the serum of early stage breast cancer patients. Ann. N. Y. Acad. Sci. **945**: 22–30.
17. LO, Y.M. *et al.* 2000. Kinetics of plasma Epstein-Barr virus DNA during radiation therapy for nasopharyngeal carcinoma. Cancer Res. **60**: 2351–2355.
18. WANG, J.Y. *et al.* 2004. Molecular detection of APC, K-ras, and p53 mutations in the serum of colorectal cancer patients as circulating biomarkers. World J. Surg. **28**: 721–726.
19. TABACK, B. & D.S. HOON. 2004. Circulating nucleic acids and proteomics of plasma/serum: clinical utility. Ann. N. Y. Acad. Sci. **1022**: 1–8.
20. HIBI, K. *et al.* 1998. Molecular detection of genetic alterations in the serum of colorectal cancer patients. Cancer Res. **58**: 1405–1407.
21. GRADY, W.M. *et al.* 2001. Detection of aberrantly methylated hMLH1 promoter DNA in the serum of patients with microsatellite unstable colon cancer. Cancer Res. **61**: 900–902.
22. TABACK, B. & D.S. HOON. 2004. Microsatellite alterations as diagnostic and prognostic molecular markers in patients with cancer. *In* Cancer Diagnostics: Current and Future Trends. R. Nakamura, *et al.*, Eds.: 395–428. Humana Press. Totowa, NJ.
23. FUJIMOTO, A. *et al.* 2004. Allelic imbalance on 12q22-23 in serum circulating DNA of melanoma patients predicts disease outcome. Cancer Res. **64**: 4085–4088.
24. HOON, D.S. *et al.* 2004. Profiling epigenetic inactivation of tumor suppressor genes in tumors and plasma from cutaneous melanoma patients. Oncogene **23**: 4014–4022.
25. SHINOZAKI, M. *et al.* 2005. Distinct hypermethylation profile of primary breast cancer is associated with sentinel lymph node metastasis. Clin. Cancer Res. **11**: 2156–2162.
26. JONES, P.A. & S.B. BAYLIN. 2002. The fundamental role of epigenetic events in cancer. Nat. Rev. Genet. **3**: 415–428.
27. HANAHAN, D. & R.A. WEINBERG. 2000. The hallmarks of cancer. Cell **100**: 57–70.
28. ANKER, P. *et al.* 1999. Detection of circulating tumour DNA in the blood (plasma/serum) of cancer patients. Cancer Metastasis Rev. **18**: 65–73.

Real-Time Quantification of Human Telomerase Reverse Transcriptase mRNA in the Plasma of Patients with Prostate Cancer

F. DASÍ,[a,b] P. MARTÍNEZ-RODES,[a] J.A. MARCH,[c] J. SANTAMARÍA,[c] J.M. MARTÍNEZ-JAVALOYAS,[c] M. GIL,[c] AND S.F. ALIÑO[a]

[a] *University of Valencia School of Medicine, Department of Pharmacology, Avda. Blasco Ibañez 15, 46010–Valencia, Spain*

[b] *University of Valencia School of Medicine, Unidad Central Investigación Medicina, Avda. Blasco Ibañez 15, 46010–Valencia, Spain*

[c] *Valencia University Clinic Hospital, Urology Department, Avda. Blasco Ibañez 17, 46010–Valencia, Spain*

ABSTRACT: The aim of this study was to evaluate the potential diagnostic value of quantitative analysis of human telomerase reverse transcriptase (hTERT) mRNA in plasma for noninvasive diagnosis of prostate cancer (PCa). Expression levels of hTERT were analyzed by real-time quantitative RT-PCR in 68 patients showing elevated prostate-specific antigen (PSA) levels and a control group of 44 healthy volunteers. Sensitivity and specificity were determined and compared to the corresponding PSA values. Median values for hTERT gene expression in the PCa patients (0.72 ng; range 0.01–12.86) were statistically significantly higher ($P <$ 0.001) than in the control group (0.13 ng; 0.02–0.35). Patients with clinically confirmed prostatitis showed lower plasma hTERT expression than PCa patients (0.29; 0.01–66.07). At a cutoff value of 0.35 sensitivity and specificity for the diagnosis of PCa were 81% and 60%, respectively. We suggest that hTERT mRNA in plasma is a very specific and sensitive method that may aid to differentiate between malignant and nonmalignant prostate tissue and may be a useful marker (in combination with PSA) for early PCa diagnosis.

KEYWORDS: plasma; serum; hTERT; RNA; prostate cancer

Address for correspondence: Dr. Salvador F. Aliño, University of Valencia School of Medicine, Department of Pharmacology, Avda. Blasco Ibañez 15, 46010–Valencia, Spain. Voice: +34-96-386-4621; fax: +34-96-386-4972.
e-mail: alino@uv.es

Ann. N.Y. Acad. Sci. 1075: 204–210 (2006).
doi: 10.1196/annals.1368.028

INTRODUCTION

Prostate cancer (PCa) is the most common male malignancy and the second leading cause of male cancer-related mortality.[1] Early detection of PCa increases the probability of successful treatment and, therefore, methods that can detect the tumor when it is still locally confined to the prostate and potentially curable are needed. Diagnostic techniques, such as serum prostate-specific antigen (PSA) testing and transrectal ultrasound have increased the sensitivity of detection, though none of the standard methods are sufficiently sensitive or specific for PCa—which makes early detection of the disease difficult.[2] Serum PSA is currently the best clinical marker for the detection of PCa and can also be used for the screening of selected populations of patients (i.e., asymptomatic men over 50 years of age), and for monitoring patients after therapy.[3] However, serum PSA levels are not specific for cancer and are elevated in men with benign prostate hyperplasia (BPH), prostatitis, and other nonmalignant prostate disorders, resulting in reduced specificity.[4] In fact, false positives for PSA result in unnecessary biopsies and other interventions. Of greater concern, 20–30% of men with PCa have serum PSA levels within the reference range, resulting in underdiagnosed disease.[5]

Therefore, new cancer-specific markers are needed that may aid early PCa diagnosis. One of the most promising tumor markers is the enzyme telomerase that has been found to be strongly associated with PCa.[6] However, although telomerase seems to be a specific marker for cancer, its clinical usefulness depends on the ability of detection in biological fluids, such us blood, urine, sputum, or bronchoalveolar lavage (BAL). Tumor-derived DNA and RNA can be found in plasma/serum of cancer patients. Some investigators have reported the detection of human telomerase reverse transcriptase (hTERT—the catalytic subunit of telomerase) mRNA in plasma/serum of patients with liver, thyroid, melanoma, or breast cancers.[7–10] In a previous report, our group showed that plasma mRNA can be detected and quantified in plasma, clearly differentiating between healthy and colorectal cancer patients.[11,12] We then wanted to expand our investigation to other cancer types, such as PCa. The aim of this study was to evaluate whether it is possible to isolate and detect circulating tumor-associated mRNA in the plasma of patients with PCa and to determine whether quantitative analysis of hTERT mRNA could serve as a reliable molecular marker of prostate malignancy.

MATERIALS AND METHODS

Patients and Sample Collection

Sixty-eight patients with elevated PSA levels and a control group of 44 healthy volunteers were studied. All samples were analyzed prospectively with no prior knowledge of patient's clinicopathological status. Patients were classified according to histopathological criteria as follows: PCa, prostatitis, and other pathologies. Venous samples were obtained from all patients and volunteers into 8-mL blood collection tubes containing sodium citrate gel and a density gradient medium (Vacutainer CPT 362761; BD, Franklin Lakes, NJ). Blood mononuclear cells and plasma were separated by centrifugation according to the manufacturer's instructions.

All the procedures described in this manuscript were in accordance with the ethical standards of the Committee on Human Experimentation (Valencia University Clinic Hospital, Spain). Written informed consent was obtained from all patients prior to their inclusion in the study.

RNA Extraction and cDNA Synthesis

RNA was isolated from 250 μL of plasma as previously described.[11,12] For reverse transcription (RT) reactions, 200 ng of the purified RNA were reverse transcribed using random hexamers with the TaqMan Reverse Transcription reagents kit (Applied Biosystems, Foster City, CA, USA) according to the manufacturer's instructions. RT conditions comprised an initial incubation step at 20°C for 10 min to allow random hexamers annealing, followed by cDNA synthesis at 42°C for 45 min, and a final inactivation step for 5 min at 95°C.

Real-Time Quantitative PCR of hTERT and GAPDH

The TaqMan Universal PCR Master Mix (Applied Biosystems) was used to detect the presence of hTERT and 18S rRNA in plasma. Five μL of the RT reaction were mixed with 45 μL of TaqMan universal PCR master mix containing 250 nM of the forward and reverse primers and 125 nM of the TaqMan probe. Primers and TaqMan probes used for hTERT and 18S rRNA have been described elsewhere.[11] PCR conditions were 10 min at 95°C for enzyme activation, followed by 40 two-step cycles (15 s at 95°C; 1 min at 60°C). The levels of 18S rRNA expression were measured in all samples to normalize hTERT expression for sample-to-sample differences in RNA input, RNA quality, and reverse transcription efficiency. A standard curve was generated using the HT-29 colon carcinoma cell line. After cDNA synthesis, dilutions were made to produce a standard curve covering a concentration range of 0.016–250 ng. qPCR reactions were performed with the ABI Prism 7700 Sequence Detection System (Applied Biosystems), and measured Ct values were transformed to ng of HT-29 hTERT and HT-29 18S rRNA using the standard curve generated in the same experiment. The ratio between ng hTERT and ng 18S rRNA represents the normalized hTERT for each sample.

PSA values were determined by routine laboratory techniques used in this hospital.

Statistical Analysis

The distribution of normalized hTERT was characterized by its median values and ranges. Differences in hTERT between PCa patients and normal volunteers were tested for statistical significance with the nonparametric Mann–Whitney U-test. $P < 0.05$ was considered statistically significant.

RESULTS

hTERT expression was measured in the plasma of 68 patients showing elevated PSA levels (PSA cutoff point ≥ 4.0 ng/mL). Of the 68 patients, 26 were diagnosed with PCa, 35 had prostatitis, and 7 other nonmalignant prostate diseases.

All patients and controls tested positive for the presence of 18S rRNA, therefore verifying the integrity of the mRNA (data not shown). FIGURE 1A shows the distribution of normalized hTERT in patients with PCa, prostatitis, and other prostate pathologies. Median values for hTERT expression in the PCa group (0.72; range 0.01–12.86) were higher than in the control group (0.13; range 0.02–0.35). The differences were statistically significant ($P < 0.001$). The highest value (0.35) of the control group was established as cutoff point. Patients with clinically confirmed prostatitis showed lower hTERT expression levels than PCa patients (0.29; 0.01–66.07).

Serum PSA was measured in all patients (FIG. 1B), and a standard cutoff value of 4.0 ng/mL was used. Serum PSA was significantly higher ($P < 0.0001$) in patients than in the control group, though there were no significant differences between patients with PCa and patients with prostatitis or other benign prostate pathologies.

The sensitivity and specificity of the hTERT assay to detect PCa were 81% and 60%, respectively, in contrast to the PSA assay that showed a higher comparable sensitivity (100%) but very poor specificity (5%). The positive predictive value (PPV) of hTERT was higher than PPV of the PSA assay (55% verses 39%). These data are summarized in TABLE 1. A strong correlation was found between hTERT mRNA and the PSA value in patients (TABLE 2). Eighty-one percent of patients with PSA values > 4.0 ng/mL showed elevated hTERT mRNA in plasma.

DISCUSSION

In this study we showed that quantification of hTERT mRNA in plasma is a very specific and sensitive method that may aid in differentiating between malignant and nonmalignant prostate tissue, and may prove to be a useful marker (in combination with PSA) for early PCa diagnosis. Recently, several

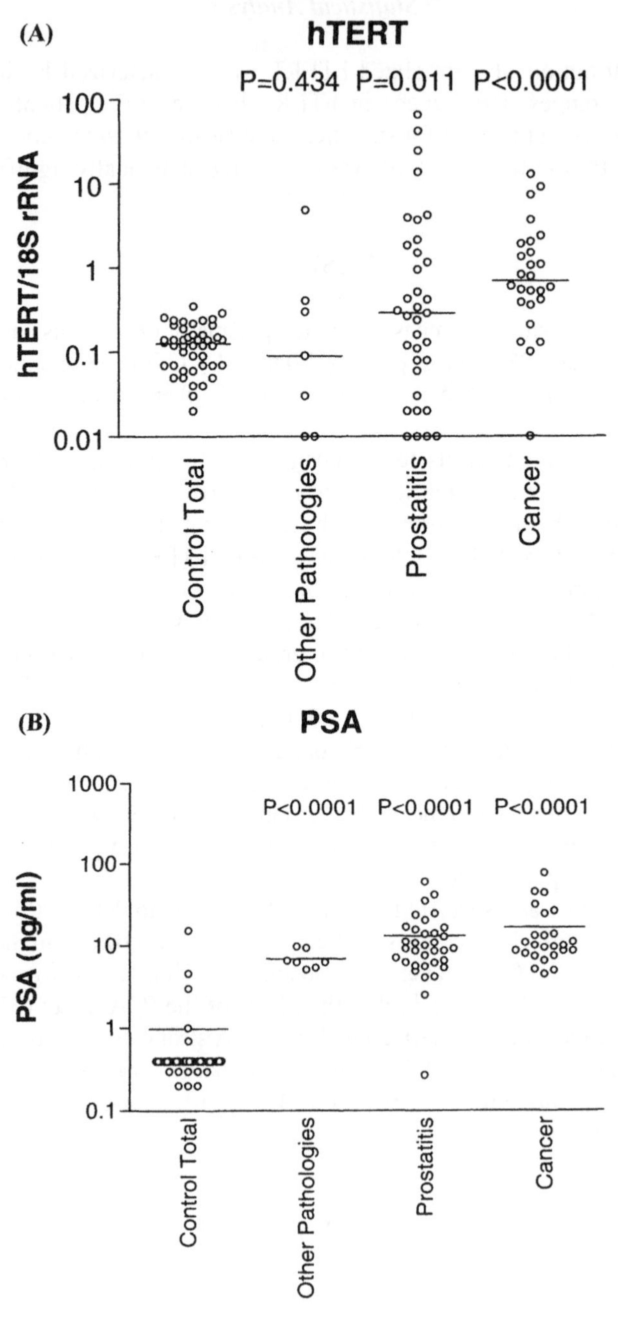

FIGURE 1. Normalized plasma hTERT mRNA (A) and serum PSA (B) in healthy volunteers, patients with PCa, patients with prostatitis, and patients with other nonmalignant prostate pathologies. Horizontal bar represents median values.

TABLE 1. Comparison of the diagnostic significance of hTERT mRNA and PSA

Cancer vs. No Cancer							
hTERT				PSA			
	Patient status				Patient status		
Blood test Results	Cancer (+)	No cancer (−)	Total	Blood test results	Cancer (+)	No cancer (−)	Total
>0.35 (+)	21	17	38	>4.0 (+)	26	40	66
<0.35 (−)	5	25	30	<4.0 (−)	0	2	2
Total	26	42	68	Total	26	42	68
	Sensitivity	81% (21/26)			Sensitivity	100% (26/26)	
	Specificity	60% (25/42)			Specificity	5% (2/42)	
PPV*		55%		PPV*		39%	

*PPV = Positive predictive value.

TABLE 2. Correlation of PSA value with hTERT mRNA in plasma

	Telomerase activity		
	Positive	Negative	Total (n)
PSA value* > 4	21	5	26
≤ 4	0	0	0
Total (n)	21	5	26

*Reference range of serum PSA is ≤ 4 ng/mL.

novel cancer-specific markers have been identified that may improve early diagnosis and can help select potentially aggressive tumors that require more aggressive treatment.[13] Telomerase activity is one of the most promising general tumor markers, and is strongly associated with PCa—showing activation in a high percentage (70–93%) of prostate tumors, whereas only low or absent activity was detected in normal and BPH tissues. However, although telomerase seems to be a very promising tumor marker, its clinical usefulness depends on the ability to detect it in nonsurgically obtained body fluids, such as urine, blood, or BAL. Previous reports showed that RNA could be extracted and amplified from the plasma/serum of patients with a variety of cancers.[7–10]

The aim of the present study was to evaluate whether quantitative analysis of plasma of hTERT mRNA could be used as a reliable molecular marker of prostate malignancy. Our study showed that median values for hTERT mRNA expression were significantly higher in the PCa group than in the controls. The sensitivity and specificity of the assay for diagnosing PCa were 81% and 60%, respectively. Serum PSA is the most widely used marker for PCa. It has been found to correlate with tumor size, and is an important predictor of tumor progression and prognosis. The sensitivity and specificity of serum PSA assay for the diagnosis of PCa were 100% and 5%, respectively. The PPV for this test was 39%. We also observed that hTERT mRNA concentration correlated with PSA levels. Taken together, these results indicate that combination of hTERT mRNA quantification in plasma and elevated PSA levels may be more specific for the presence of PCa than PSA levels alone. This is particularly important for the group of patients with PSA levels of 4 to 10 ng/mL, since although <25% have PCa, most undergo prostate biopsy.[14]

In summary, we suggest that the quantitative analysis of hTERT mRNA in plasma can be useful for the selection of PCa and cancer-free cases. However, additional studies including more patients and exploring additional laboratory procedures will be required to validate these results for future clinical applications.

REFERENCES

1. JEMAL, A. *et al.* 2003. Cancer statistics, 2003. CA. Cancer J. Clin. **53:** 5–26.
2. BOTCHKINA, G.I. *et al.* 2005. Noninvasive detection of prostate cancer by quantitative analysis of telomerase activity. Clin. Cancer Res. **11:** 3243–3249.
3. GOHAGEN, J.K. *et al.* 1994. Prostate cancer screening in the prostate, lung, colorectal and ovarian cancer screening of the National Cancer Institute. J. Urol. **152:** 1905–1909.
4. PANNEK, J. & A.W. PARTIN. 1997. Prostate-specific antigen: what's new in 1997. Oncology **11:** 1273–1278.
5. WANG, M.D. *et al.* 2001. Detection of telomerase activity in prostatic fluid specimens. Urol. Oncol. **6:** 4–9.
6. TRICOLI, J.V. *et al.* 2004. Detection of prostate cancer and predicting progression: current and future diagnostic markers. Clin. Cancer Res. **10:** 3943–3953.
7. MIURA, N. *et al.* 2005. Serum human telomerase reverse transcriptase messenger RNA as a novel tumor marker for hepatocellular carcinoma. Clin. Cancer Res. **11:** 3205–3209.
8. NOVAKOVIC, S. *et al.* 2004. Detection of telomerase RNA in the plasma of patients with breast cancer, malignant melanoma or thyroid cancer. Oncol. Rep. **11:** 245–252.
9. KOPRESKI, M.S. 1999. Detection of tumor messenger RNA in the serum of patients with malignant melanoma. Clin. Cancer Res. **5:** 1961–1965.
10. CHEN, X. *et al.* 2000. Telomerase RNA as detection marker in the serum of breast cancer patients. Clin. Cancer. Res. **6:** 3823–3826.
11. DASI, F. *et al.* 2001. Real-time quantification in plasma of human telomerase reverse transcriptase (hTERT) mRNA: a single blood test to monitor disease in patients. Lab. Invest. **81:** 767–769.
12. LLEDO, S. *et al.* 2004. Real-time quantification in plasma of human telomerase reverse transcriptase (hTERT) mRNA in patients with colorectal cancer. Color Dis. **6:** 236–242.
13. DE KOK, J.B. *et al.* 2002. DD3PCA3, a very sensitive and specific marker to detect prostate tumors. Cancer Res. **62:** 2695–2698.
14. CATALONA, W.J. *et al.* 1991. Measurement of prostate specific antigen in serum as screening test for prostate cancer. N. Engl. J. Med. **324:** 1156–1161.

Epigenetic Analysis of Body Fluids and Tumor Tissues

Application of a Comprehensive Molecular Assessment for Early-Stage Breast Cancer Patients

BRET TABACK,[a] ARMANDO E. GIULIANO,[b] RON LAI,[c] NORA HANSEN,[b] FREDERICK R. SINGER,[d] KLAUS PANTEL,[e] AND DAVE S.B. HOON[c]

[a]*Division of Surgical Oncology, Columbia University Medical Center, New York, New York 10032, USA*

[b]*Joyce Eisenberg-Keefer Breast Center, John Wayne Cancer Institute, Santa Monica, California 90404, USA*

[c]*Department of Molecular Oncology, John Wayne Cancer Institute, Santa Monica, California 90404, USA*

[d]*Department of Skeletal Biology, John Wayne Cancer Institute, Santa Monica, California 90404, USA*

[e]*Institute of Tumor Biology, University Hospital Hamburg-Eppendorf, Hamburg D-20246, Germany*

ABSTRACT: Breast cancer recurrence is a result of undetected metastasis present at the time of primary patient treatment. More sensitive methods are needed to identify subclinical disease progression to better accompany those increasing advances in early breast cancer screening. Aberrant hypermethylation of tumor-suppressor genes is found frequently in primary breast tumors and has been implicated in disease initiation and progression. Epigenetic characterization of tumor cells may provide highly specific and sensitive molecular surrogates for surveillance. We evaluated whether tumor-associated methylated DNA markers could be identified circulating in bone marrow (BM) aspirates and paired serum samples from 33 early-stage patients undergoing surgery for breast cancer. Quantitative methylation-specific PCR (qMSP) was performed using a selected tumor-related gene panel for RAR-ß2, MGMT, RASSF1A, and APC. Tumor-associated hypermethylated DNA was identified in 7 (21%) of 33 BM aspirates and 9 (27%) serum samples. In three patients both

Address for correspondence: Dr. Dave S.B. Hoon, Department of Molecular Oncology, John Wayne Cancer Institute, 2200 Santa Monica Blvd, Santa Monica, CA 90404. Voice: 310-449-5267; fax: 310-449-5282.

e-mail: Hoon@jwci.org

Ann. N.Y. Acad. Sci. 1075: 211–221 (2006). © 2006 New York Academy of Sciences.
doi: 10.1196/annals.1368.029

BM and serum were positive for hypermethylation. The most frequently detected hypermethylation marker was RASSF1A occurring in 7 (21%) patients. Concordance was present between gene hypermethylation detected in BM or serum samples, and matched-pair primary tumors. Advanced AJCC stage was associated with an increased incidence of circulating gene hypermethylation. In addition, methylation patterns in the sentinel lymph node (SLN) metastasis corresponded with that of the primary tumor, confirming epigenetic clonality is associated with early tumor dissemination. This study demonstrates the novel finding of tumor-associated epigenetic markers in BM aspirates/blood and their potential role as targets for molecular detection.

KEYWORDS: **methylation; breast cancer; bone marrow**

INTRODUCTION

A variety of genetic alterations including microsatellite instability, allelic loss, mutation, and amplification have been described in primary breast cancers. These events result in loss of gene function and have been implicated in tumor development and progression. Clinical tools (i.e., radiographic) used to detect breast cancer progression have been limited particularly in this current era of earlier disease diagnosis. The most sensitive method for the identification of breast cancer progression at the time of patient diagnosis is histopathologic lymph node evaluation. However, 20–30% of node-negative breast cancer patients will subsequently develop recurrent disease.[1,2] Therefore, recurrence may be considered a consequence of systemic occult metastasis not detected at the time of patient diagnosis and treatment. The most frequent site of breast cancer metastasis is bone.[3] Identification of patients at increased risk for systemic metastasis may improve prognostic staging and provide selection for additional therapy that may have a significant impact on disease outcome. Direct assessment of small amounts of body fluids for circulating tumor cells using microscopy has shown to be tedious.[4] This technique is labor-intensive, insensitive, and subjective. Furthermore, the rapid circulation and turbulent environment of blood may contribute to the low yield. In contrast, detection of occult tumor cells in bone marrow (BM) using immunohistochemistry has been associated with the subsequent development of systemic metastasis and shows promise as a marker of a poorer prognostic outcome.[5] Regardless, identification of a few tumor cells among a background of one million normal BM cells can be difficult and tedious. More sensitive techniques, such as RT-PCR can facilitate identification but may have diminished specificity as tumor cell-specific mRNA markers are uncommon and expression levels may vary substantially affecting results.

Recently, cell-free DNA has been identified in the serum and plasma from patients with various cancers.[6] These circulating nucleic acids have demonstrated similar genetic alterations and characteristics as those found in the primary tumor. Their presence in blood can be readily identified using common PCR techniques and appear to be elevated during disease progression.[7–9]

Alternatively, promotor region hypermethylation has been described as a common genetic abnormality occurring in various cancers. Aberrant methylation of CpG islands in promotor regions of putative tumor-suppressor and related genes resulting in their silencing has been implicated in oncogenesis and tumor progression. Identification of these additional genetic events may offer a more accurate molecular portrait accounting for a tumor's metastatic potential and provide unique tumor-specific surrogate markers for monitoring occult disease progression. Quantitative methylation-specific PCR (qMSP) provides a highly sensitive DNA-based assay for the detection of methylated alleles associated with breast cancer.[10–12]

Because BM is the most common site for systemic relapse following breast cancer diagnosis, we attempted to determine whether BM aspirate plasma could provide a viable source to detect tumor-specific epigenetic alterations associated with systemic metastasis from early-stage breast cancer patients. Identification of circulating DNA markers in BM may serve as a surrogate for occult tumor cells in BM and may provide an additional source (along with blood) for a more comprehensive patient evaluation.

MATERIALS AND METHODS

BM aspirates were collected prospectively in 4.5-mL sodium citrate tubes (Becton Dickinson, Franklin Lakes, NJ) through bilateral anterior iliac approach from 33 consecutive patients: 17 American Joint Committee on Cancer (AJCC) stage I patients, 14 AJCC stage II patients, and 2 AJCC stage III patients. All were undergoing surgical resection of their primary breast cancer at the Saint John's Health Center/John Wayne Cancer Institute. In addition, BM aspirates were obtained from 5 healthy female donors (Cambrex Bio Science, Walkersville, MD) and peripheral blood from 10 healthy female donors that served as respective negative controls. Institutional Review Board approved consent forms were signed by all patients prior to participation in the study. BM was drawn and (cell-free supernatant) plasma was separated, filtered, and cryopreserved as previously described.[13]

In addition, match-paired peripheral venous blood was drawn preoperatively. DNA was extracted from 1 mL of both peripheral blood serum and BM aspirate plasma using QIAamp extraction kit (Qiagen, Valencia, CA) as previously described.[8] Additionally, each BM aspirate was assessed for the presence of occult tumor cells by standard histologic staining methods using hematoxylin and eosin (H&E).

To determine the correlation of gene hypermethylation found in the BM and that of the primary breast tumor, DNA was isolated from ten 10-μm sections cut from paraffin-embedded tissue blocks as previously described.[14] Samples were deparaffinized, microdissected using laser capture microscopy (LCM) (Arcturus, Mountain View, CA), and incubated in lysis buffer (50 mM Tris-HCl, 1 mM EDTA, and 0.5% Tween 20) and proteinase K at 37°C overnight and then heated at 95°C for 10 min as described previously.[14]

Among the 33 patients in this study 10 had sentinel lymph nodes (SLN) positive for metastasis by conventional H&E staining. DNA was isolated from the SLN metastasis of these 10 patients as follows: 10-μm-thick sections were cut from each SLN and after deparaffinization the metastasis were microdissected using LCM followed by DNA extraction as described above. The purpose of this evaluation was to determine the concordance between primary tumor methylation status and that found in its corresponding earliest readily detectable metastasis (i.e., the SLN). For control, SLNs obtained from five patients with ductal carcinoma *in situ* (DCIS) and no evidence of metastasis following H&E and immunohistochemical analysis were processed for DNA, bisulfite treated and assayed by qMSP.

Sodium bisulfite modification was performed on 1 μg of genomic DNA obtained from paraffin-embedded tissues and 50 ng of genomic DNA collected from paired blood and BM aspirates, as described previously.[14] Primer sets were used for the detection of four genes frequently hypermethylated in breast cancer: RAS association domain family protein 1 A protein (RASSF1A), adenomatous polyposis coli (APC), retinoic acid-binding receptor-β2 (RAR-β2), and O^6-methylguanine-DNA methyltransferase (MGMT). In addition, MYOD was assessed as an internal control to confirm DNA presence in the PCR reaction. qMSP was performed with an initial incubation for 15 min at 95°C followed by 35 cycles (40 cycles for BM aspirate and serum samples) of denaturation at 94°C for 30 s, annealing at 50–56°C, and extension for 90 s at 72°C, followed by a final extension step of 72°C for 5 min. For each methylation-specific (MSP) reaction normal donor lymphocyte DNA served as a negative control, *Sss*I treated lymphocyte served as a positive control for methylated DNA markers and water served as a reagent control for contamination. Respective negative controls for marker validation consisted of peripheral blood and BM aspirates from healthy female donors and histopathologically negative SLNs from patients with DCIS. The standard curve for quantifying methylated DNA copy number was constructed using templates with known numbers of the cloned DNA template. Standard curves for each marker were generated by using the threshold cycle of dilutions of known number of DNA templates and sample results were considered positive if at least 10 copies were detected on final analysis.

Clinical and pathologic data was obtained from patient chart review and the John Wayne Cancer Institute's Breast Database. Chi-square and Wilcoxon Rank Sum tests were performed for statistical evaluation for the association of

TABLE 1. Frequency of gene hypermethylation in patients' serum and BM

	Frequency in patients' body fluid ($n = 33$)	
Marker	Serum	Bone marrow
RASSF1A	5 (15%)	7 (21%)
MGMT	2 (6%)	2 (6%)
RAR-β2	1 (3%)	2 (6%)
APC	1 (3%)	0

TABLE 2. Gene hypermethylation detection in breast cancer patients' serum and BM according to AJCC stage

	Patients with hypermethylation	
AJCC stage	Serum	Bone marrow
I ($n = 17$)	4 (24%)	3 (18%)
II ($n = 14$)	4 (29%)	3 (21%)
III ($n = 2$)	1 (50%)	1 (50%)

BM and serum methylation status and known prognostic parameters in breast cancer.

RESULTS

We identified several tumor-suppressor genes as targets for body fluid assessment that have been reported to be frequently hypermethylated in primary breast cancers as compared to normal tissue counterparts.[11,15–17] In this study, circulating tumor DNA containing gene promotor hypermethylation for any one marker was identified in the BM of 7 (21%) of 33 patients. The most frequently detected hypermethylated gene was RASSF1A occurring in 5 (15%) patients' BM, followed by MGMT in 2 (6%) patients, and RAR-β2 and APC in 1 (3%) patient each (TABLE 1). Five patients demonstrated one hypermethylated gene in their BM, whereas two patients had two hypermethylated genes identified and in 26 patients no hypermethylated DNA sequences could be detected for any of the genes assessed. No hypermethylated markers were detected in the BM from five healthy female donors.

There was an increased association between the presence of gene hypermethylated markers detected in the BM and advanced disease stage. Three (18%) of 17 AJCC stage I patients demonstrated hypermethylated DNA for at least one marker in their BM, in contrast to 3 (21%) of 14 AJCC stage II patients, and 1 of 2 AJCC stage III patients (TABLE 2).

Hypermethylation was detected in paired peripheral blood serum in 9 (27%) of 33 patients. Again RASSF1A was most frequently identified occurring in 7 (21%) patients' serum samples followed by RAR-β2 (6%) and MGMT (6%) (TABLE 1). Eight patients demonstrated hypermethylated genes in serum for

TABLE 3. Hypermethylation status: Concordance between patients' serum and BM

		Serum	
		Yes	No
Bone marrow	Yes	3	4
	No	6	20

Yes: presence of hypermethylation detected for any one gene.
No: absence of hypermethylation detected for any gene.

one gene and one patient for three genes. No hypermethylated genes were detected in serum from 10 healthy female donors. Similarly, there was an increased association between the presence of gene hypermethylation in serum and advanced AJCC stage. Hypermethylation for any one gene was identified in 4 (24%) of 17 AJCC stage I patients' serum, whereas 4 (29%) of 14 AJCC stage II patients, and 1 of 2 AJCC stage III patients had these findings (TABLE 2).

Twelve clinicopathologic prognostic factors were assessed for correlation with BM methylation status: patient age, histologic tumor type, size, grade, Bloom–Richardson score, lymph node involvement, AJCC stage, receptor status (estrogen [ER], progesterone [PR], HER2), Ki-67, and p53 status. A trend toward increased circulating methylated DNA in serum from patients with PR negative tumors was identified: 5 (50%) of 10 patients as compared to 4 (18%) of 23 patients with PR positive tumors. In multivariate analysis, patients with PR positive tumors were less likely to have methylation markers in their BM and serum, odds ratio 0.04, 95% CI: 0.00–0.82 ($P < 0.04$). However, due to the limited sample size as a result of few comprehensively collected paired specimens available for analysis no other correlations could be identified with BM or serum methylation status. These results need to be verified with a larger sample size.

Concordance between the presence of serum and/or BM hypermethylation status among patients is shown in TABLE 3. However, identification of the same gene hypermethylated between the BM and serum occurred in only two patients, with one additional patient having the same gene hypermethylation profile in BM for two of the three genes detected in serum.

To determine whether a correlation existed between gene hypermethylation detected in patients' BM and their primary tumor, DNA was isolated from primary tumors and evaluated with the same hypermethylated gene. Of 13 patients with BM and/or serum positive for hypermethylated markers, 8 had primary tumor blocks available for assessment. In all eight patients the hypermethylated gene(s) identified in the BM/serum was also hypermethylated, respectively, in the primary tumor. Conventional histologic analysis of all specimens using standard H&E staining did not demonstrate occult tumor cells in any of the BM samples.

TABLE 4. Methylation status of primary tumor and corresponding SLN metastasis (10 pairs)

Methylation status of primary tumor → SLN metastasis	APC	RAR-β2	RASSF1A	MGMT
M → M	5	3	3	0
M → U	1	2	5	0
U → M	0	2	0	0
U → U	4	3	2	0

M = methylated; U = unmethylated; → = methylation status of the primary tumor and its corresponding SLN metastasis.

Eleven patients had SLN metastasis of which 10 had paraffin blocks available for analysis. To evaluate whether SLN metastasis exhibited a similar methylation profile as that of the primary tumor, MSP analysis was performed on the paired SLNs. APC methylation was most frequent occurring in the SLNs in 5 (83%) of the 6 patients whose primary tumors demonstrated hypermethylation of this gene (TABLE 4). In the remaining four patients both the SLN and primary tumors were unmethylated at this site. Three (60%) of five patients with primary tumors containing methylated RAR-β2 had the same finding in the corresponding SLNs and RASSF1A methylation was demonstrated in the SLNs from 3 of 8 patients with this finding in their primary tumors. MGMT was not detected in any of the SLN primary tumor pairs. In general, the methylation status for the paired SLN metastasis and corresponding primary tumor correlated as follows: 90% for APC, 60% for RAR-β2, and 50% for RASSF1A. None of the methylated genes could be detected in the SLNs from five patients with DCIS, except for one case that was positive for RAR-β2 methylation.

DISCUSSION

Advances in breast imaging modalities and greater awareness for breast cancer screening have resulted in a dramatic increase in the detection of smaller primary breast cancers. Concurrently, these tumors are less likely to be associated with readily identified metastasis. Therefore newer techniques are needed to identify the presence of occult disease progression in this current era of early breast cancer diagnosis. Cytologic evaluation for the presence of circulating tumor cells remains cumbersome and subjective. RT-PCR assays have proven more facile and sensitive, however specificity remains an issue. More recently epigenetic events involving gene promoter region hypermethylation have been shown to occur more frequently in a variety of primary tumors suggesting a role in cancer development and progression.[11,18] Furthermore, circulating nucleic acids have been identified in cancer patient's serum/plasma that harbor these similar alterations providing a potential surrogate marker for monitoring disease.[7,8] MSP provides a rapid, straightforward method to assess their presence in a quantitative manner.[19] In this study we applied this technique to determine whether tumor-associated hypermethylated DNA could be identified in circulation from early-stage breast cancer patients. Genes comprising the panel were selected on the basis of demonstrating increased methylation in primary breast cancers and/or having potential prognostic significance.[12,20,21] Our findings characterized tumor-associated DNA circulating in breast cancer patient's blood. Recently, investigators have shown RASSF1A and/or APC hypermethylation identified in breast cancer patient's sera was associated with a worse outcome.[12] Interestingly at a median follow-up of 3.7 years, 20% of these patients had already died suggesting a cohort studied with more advanced disease. Not only did we also find methylated RASSF1A to be more frequent in blood than other methylated markers but we have identified its presence in BM, a common site for tumor cell metastasis. Our report demonstrates that circulating methylated tumor DNA can be detected in the blood and BM from patients with earlier stage breast cancers and thus may have clinical implications for surveillance of patients sooner in their disease course.

In this study, to permit a comprehensive evaluation of early-stage breast cancer patients' tumor tissues and body fluids, we characterized the methylation patterns in paired primary tumors and their corresponding SLN metastasis. We found that although, in general, gene hypermethylation status in the SLN corresponded with that of the primary tumor there was no significant increase in overall frequency of the number of hypermethylated genes in SLN metastasis when compared with the primary tumor. This is in contrast to a previous report[22]; however, in our study we evaluated the SLN that represents the first site of metastasis and therefore genotypically may more closely resemble the primary tumor. These findings provide further evidence that the earliest form of metastasis has a clonal epigenetic relation to the primary tumor. On the contrary, larger non-SLN metastasis may contain a greater degree of heterogeneity and/or develop additional methylation events that occur further along during tumor progression.

Methylated DNA genes have been used to detected breast cancer epithelial cells in ductal lavage fluid and nipple aspirates.[23,24] In this study we demonstrate for the first time this technique for identifying tumor-associated hypermethylated DNA in BM, aspirates from patients with early-stage breast cancer. We detected gene hypermethylation in the BM and blood from 21% and 27% of patients, respectively. As bone is the most frequent location for systemic relapse following a diagnosis of breast cancer, assessing this site provides a logical approach for identifying molecular surrogates for occult metastasis. BM aspirates, which can be performed during surgery for the primary treatment of breast cancer, may prove complementary to blood for the identification of patients at increased risk for recurrence. These early findings will need to be validated in larger patient populations with substantially longer-term follow-up with adequate outcome data to determine clinical utility. In contrast to the identification of methylated DNA in BM, we did not detect the presence of circulating tumor cells in any of these samples with H&E—many of these patients were participants in the American College of Surgeons Oncology Group Z10 and therefore BM immunohistochemical results are currently blinded. These findings reveal the relative inadequacy of microscopic BM H&E assessment as compared with molecular diagnostic techniques. To our knowledge the only other report evaluating circulating tumor cell detection with that of circulating tumor-associated DNA was performed on peripheral blood and corroborates our results in BM.[25] In their study, which evaluated a majority of patients with stage IV breast cancer, circulating DNA containing loss of heterozygosity (LOH) could readily be found in patients' serum whereas almost in none of the cases could circulating tumor cells be identified.

The current findings provide a most comprehensive and quantitative assessment of early-stage breast cancer patients' body fluids for the presence of tumor-related DNA. The objective is that a more complete evaluation of sources and potential sites of subsequent metastasis will prove more informative for patient care decisions. The demonstration of circulating tumor-associated methylated DNA in blood and BM of early-stage patients is of considerable interest and may be of particular clinical utility in view of the fact that bone and blood are a frequent site and route for systemic recurrence and disease spread, respectively. The results are consistent with our previous study assessing circulating microsatellites for LOH in breast cancer patients.[13] However, the evaluation of LOH in body fluids is associated with a number of limitations including contamination from normal DNA, inability to readily quantitate, and the fundamentally inherent difficulty associated with an assay that assesses for the loss of a marker (as in LOH) as compared to its gain (as in methylation). Larger studies with prospectively collected paired tumor tissues and body fluids are currently under way. At present the evaluation of tissues and body fluids in patients with early-stage breast cancer provides the most promise for assessing the earliest stages of disease progression. The identification of tumor-specific genetic alterations in cancer patient's tumor tissues provides unique and readily assessable molecular targets for establishing patient-specific disease surveillance programs. Innovative techniques are needed that detect and characterize subclinical disease in this new era of early breast cancer diagnosis.

ACKNOWLEDGMENTS

This work was supported in part by the California Breast Cancer Research Program Grant 7WB-0021 (DSBH), the Leslie and Susan Gonda (Goldschmied) Foundation, and the Ben B., Joyce E. Eisenberg Foundation (Los Angeles), Avon Foundation, and the Fashion Footwear Association of New York.

REFERENCES

1. FISHER, B. *et al.* 1989. A randomized clinical trial evaluating tamoxifen in the treatment of patients with node-negative breast cancer who have estrogen-receptor-positive tumors. N. Engl. J. Med. **320:** 479–484.
2. ROSEN, P.R. *et al.* 1989. A long-term follow-up study of survival in stage I (T1N0M0) and stage II (T1N1M0) breast carcinoma. J. Clin. Oncol. **7:** 355–366.
3. GOLDHIRSCH, A. *et al.* 1988. Relapse of breast cancer after adjuvant treatment in premenopausal and perimenopausal women: patterns and prognoses. J. Clin. Oncol. **6:** 89–97.
4. MOLINO, A. *et al.* 1991. A comparative analysis of three different techniques for the detection of breast cancer cells in bone marrow. Cancer **67:** 1033–1036.
5. BRAUN, S. *et al.* 2000. Cytokeratin-positive cells in the bone marrow and survival of patients with stage I, II, or III breast cancer. N. Engl. J. Med. **342:** 525–533.
6. SIDRANSKY, D. 1997. Nucleic acid-based methods for the detection of cancer. Science **278:** 1054–1059.
7. SILVA, J.M. *et al.* 1999. Presence of tumor DNA in plasma of breast cancer patients: clinicopathological correlations. Cancer Res. **59:** 3251–3256.
8. TABACK, B. *et al.* 2001. Microsatellite alterations detected in the serum of early stage breast cancer patients. In circulating nucleic acids in plasma/serum ii. Ann. N. Y. Acad. Sci. **945:** 22–30.
9. SILVA, J.M. *et al.* 2002. Tumor DNA in plasma at diagnosis of breast cancer patients is a valuable predictor of disease-free survival. Clin. Cancer Res. **8:** 3761–3766.
10. LO, Y.M. *et al.* 1999. Quantitative analysis of aberrant p16 methylation using real-time quantitative methylation-specific polymerase chain reaction. Cancer Res. **59:** 3899–3903.
11. ESTELLER, M. *et al.* 2001. A gene hypermethylation profile of human cancer. Cancer Res. **61:** 3225–3229.
12. MULLER, H.M. *et al.* 2003. DNA methylation in serum of breast cancer patients: an independent prognostic marker. Cancer Res. **63:** 7641–7645.
13. TABACK, B. *et al.* 2003. Detection of tumor-specific genetic alterations in bone marrow from early-stage breast cancer patients. Cancer Res. **63:** 1884–1887.
14. HOON, D.S. *et al.* 2004. Profiling epigenetic inactivation of tumor suppressor genes in tumors and plasma from cutaneous melanoma patients. Oncogene **23:** 4014–4022.
15. LEHMANN, U. *et al.* 2002. Quantitative assessment of promoter hypermethylation during breast cancer development. Am. J. Pathol. **160:** 605–612.
16. DAMMANN, R. *et al.* 2001. Hypermethylation of the cpG island of Ras association domain family 1A (RASSF1A), a putative tumor suppressor gene from the 3p21.3 locus, occurs in a large percentage of human breast cancers. Cancer Res. **61:** 3105–3109.
17. JIN, Z. *et al.* 2001. Adenomatous polyposis coli (APC) gene promoter hypermethylation in primary breast cancers. Br. J. Cancer. **85:** 69–73.
18. NASS, S.J. *et al.* 2000. Aberrant methylation of the estrogen receptor and E-cadherin 5′ CpG islands increases with malignant progression in human breast cancer. Cancer Res. **60:** 4346–4348.
19. HERMAN, J.G. *et al.* 1996. Methylation-specific PCR: a novel PCR assay for methylation status of CpG islands. Proc. Natl. Acad. Sci. USA **93:** 9821–9826.

20. WIDSCHWENDTER, M. *et al.* 2000. Methylation and silencing of the retinoic acid receptor-beta2 gene in breast cancer. J. Natl. Cancer Inst. **92:** 826–832.
21. WIDSCHWENDTER, M. & P.A. JONES. 2002. DNA methylation and breast carcinogenesis. Oncogene **21:** 5462–5482.
22. MEHROTRA, J. *et al.* 2004. Very high frequency of hypermethylated genes in breast cancer metastasis to the bone, brain, and lung. Clin Cancer Res. **10:** 3104–3109.
23. EVRON, E. *et al.* 2001. Detection of breast cancer cells in ductal lavage fluid by methylation-specific PCR. Lancet **357:** 1335–1336.
24. KRASSENSTEIN, R. *et al.* 2004. Detection of breast cancer in nipple aspirate fluid by CpG island hypermethylation. Clin. Cancer Res. **10:** 28–32.
25. SCHWARZENBACH, H. *et al.* 2004. Detection and characterization of circulating microsatellite-DNA in blood of patients with breast cancer. Ann. N. Y. Acad. Sci. **1022:** 25–32.

Comparison of Genetic Alterations Detected in Circulating Microsatellite DNA in Blood Plasma Samples of Patients with Prostate Cancer and Benign Prostatic Hyperplasia

IMKE MÜLLER, KAROLINE URBAN, KLAUS PANTEL, AND HEIDI SCHWARZENBACH

Institute of Tumor Biology, University Medical Center Hamburg-Eppendorf, Hamburg, Germany

ABSTRACT: Prostate cancer is the most frequent malignant disease and the second most frequent cause of death due to cancer in men in the Western world. Since serum prostate-specific antigen (PSA) and its subforms show poor specificity in clinical practice, a molecular marker for the detection and discrimination of prostate cancer (PCa) could be of great interest. To investigate the potential significance of genetic aberrations, such as loss of heterozygosity (LOH), in PCa we identified and characterized allelic losses in circulating tumor-associated DNA in blood from patients with localized PCa. Genomic DNA extracted from cell-free plasma of blood samples drawn from 65 PCa patients was analyzed using a panel of 15 polymorphic microsatellite markers mapping to known tumor-suppressor genes. Comparative analyses were performed with a control group of 36 patients with benign prostatic hyperplasia (BPH). In the current study, we demonstrate that PCa patients had higher DNA concentrations in their blood circulation than BPH patients. In the marker panel studied, LOH was more frequently detected in PCa patients (34%) than in BPH patients (22%). The incidence of LOH in the plasma DNA of PCa patients was highest at chromosomal regions 3p24 (THRB, 22%) and 8p21 (D8S360, 22%) in comparison to the BPH control cohort, which frequently showed LOH at loci 8q21, 8p21, 9p21, and 11q22 (D8S286, D8S360, D9S1748, and D11S898, each 6%). These results indicate that microsatellite analysis using plasma DNA may be an interesting tool for molecular screening of PCa patients.

KEYWORDS: circulating tumor DNA; microsatellite aberrations; loss of heterozygosity (LOH); prostate cancer; BPH

Address for correspondence: Dr. Heidi Schwarzenbach, Institute of Tumor Biology, University Medical Center Hamburg-Eppendorf, Martinistrasse 52, 20246 Hamburg, Germany. Voice: +49-40-42-803-7494; fax: +49-40-42-803-6546.
e-mail: hschwarz@uke.uni-hamburg.de

Ann. N.Y. Acad. Sci. 1075: 222–229 (2006).
doi: 10.1196/annals.1368.030

INTRODUCTION

Serum prostate-specific antigen (PSA) together with digital rectal examination (DRE), transrectal ultrasound (TRUS), and needle biopsy are currently used for the early detection of prostate cancer (PCa).[1] Despite lack of specificity, serum PSA and its subforms (e.g., percentage free PSA), age-related PSA cut-off levels, PSA kinetics (PSA velocity), and PSA density are currently recognized as the most useful tumor markers available in clinical diagnosis. Because of the low specificity of PSA, however, a great need exists for novel tumor markers that are able to discriminate malignant PCa from benign prostatic hyperplasia (BPH).[2]

The development of PCa is associated with numerous genetic alterations, such as changes in genes encoding molecules involved in cell–cell adhesion, cell cycle control, and apoptosis.[3–5] A number of publications have shown that tumor-associated DNA from PCa patients exhibits specific microsatellite alterations, like loss of heterozygosity (LOH).[6–9] Moreover, higher plasma DNA concentrations in PCa patients have been described.[10–12] The origin of circulating DNA in blood is still debated and it has been suggested that apoptotic and necrotic as well as active processes are involved in the release of DNA into the blood stream early during tumorigenesis.[13,14] However, so far no study has focused on detection of LOH in the circulating plasma DNA in PCa patients. A high incidence of LOH has been observed in the serum or plasma of patients with other tumors, such as breast, lung, kidney, gastrointestinal, head and neck carcinomas, and melanoma.[15–22] Peripheral blood from PCa patients could therefore be an important source for the detection of genetic alterations and has the additional benefit that this is less invasive and more convenient than biopsies.

In this study, we demonstrate that the median DNA levels and the frequency of LOH were higher in the peripheral blood of PCa than in BPH plasma samples. Furthermore, we show that the panel of polymorphic markers used in this study could be subdivided into two groups: one group that shows LOH in both PCa and BPH samples and the other showing LOH only in PCa patients.

MATERIALS AND METHODS

The current study on LOH in circulating plasma DNA included 101 men (65 with localized PCa and 36 BPH patients) and was approved by the local ethics review board. All patients gave their written consent prior to blood collection. PCa patients underwent DRE, TRUS, and ultrasound-guided 10-core biopsy. BPH was diagnosed when either the prostate saturation biopsy or two series of a systematic 10-core prostate biopsy were negative.

Circulating DNA was isolated from 3–4 mL of plasma using the QIAamp DNA Mini Kit and a vacuum chamber (Qiagen QIAvac24) according to the

TABLE 1. Microsatellite markers used

Microsatellite marker	Chromosomal locus	Tumor-suppressor gene
D3S3703	3q13	n.a.
THRB	3p24	thyroid hormone receptor β
D6S474	6q21–22	n.a.
D6S1631	6q16	TAK1
D7S522	7q31	Caveolin1
D8S87	8p12	Neuregulin 1
D8S137	8p21.1	Dematin
D8S286	8q21.3	HNF-4γ
D8S360	8p21	n.a.
D9S171	9p21–22	CDKN2/p16
D9S1748	9p21	CDKN2/p16
D10S1765	10q23	PTEN
D11S898	11q22	n.a.
D11S1313	11q11–11p11	n.a.
TP53.6	17p13	tumor suppressor p53

n.a.: tumor suppressor gene is not analyzed.

manufacturer's instructions (Qiagen, Hilden, Germany). Quantification of the isolated DNA was performed spectrophotometrically using a biophotometer (Eppendorf, Hamburg, Germany).

The fluorescence-labeled PCR products amplified with a panel of 15 polymorphic microsatellite markers (Sigma, Taufkirchen, Germany, TABLE 1) and the AmpliTaq Gold DNA-Polymerase (Applied Biosystems, Mannheim, Germany) were separated by capillary gel electrophoresis on the Genetic Analyzer 310 (Applied Biosystems). The presence of LOH was defined by a >40% reduction in the fluorescence intensity. Statistical analysis was performed using the SPSS software (version 11.0, SPSS Inc, Chicago, IL, USA).

RESULTS

The results of the spectrophotometric quantification showed a wide range of DNA concentrations in the plasma of BPH and PCa patients (FIG. 1). However, median plasma DNA levels in PCa patients were higher than in BPH patients ($P < 0.05$). In BPH patients, the median value was 481 ng/mL, with a range of 55 to 2,908 ng/mL. In comparison, in PCa patients the median DNA concentration was 590 ng/mL, with a range of 81 to 2,792 ng/mL.

To assess the genetic alterations in the plasma DNA derived from PCa and BPH patients we used a set of 15 microsatellite primers binding to chromosomes 3, 6, 7, 8, 9, 10, 11, and 17, as described in TABLE 1. In the PCa group 34% (22 of 65) of the patients showed at least one LOH at any marker. In the BPH group 22% (8 of 36) had at least one LOH in their circulating plasma DNA.

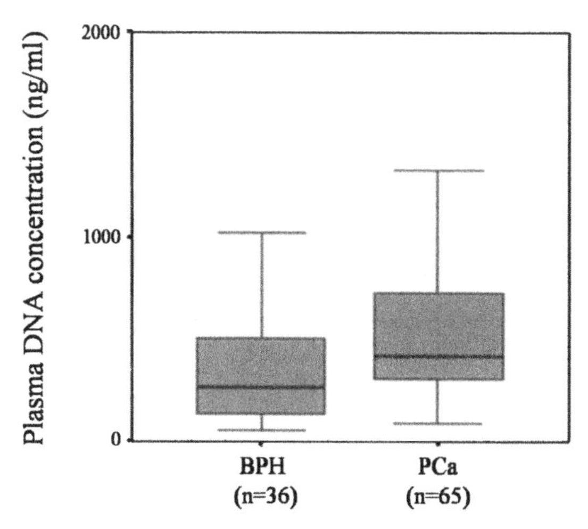

FIGURE 1. Box plot comparison of plasma DNA concentrations in patients with localized PCa and BPHs.

FIGURE 2 shows a representative LOH result found in the circulating DNA of a PCa patient. The right allele in the plasma DNA is significantly reduced compared to the respective allele in the reference DNA.

Most LOH was observed at the markers near thyroid hormone receptor beta (THRB) (22%) and D8S360 (22%) in PCa patients (FIG. 3, TABLE 2). In the BPH group LOH could only be detected at D8S360 (6%), but not at THRB. LOH was also seen in D8S137 quite frequently in PCa patients (14%), but not in BPH patients. In addition to D8S360 the BPH group demonstrated LOH at D8S286, D9S1748, and D11S898 (6% each). The frequencies of LOH in the PCa group were similar (9%, 3%, and 10%, respectively).

DISCUSSION

Development of suitable noninvasive, novel tumor markers is crucial for an improved early detection of PCa since the established serum marker PSA lacks specificity. Techniques, such as fluorescence *in situ* hybridization (FISH), comparative genomic hybridization (CGH), and microsatellite-based PCR have demonstrated amplifications and deletions in chromosomes associated with PCa.[23] We investigated the incidence of allelic losses in plasma DNA derived from PCa patients by a PCR-based fluorescence microsatellite analysis using a DNA Analyzer.

Our studies showed that 22 of 65 patients (34%) with localized PCa and 8 of 36 BPH patients (22%) showed at least one LOH in the circulating DNA. Most LOH in PCa patients was detected at markers THRB (22%) and

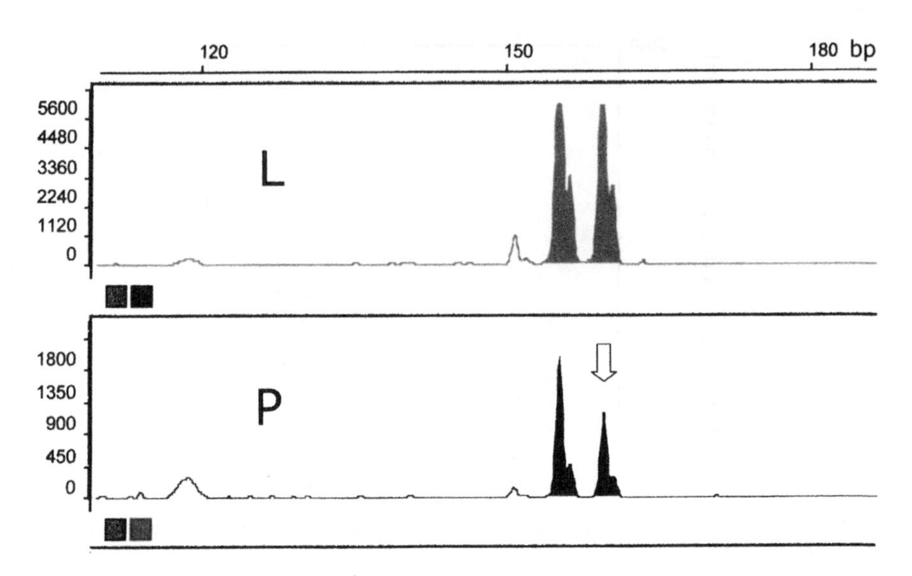

FIGURE 2. Representative example of LOH. The fluorescence-labeled PCR products amplified with marker D6S474 were analyzed after capillary gel electrophoresis by the Gene Scan software on a Genetic Analyzer. The vertical axis shows the fluorescent intensity of the PCR product and the horizontal axis represents the fragment size of the amplicons. The arrow indicates the LOH. L refers to lymphocyte and P refers to plasma.

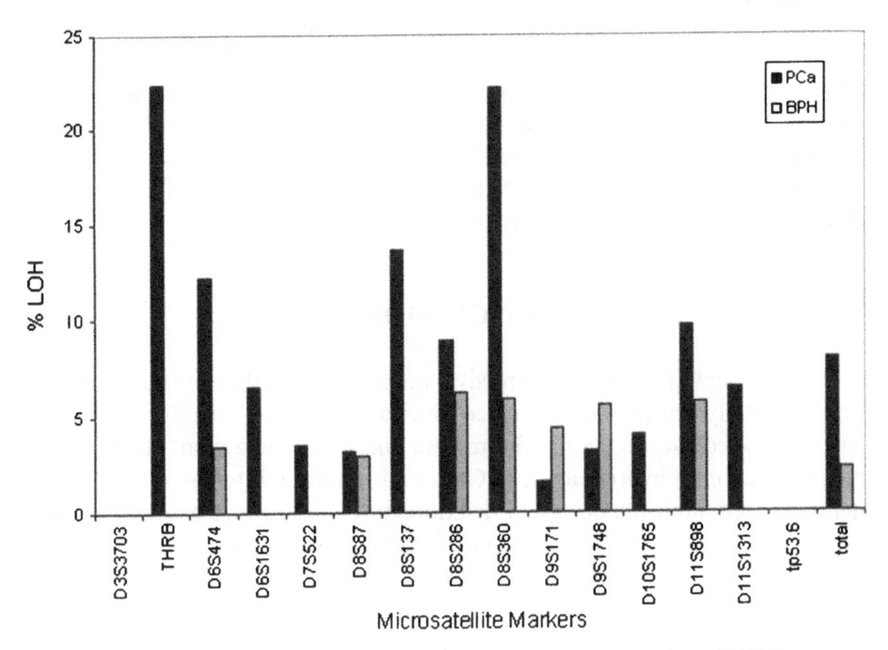

FIGURE 3. Incidence of LOH at 15 different microsatellite markers. LOH frequency (%) in the circulating plasma DNA in patients with PCa (black) and BPH (gray). No LOH was found at markers D3S3703 and TP53.6.

TABLE 2. Frequency of LOH in plasma of patients with BPH and PCa

	BPH ($n = 36$)			PCa ($n = 65$)		
Marker	LOH	Informative	%	LOH	Informative	%
D3S3703	0	22	0	0	18	0
THRB	0	20	0	13	58	22
D6S474	1	29	3	5	41	12
D6S1631	0	30	0	4	61	7
D7S522	0	32	0	1	28	4
D8S87	1	34	3	2	63	3
D8S137	0	34	0	3	22	14
D8S286	1	16	6	4	45	9
D8S360	1	17	6	4	18	22
D9S171	1	23	4	1	61	2
D9S1748	2	36	6	2	63	3
D10S1765	0	24	0	1	25	4
D11S898	2	35	6	6	62	10
D11S1313	0	33	0	4	62	7
tp53.6	0	18	0	0	0	0

D8S360 (22%). The control group of BPH patients revealed most LOH with markers D8S286, D8S360, D9S1748, and D11S898 (6% each).

The incidence of LOH at the markers D6S474, D8S87, D8S286, D8S360, D9S171, D9S1748, and D11S898 was not able to discriminate between PCa and BPH. The association of these markers with genes involved in cell proliferation is supported by the fact that the chromosomal regions 6q21–22, 8p12, 8q21, 8p21, 9p21, and 11q22 encode tumor-suppressor proteins, which are involved in cell cycle control, apoptosis, cell–cell, and cell–matrix interactions.[24–26] Loss of expression of these proteins might lead to a higher cell proliferation activity playing a role in both PCa and BPH.

LOH at chromosomal region 3p24 (marker THRB) only occurred in patients with localized PCa, but not in BPH patients, suggesting that deletion of this area seems to be tumor specific and may be involved in tumorigenesis. Kato *et al.* reported that mice carrying a mutant THRB gene spontaneously developed follicular thyroid carcinomas.[27] LOH was also specifically found with marker D8S137 in PCa patients, suggesting that it appears to play a role in early tumorigenesis. Although LOH with markers D6S1631, D7S522, D10S1765, and D11S1313 was only detected in PCa, the frequencies (7%, 4%, 4%, and 7%, respectively) were relatively low.

In conclusion, LOH is more frequently detected in the peripheral blood of patients with PCa than in BPH patients. Among the markers tested some showed allelic losses in both BPH and PCa patients, whereas DNA deletions of the markers THRB and D8S137 were tumor specific and might be involved in neoplastic processes.

REFERENCES

1. CATALONA, W.J. *et al.* 1994. Comparison of digital rectal examination and serum prostate specific antigen in the early detection of prostate cancer: results of a multicenter clinical trial of 6630 men. J. Urol. **151:** 1283–1290.
2. STAMEY, T.A. *et al.* 2004. The prostate specific antigen era in the United States is over for prostate cancer: what happened in the last 20 years? J. Urol. **172:** 1297–1301.
3. ELO, J.P. & T. VISAKORPI. 2001. Molecular genetics of prostate cancer. Ann. Med. **33:** 130–141.
4. SARIC, T. *et al.* 1999. Genetic pattern of prostate cancer progression. Int. J. Cancer **81:** 219–224.
5. NUPPONEN, N.N. & T. VISAKORPI. 2000. Molecular cytogenetics of prostate cancer. Microsc. Res. Tech. **51:** 456–463.
6. VON KNOBLOCH, R. *et al.* 2004. Genetic pathways and new progression markers for prostate cancer suggested by microsatellite allelotyping. Clin. Cancer Res. **10:** 1064–1073.
7. DONG, J.T. 2001. Chromosomal deletions and tumor suppressor genes in prostate cancer. Cancer Metastasis Rev. **20:** 173–193.
8. DONG, J.T. *et al.* 2001. Loss of heterozygosity at 13q14 and 13q21 in high grade, high stage prostate cancer. Prostate **49:** 166–171.
9. JENKINS, R. *et al.* 1998. Prognostic significance of allelic imbalance of chromosome arms 7q, 8p, 16q, and 18q in stage T3N0M0 prostate cancer. Genes Chromosomes Cancer **21:** 131–143.
10. JUNG, K. *et al.* 2004. Increased cell-free DNA in plasma of patients with metastatic spread in prostate cancer. Cancer Lett. **205:** 173–180.
11. WU, T.L. *et al.* 2002. Cell-free DNA: measurement in various carcinomas and establishment of normal reference range. Clin. Chim. Acta **321:** 77–87.
12. BODDY, J.L. *et al.* 2005. Prospective study of quantitation of plasma DNA levels in the diagnosis of malignant versus benign prostate disease. Clin. Cancer Res. **11:** 1394–1399.
13. STROUN, M. *et al.* 2001. About the possible origin and mechanism of circulating DNA apoptosis and active DNA release. Clin. Chim. Acta **313:** 139–142.
14. JAHR, S. *et al.* 2001. DNA fragments in the blood plasma of cancer patients: quantitations and evidence for their origin from apoptotic and necrotic cells. Cancer Res. **61:** 1659–1665.
15. SCHWARZENBACH, H. *et al.* 2004. Detection and characterization of circulating microsatellite-DNA in blood of patients with breast cancer: circulating nucleic acids in plasma/serum proteomics III. Ann. N. Y. Acad. Sci. **1022:** 25–32.
16. SILVA, J.M. *et al.* 2002. Tumor DNA in plasma at diagnosis of breast cancer patients is a valuable predictor of disease-free survival. Clin. Cancer Res. **8:** 3761–3766.
17. CHEN, X.Q. *et al.* 1999. Detecting tumor-related alterations in plasma or serum DNA of patients diagnosed with breast cancer. Clin. Cancer Res. **5:** 2297–2303.
18. CHEN, X.Q. *et al.* 1996. Microsatellite alterations in plasma DNA of small cell lung cancer patients. Nat. Med. **2:** 1033–1035.
19. ANKER, P. *et al.* 1997. K-ras mutations are found in DNA extracted from the plasma of patients with colorectal cancer. Gastroenterology **112:** 1114–1120.
20. GOESSL, C. *et al.* 1998. Microsatellite analysis of plasma DNA from patients with clear cell renal carcinoma. Cancer Res. **58:** 4728–4732.

21. NAWROZ, H. *et al.* 1996. Microsatellite alterations in serum DNA of head and neck cancer patients. Nat. Med. **2:** 1035–1037.
22. TABACK, B. *et al.* 2001. Prognostic significance of circulating microsatellite markers in the plasma of melanoma patients. Cancer Res. **61:** 5723–5726.
23. OZEN, M. & S. PATHAK. 2000. Genetic alterations in human prostate cancer: a review of current literature. Anticancer Res. **20:** 1905–1912.
24. TAL-OR, P. *et al.* 2003. Neuregulin promotes autophagic cell death of prostate cancer cells. Prostate **55:** 147–157.
25. FROMONT, G. *et al.* 2003. Allelic losses in localized prostate cancer: association with prognostic factors. J. Urol. **170:** 1394–1397.
26. PERINCHERY, G. *et al.* 1999. High frequency of deletion on chromosome 9p21 may harbour several tumor-suppressor genes in human prostate cancer. Int. J. Cancer **83:** 610–614.
27. KATO, Y. *et al.* 2004. A tumor suppressor role for thyroid hormone beta-receptor in a mouse model of thyroid carcinogenesis. Endocrinology **145:** 4430–4438.

Quantification of Total Plasma Cell-Free DNA in Ovarian Cancer Using Real-Time PCR

APARNA A. KAMAT,[a] ANIL K. SOOD,[a,b] DIANNE DANG,[b]
DAVID M. GERSHENSON,[a] JOE L. SIMPSON,[b]
AND FARIDEH Z. BISCHOFF[b]

[a]*Department of Gynecologic Oncology, University of Texas, M. D. Anderson Cancer Center, Houston, Texas 77030, USA*

[b]*Cancer Biology, University of Texas, M. D. Anderson Cancer Center, Houston, Texas 77030, USA*

[c]*Department of Obstetrics and Gynecology, Baylor College of Medicine, Houston, Texas, USA*

ABSTRACT: Our objective was to compare the levels of total circulating plasma cell-free DNA (CfDNA) using real-time PCR in patients with late-stage ovarian cancer with those in unaffected controls. Following IRB consent, DNA was extracted from archived frozen plasma of 19 patients with primary ovarian carcinoma and 12 age-matched controls using Qiagen DNA Isolation Kits. Quantification of total CfDNA was performed using real-time PCR with the TaqMan Assay for GAPDH, β-actin and β-globin and the number of genome equivalents (GE/mL) were determined from a standard curve. CfDNA levels of these loci were compared between the groups with Student's *t*-test, with $P < 0.05$ being statistically significant. The mean age of the patients was 61.6 years (±9.6) and of the controls was 54 years (±12.2). All patients had high-grade, advanced stage (III or IV) serous ovarian carcinomas. Preoperative CA-125 levels ranged from 43 to 15,626 IU/mL (mean 2487.2 ± 3686 IU/mL). Total CfDNA in ovarian cancer was higher among patients with ovarian cancer as compared to controls at all three loci: GAPDH ($P = 0.022$), β-actin ($P = 0.025$), and β-globin ($P = 0.0089$). CfDNA is elevated in advanced stage disease compared to controls. These preliminary results suggest that total CfDNA in the plasma of patients with ovarian cancer may be useful for noninvasive screening and disease surveillance.

KEYWORDS: cell-free DNA; real-time PCR; quantification; ovarian cancer

Address for correspondence: Farideh Z. Bischoff, Department of Obstetrics and Gynecology, Baylor College of Medicine, 6550 Fannin, Houston, Texas 77030. Voice: 713-798-7559; fax: 713-798-5575. e-mail: bischoff@bcm.tmc.edu

**Ann. N.Y. Acad. Sci. 1075: 230–234 (2006). © 2006 New York Academy of Sciences.
doi: 10.1196/annals.1368.031**

INTRODUCTION

Ovarian cancer remains the leading cause of death from a gynecological malignancy in the United States.[1] Most patients with ovarian cancer generally present with widely metastatic disease at diagnosis, contributing to the high mortality associated with this disease.[2] In stark contrast, survival rates for women whose tumors are confined to the ovary are nearly 90%.[3] Although improvements in ovarian cancer therapy, such as the use of combination regimens of platinum- and taxane-based agents, have improved the frontline response rate of patients undergoing treatment for ovarian cancer, the key to successful curative treatment of this disease remains early diagnosis.

It is well established that "cell-free" DNA (CfDNA) circulates in plasma of normal healthy individuals.[4] It is also known that tumor-specific DNA can coexist with normal CfDNA in the plasma of cancer patients[4,5] and that most of this DNA is derived from the tumor itself with correlation between tumor load and the quantity of CfDNA detected.[4,5] This has been observed for a wide range of malignancies, including ovarian cancer.[6–8] Thus, strategies for analysis of tumor-specific DNA in plasma have been proposed using quantification of either total (normal and tumor) plasma DNA amounts based on detection of a ubiquitous locus (e.g., betaglobin) or tumor-specific (genetic and epigenetic) sequence alterations. These methods could be used noninvasively to screen patients for the presence and severity of cancer as well as to monitor response to treatment. Given availability of highly sensitive real-time PCR assays that enable detection of single copy sequences, optimal methods of plasma DNA isolation followed by a plasma-based molecular diagnostic test would perhaps identify asymptomatic patients with early and clinically curable ovarian cancer. Such a technique could potentially be used as a surrogate marker to monitor disease progression as well as aid in early diagnosis. Thus, the goal of this study was to quantify total plasma CfDNA using real-time PCR from patients with ovarian cancer and compare these levels with unaffected age-matched healthy controls.

METHODS

Patient Plasma Samples and Clinical Information

Following IRB approval, archived frozen plasma samples were obtained from the Gynecologic Oncology Tumor Bank at M.D. Anderson Cancer Center from the year 2000 to 2002. Plasma samples from 19 patients with epithelial ovarian carcinoma (EOC) and 12 patients who had no personal or family history of cancer (controls) were selected. All tumors were surgically staged on the basis of the International Federation of Gynecology and Obstetrics (FIGO) staging system.

All the blood samples had been collected immediately prior to surgery and plasma was separated by centrifugation at 1,000 × g, followed by storage at −80°C until use. The samples were then thawed at room temperature and were subjected to a second round of centrifugation at 1,000 × g to remove any cellular contamination; the supernatant was separated and used for DNA extraction.

Clinical information including demographic information, CA-125 level, tumor stage, and histology was obtained from the patients' medical records. All patients underwent surgical exploration and primary surgical cytoreduction as the initial treatment. Pathologic diagnosis and tumor grading were confirmed by reviewing the surgical pathology reports.

Extraction and Quantification of Total Plasma Cell-Free DNA

DNA extraction was performed using the Qiagen DNA extraction Mini Kit (QIAGEN Sciences, MD, USA) according to manufacturer's instructions. All samples were processed using the same protocol. Quantification of total plasma DNA was performed using real-time PCR with the TaqMan Assay (Applied Biosystems, Foster City, CA, USA) with primers directed to endogenous control genes (GAPDH, β-actin and β-globin,[9,10] Applied Biosystems, Foster City, CA, USA). A standard curve consisting of known DNA concentrations was created to quantify the number of genome equivalents (GE/mL) in each of the unknown samples. All samples were run in triplicate.

Statistical Analysis

The mean values of total CfDNA were compared among the groups using the Student's t-test. A P-value <0.05 was considered statistically significant.

RESULTS

The mean age (±SD) was 61.6 years (±9.6) for patients, and 54 years (±12.2) for controls. All patients had high-grade, advanced stage (III or IV) serous ovarian carcinomas. Preoperative CA-125 levels ranged from 43 to 15,626 IU/mL (mean 2487.2 ± 3686). None of the controls had a history of proven cancer, including ovarian cancer.

In order to optimize quantification of CfDNA from plasma, real-time PCR was performed, using three endogenous loci in order to correct for efficiency of amplification. The mean levels of CfDNA (±SE) in controls and ovarian cancer patients using primers directed to GAPDH, β-actin and β-globin were higher in the ovarian cancer patients compared to controls. These data are shown in TABLE 1.

TABLE 1. Plasma total CFDNA levels in the plasma of patients with ovarian cancer and healthy controls

Mean CFDNA (GE/mL)	Controls ($n = 12$)	Ovarian Cancer ($n = 19$)	P-value
GAPDH	5266 ± 796	9581 ± 6723	0.022
β-actin	6608 ± 985	12994 ± 6723	0.025
β-globin	10105 ± 1366	20018 ± 14073	0.0089

Results are mean ± SE.

While the sample size of this pilot study is small, it appears that patients with EOC had higher levels of CfDNA compared to controls at all three loci. However, amplification efficiency was found to be the best for the β-globin locus as seen in TABLE 1.

DISCUSSION

Our results demonstrate that the levels of total CfDNA are elevated in patients with invasive EOC. These results were validated using three different endogenous loci, all of which showed elevated levels among EOC samples. Among the three loci, amplification efficiency was greatest with β-globin. This response may be the result of varying primer sequence or the presence of differential representation of annealing sites in the CfDNA fragments in the plasma. Nevertheless, mean CfDNA levels were significantly higher among ovarian cancer patients, with greatest differences seen with β-globin (TABLE 1). These results are promising and could provide an additional marker to identify patients with malignant ovarian neoplams.

The amount of DNA isolated from plasma is dependent on the specific method of DNA isolation employed. Factors that can potentially affect the rate of detection of CfDNA from the plasma are storage of the blood specimen (EDTA or citrate) or centrifugation speed. Both can also affect the amount of CfDNA recovered.[11,12] Therefore, in order to use CfDNA levels for diagnostic or prognostic purposes, it is necessary to standardize sample collection and processing techniques as was done in this study. Total CfDNA measurements detect both normal and tumor-specific DNA. Historically, while a large fraction of this CfDNA is tumor derived, it is possible for other conditions, and thus cell types, to cause elevations in these levels. Thus, it is important to have pathologic confirmation of disease status in the study groups. Moreover, accurate designation of control groups and consideration of demographic variables are key to accurate and reproducible results.

In summary, these preliminary results point to a potential role for using total CfDNA levels as an additional noninvasive technique to identify women with malignant ovarian disease. Our data indicate that total CfDNA can be quantified reliably using real-time PCR in the plasma of patients with ovarian cancer and that amplification efficiency is greatest using β-globin as the endogenous control locus. These results are promising, but would have to be validated using a larger sample size before being considered for clinical application.

REFERENCES

1. JEMAL, A. *et al.* 2005. Cancer statistics. CA Cancer J. Clin. **55:** 10–30.
2. BEREK, J. & N. HACKER. 2000. Practical Gynecologic Oncology. Third edition. Lippincott Williams & Wilkins, Philadelphia, PA.
3. AHMED, F.Y. *et al.* 1996. Natural history and prognosis of untreated stage I epithelial ovarian carcinoma. J. Clin. Oncol. **14:** 2968–2975.
4. STROUN, M. *et al.* 2000. The origin and mechanism of circulating DNA. Ann. N. Y. Acad. Sci. **906:** 161–168.
5. ANKER, P. *et al.* 1999. Detection of circulating tumor DNA in the blood (plasma/serum) of cancer patients. Cancer Metastasis Rev. **18:** 65–73.
6. NAWROZ, H. *et al.* 1996. Microsatellite alterations in serum DNA of head and neck cancer patients. Nat. Med. **2:** 1035–1037.
7. WU, T.L. *et al.* 2002. Cell-free DNA: measurement in various carcinomas and establishment of normal reference range. Clin. Chim. Acta **321:** 77–87.
8. CHANG, H.W. *et al.* 2002. Assessment of plasma DNA levels, allelic imbalance, and CA 125 as diagnostic tests for cancer. J. Natl. Cancer Inst. **94:** 1697–1703.
9. LO, Y.M. *et al.* 1998. Quantitative analysis of fetal DNA in maternal plasma and serum: implications for noninvasive prenatal diagnosis. Am. J. Hum. Genet. **62:** 768–775.
10. BISCHOFF, F.Z. *et al.* 2003. Detecting fetal DNA from dried maternal blood spots: another step towards broad scale non-invasive prenatal genetic screening and feasibility testing. Reprod. Biomed. Online **6:** 349–351.
11. CHIU, R. *et al.* 2001. Effects of blood-processing protocols on fetal and total DNA quantification in maternal plasma. Clin. Chem. **47:** 1607–1613.
12. LUI, Y.Y.N. *et al.* 2002. Predominant hematopoietic origin of cell-free DNA in plasma and serum after sex-mismatched bone marrow transplantation. Clin. Chem. **48:** 421–427.

Cell-free DNA and RNA in Plasma as a New Molecular Marker for Prostate and Breast Cancer

EIRINI PAPADOPOULOU,[a,d] ELIAS DAVILAS,[b] VASILIOS SOTIRIOU,[b] ELEFTHERIOS GEORGAKOPOULOS,[c] STAVROULA GEORGAKOPOULOU,[d] ALEXANDER KOLIOPANOS,[e] FILIPOS AGGELAKIS,[f] KONSTANTINOS DARDOUFAS,[f] NIKI J. AGNANTI,[g] IRINI KARYDAS,[h] AND GEORGIOS NASIOULAS[a]

[a]*Molecular Biology Research Center HYGEIA 151 23 Marousi, Athens, Hellas, Greece*

[b]*Urologic Clinic, Diagnostic and Therapeutic Center of Athens, "HYGEIA" Kiffisias Ave. and 4 Erythrou Stavrou Str., 151 23 Marousi, Athens, Hellas, Greece*

[c]*Acrongenomics Inc., Calgary-Alberta, T2G 0T7, Canada*

[d]*Eurogenet Laboratories S.A., 27 Marathonos Ave Pallini, Greece*

[e]*Surgical Clinic, General State Hospital of Athens "G. Gennimatas" Mesogeion Avenue, Athens, Greece*

[f]*Centre of Radiation Oncology, Diagnostic and Therapeutic Center of Athens, "HYGEIA" Kiffisias Ave. and 4 Erythrou Stavrou Str., 151 23 Marousi, Athens, Hellas, Greece*

[g]*Department of Pathology, Medical School, University of Ioannina, Hellas, Greece*

[h]*Breast Center, Diagnostic and Therapeutic Center of Athens, "HYGEIA" Kiffisias Ave. and 4 Erythrou Stavrou Str., 151 23 Marousi, Athens, Hellas, Greece*

ABSTRACT: **In this study, we examined several molecular markers in prostate and breast cancer patients and in normal individuals. The markers tested were: variations in the quantity of plasma DNA, glutathione-*S*-transferase P1 gene (GSTP1), Ras association domain family 1A (RASSF1A), and ataxia telangiectasia mutated (ATM) methylation status in plasma, carcinoembryonic antigen (CEA) and prostate-specific membrane antigen (PSMA) mRNA in peripheral blood mononuclear**

Address for correspondence: G. Nasioulas, Ph.D., Molecular Biology Research Center HYGEIA Antonis Papayiannis, Kifissias Ave. and Erythrou Stavrou 4 Str., 15123 Maroussi, Athens, Greece. Voice: +30-210-686-7932; fax: +30-210-685-2050.
e-mail: g.nasioul@hygeia.gr

Ann. N.Y. Acad. Sci. 1075: 235–243 (2006). © 2006 New York Academy of Sciences.
doi: 10.1196/annals.1368.032

cells (PBMC) and plasma samples from prostate cancer patients. DNA quantification in plasma was performed using real-time PCR (RT-PCR). We assessed the methylation status of GSTP1 in plasma DNA using methylation-specific PCR (MSP) assay, while the methylation status of RASSF1A and ATM genes was examined by the MethyLight technology. RT-PCR analysis was used for the detection of mRNA, PSMA, and CEA. In 58.3% of newly diagnosed prostate cancer patients and 26.7% of prostate cancer patients under therapy, plasma DNA levels were increased. Additionally, 48.5% of breast cancer patients showed plasma DNA levels above the cutoff limit. GSTP1 Promotor hypermethylation was detectable in 75% of plasma samples obtained from patients with newly diagnosed prostate cancer and in 36.8% of patients under therapy, whereas 26% and 14% of the breast cancer patients tested were positive for RASSF1A and ATM methylation, respectively. The combination of DNA load and promotor methylation status identified 88% of prostate cancer patients and 54% of breast cancer patients. This study shows that free-circulating DNA can be detected in cancer patients compared with disease-free individuals, and suggests a new, noninvasive approach for early detection of cancer.

KEYWORDS: methylation; DNA quantification; methylation-specific PCR; methyLight

INTRODUCTION

Improvements in molecular approaches have permitted the identification of tumor-derived nucleic acids (DNA/RNA) in the nucleated blood cells, as well as in the blood plasma of cancer patients. In clinical oncology, the use of blood samples for the detection of sporadic cancer is considered as a possible objective for the near future.

To date, several studies have reported that cancer patients exhibit increased amounts of DNA in plasma or serum, and molecular studies have provided evidence that the origin of this DNA is the tumor tissue.[1–5] In addition, the investigation of circulating tumor cells in minimal residual disease demonstrated the presence of free tumor mRNA in the plasma of patients with cancer.[6–9] Following these observations, it has been proposed that circulating tumor DNA and RNA can be used as a diagnostic or prognostic marker for cancer.

In the present study, we measured the quantity of DNA in plasma of primary breast cancer and prostate cancer patients, as well as in healthy controls. We also examined the promotor methylation status of the GSTP1 gene for prostate cancer, as well as RASSF1A and ATM genes for breast cancer. Furthermore, the presence of mRNA for CEA and PSMA in PBMC and plasma samples of prostate cancer patients was also assessed.

MATERIALS AND METHODS

Samples Collection and Nucleic Acids Isolation

Blood samples were collected from patients with breast and prostate cancer before therapy, as well as from prostate cancer patients undergoing therapy, right after the completion of the treatment. Healthy volunteers were also included in the study. Plasma was separated from the cellular fraction by centrifugation twice at 2000 rpm for 10 min at 4°C. DNA was isolated using the QIAamp DNA Blood Mini Kit (QIAGEN, Valencia, CA).

DNA Quantification

Quantification of the circulating plasma DNA was performed by real-time quantitative PCR analysis, using a Chromo4™ continuous fluorescence detection system (MJ Research Inc., Waltham, MA). Purified DNA was quantified by a real-time quantitative PCR for the β-globin gene as described previously.[10] For each reaction, 6 μL of the extracted plasma DNA was used and analyzed in triplicate.

Methylation-Specific PCR (MSP)

DNA methylation patterns in the CpG island of the GSTP1, RASSF1, and ATM genes were determined by chemical modification of the unmethylated and not the methylated cytosines to uracil. GSTP1 primers were fluorescently labeled and were specific for the methylated and the unmethylated GSTP1 promotor. Fluorescent MSP products were separated electrophoretically on a 2% agarose gel and analyzed by laser fluorescence using an automated gene sequencer (ABI Prism™ 310, AME Bioscience Ltd. Toroed, Norway) and the GeneScan Analysis program.[11]

RASSF1A and ATM methylation was analyzed by the MethyLight, a fluorescence-based, real-time PCR (RT-PCR) assay, as described previously.[4,12] Briefly, two sets of primers and probes, designed specifically for bisulfite-converted DNA, were used: a methylated set for RASSF1A or ATM and a reference set, β-actin (ACTB), to normalize for input DNA. Specificity of the reactions for methylated DNA was confirmed using site-specific DNA methyl transferase (SssI New England Biolabs, Beverly, MA). The percentage of fully methylated molecules at a specific locus was calculated by dividing the GENE:ACTB ratio of a sample by the GENE:ACTB ratio of SssI-treated DNA and multiplying by 100. For each MethyLight reaction 10 μL of bisulfite-treated genomic DNA was used. A gene was considered methylated if the percentage of fully methylated reference value was >0.

Reverse Transcription PCR

CEA and PSMA mRNA were amplified using reverse transcription PCR as described previously.[13,14]

Statistical Analysis

To assess whether circulating DNA might discriminate between prostate cancer patients versus healthy individuals, we computed sensitivity and specificity estimates for different DNA thresholds, and the area under the curve for receiver operates curves (ROC) according to Hanley and McNeil.[15] Statistical analysis was performed with the software MedCalc 7.2.0.2 (MedCalc´ Software, Mariakerke, Belgium). The probability (*P* values) was detected by a nonpaired two-tail type 2 Student's *t*-test. We also used the Spearman's rank correlation coefficients (r_s). $P < 0.05$ was considered statistically significant.

RESULTS

DNA Quantification in Plasma of Prostate Cancer Patients

DNA levels were measured by RT-PCR in 12 patients with newly diagnosed prostate cancer and in 15 prostate cancer patients subjected to radiotherapy or/and chemotherapy and/or radical prostatectomy. Thirteen healthy controls were also examined. Mean plasma DNA was 20.2 ± 18.7 ng/mL for the newly diagnosed prostate cancer patients. This was significantly higher than in the controls ($P = 0.0097$). In patients subjected to therapy the mean value of plasma DNA concentration was 9.7 ± 14.4 ng/mL and was not significantly different to that of the control group (4.7 ± 4 ng/mL) ($P = 0.2494$). A cut-off level of 10 ng/mL was used to distinguish between healthy donors and patients. Using this cutoff level 7 out of 12 newly diagnosed prostate cancer patients had high plasma DNA values (sensitivity 58.3%), whereas 4 out of 15 prostate cancer patients under therapy had high plasma DNA (sensitivity 26.7%) (TABLE 1).[16] In addition, only 1 of the 13 healthy controls showed levels above cutoff (specificity 92.3%). The area under the ROC curve was 0.708 (95% confidence interval, 0.494–0.871) for the newly diagnosed prostate cancer patients and 0.528 (95% confidence interval, 0.332–0.718) for those under therapy.

In newly diagnosed prostate cancer patients, DNA concentration in plasma did not show any statistical association with Gleason score ($rs = 0.283$, $P = 0.3476$). Similarly, in prostate cancer patients under therapy, Gleason score did not associate with plasma DNA ($rs = 0.141$, $P = 0.5489$).

TABLE 1. Molecular markers in blood samples of prostate cancer patients and normal volunteers

Patients	Increased RT-PCR	GST‡	PSMA* plasma	PSMA* PBMC	CEA* plasma	CEA* PBMC
P.Ca	7/12	9/12	4/12	2/12	1/12	5/12
P.Ca under treatment	4/15	7/19	3/19	1/19	0/19	3/19
Normal	1/13	0/9	0/9	0/9	0/9	0/9

Abbreviations: P.Ca = prostate cancer.
Quantification of the circulating plasma DNA was performed by RT-PCR for the β-globin gene.
‡GSTP1 methylation was detected by MSP in plasma samples.
*The presence of PSMA and CEA mRNA was studied by RT-PCR in PBMC and plasma samples.

TABLE 2. Comparison of plasma DNA of breast cancer patients and normal volunteers

Patients	Increased DNA in plasma	RASSF1A‡	ATM‡
B.Ca	33/68	13/50	7/50
Normal	51/54	0/14	0/9

Abbreviations: B.Ca = Breast cancer.
Quantification of the circulating plasma DNA was performed by RT-PCR for the β-globin gene.
‡RASSF1A and ATM methylation was detected by MethyLight in plasma samples.

DNA Quantification in Plasma of Breast Cancer Patients

Mean plasma DNA in breast cancer patients was 45.8 ± 8.5 ng/mL, which was higher compared to that of the control group (11.4 ± 9.5 ng/mL). A cutoff level of 10.8 ng/mL was used to distinguish between healthy donors and patients. Using this cutoff level 33 out of 68 newly diagnosed breast cancer patients had high plasma DNA values (sensitivity 48.5%), whereas only 3 of the 54 healthy controls showed levels above cutoff (specificity 94.4%) (TABLE 2). The value of the AUC–ROC was 0.827 (95% confidence interval 0.748–0.889), indicating a good discrimination power of the test. The mean value of plasma DNA in patients with lymph node metastasis was higher (69.9 ± 285.8 ng/mL) compared to patients without metastasis (11.3 ± 7 ng/mL).

GSTP1 Methylation Status in Plasma

GSTP1 promotor hypermethylation was detectable in 9 of 12 (75%) of plasma samples obtained from patients with newly diagnosed prostate cancer and in 7 of 19 (36.8%) patients under therapy whereas none of the 9 normal volunteers examined exhibited GSTP1 hypermethylation in plasma (TABLE 1).[16]

RASSF1A and ATM Methylation Status in Plasma

RASSF1A promotor hypermethylation was detectable in 13 out of 50 (26%) plasma DNA samples obtained from patients with breast cancer, while ATM hypermethylation was detected in 7 out of 50 (14%) patients tested. Furthermore, 36% of the breast cancer patients examined showed Promotor hypermethylation of at least one of the two genes tested. None of the 14 normal volunteers examined exhibited hypermethylation of either of these genes in plasma (TABLE 2).

Detection of PSMA mRNA in Plasma and PBMC

Presence of mRNA for PSMA in plasma and blood samples was assessed. Four (33.3%) of the 12 patients with newly diagnosed disease were positive mRNA for PSMA in plasma, whereas only 2 (16.7%) of them were positive for PSMA mRNA in PBMC (TABLE 1).[16] Among the prostate cancer patients under therapy, three (15.8%) were positive for PSMA mRNA in plasma, while only one (5.3%) was positive in PBMC. In addition, none of the nine normal volunteers had detectable amounts of PSMA mRNA in plasma or blood cells.

Detection of mRNA for CEA in Plasma and PBMC

The analysis of mRNA for CEA in plasma and blood samples in the group of patients with untreated prostate cancer revealed that 1 of the 12 patients was positive for CEA mRNA in plasma and 5 (41.67%) patients were positive for CEA mRNA in PBMC (TABLE 1).[16]

Only 3 (15.8%) patients with prostate cancer under therapy were positive for CEA mRNA in PBMC, while it was not possible to detect extracellular-free CEA mRNA in this group. In addition, none of the normal volunteers presented detectable amounts of CEA mRNA in plasma or blood cells.

DISCUSSION

DNA Quantification in Plasma

The present study provides evidence of an increased amount of plasma DNA with features of tumor DNA in prostate and breast cancer patients. This result was confirmed by the RT-PCR for the quantification of plasma DNA as well as by the detection of hypermethylation in GSTP1, RASSF1A, and ATM genes in plasma DNA.

We found that approximately 60% of prostate cancer patients and 48% of breast cancer patients had elevated levels of cell-free DNA. The measurement of cell-free DNA probably can complement other tests to indicate the presence of cancer. On the other hand the mean amount of cell-free DNA measured in prostate cancer patients after the completion of the therapy was close to those observed in the control group. This is in accordance with previous observations suggesting that plasma DNA can undergo quantitative changes in cancer patients after radiation therapy.[17] However, its utility in monitoring chemotherapy and radiotherapy has yet to be fully defined.

Several studies have reported that cancer patients exhibit increased amounts of DNA in plasma or serum, and molecular studies, such as promotor hypermethylation have provided evidence of the tumor origin of this DNA.[1–5]

Promotor Methylation

GSTP1

GSTP1 Promotor hypermethylation is the most frequent prostate cancer-specific event.[13] Our results indicate that GSTP1 hypermethylation can be detected in the majority of newly diagnosed prostate cancer patients (75%) but the percentage of detection is reduced (36.84%) in patients under therapy. All patients with increased amounts of DNA in plasma also presented aberrant GSTP1 Promotor methylation in plasma. The association of GSTP1 methylation status with the response to treatment and tumor-free survival is yet to be established. The majority of patients with newly diagnosed prostate cancer exhibited either elevated DNA levels or aberrant GSTP1 methylation in plasma, while 88% of them exhibited both DNA alterations.

RASSF1A and ATM

Promotor hypermethylation of RASSF1A or ATM genes was detectable in 36% (18 out of 50) of the breast cancer patients examined. Additionally, the majority of patients with breast cancer exhibited either elevated DNA levels or aberrant Promotor methylation in plasma, while 54% of them (27 out of 50) exhibited both DNA alterations. The present report is the first description of ATM gene hypermethylation in plasma DNA.

Several studies have reported that hypermethylation of the CpG island Promotor of RASSF1A may play an important role in breast cancer pathogenesis and hypermethylation of this gene was detected in plasma and serum samples of breast cancer patients.[5,6] Dulaimi *et al.* detected RASSF1A hypermethylation in 65% of breast tumor DNA and in 56% of the serum DNA tested,[5] while Muller *et al.* detected RASSF1A hypermethylation in 23% of serum DNA extracted from patients with primary breast cancer and in 80% of those with recurrent breast cancer.[4] Quynh reported epigenetic silencing of ATM expression in 78% of locally advanced breast tumors as judged by methylation-specific PCR.[18]

Analysis of the methylation status of different genes seems to be specific, since it has been reported that a gene unmethylated in the tumor DNA is always found to be unmethylated in the matched serum DNA (100% specificity). Although the method is specific, its sensitivity seems to be reduced in the body fluids since only a percentage of patients with methylated genes in tumor DNA have the same aberrant Promotor methylation pattern in the paired serum DNA.[6] Additionally, all the genes known are hypermethylated in only a proportion of breast cancer patients. For these reasons, it is necessary to analyze a panel of genes in plasma or serum so as to increase the sensitivity of the method.

PSMA and CEA mRNA

Cell-free PSMA mRNA could be detected even in the absence of cell-derived mRNA. All patients positive for PSMA mRNA in PBMC were positive for plasma PSMA mRNA. On the contrary, CEA mRNA was detected in only 1 of the 12 patients with untreated tumors examined, while cell-derived CEA mRNA was detected in 41.67% of the cases.

Chen *et al.* showed that it was possible to amplify telomerase RNA template and telomerase reverse transcriptase protein mRNA in the serum of breast cancer patients.[7] Silva *et al.* also detected epithelial tumor RNA in plasma from colon cancer patients. In this study, CEA mRNA was detected in the 32% of plasma samples. Advanced stages and soluble CEA status were associated with the presence of CEA, CK19, or both RNAs in plasma.[19]

Further studies are needed to understand the correlation of these new molecular markers with cancer diagnosis, outcome of disease, and eventually treatment response.

In this study, several molecular markers were examined for their usefulness in the detection of breast and prostate cancer (TABLES 1 and 2). So far our results show that the most promising is the combination of DNA load and Promotor methylation in plasma, leading to the detection of 88% of the cases with untreated prostate cancer and 54% of the cases with untreated breast cancer.

REFERENCES

1. SOZZI, G. *et al.* 2003. Quantification of free circulating DNA as a diagnostic marker in lung cancer. J. Clin. Oncol. **21:** 3902–3908.
2. GAL, S. *et al.* 2004. Quantitation of circulating DNA in the serum of breast cancer patients by real-time PCR. Br. J. Cancer **90:** 1211–1215.

3. SILVA, J.M. *et al.* 1999. Presence of tumour DNA in plasma of breast cancer patients: clinicopathological correlations. Cancer Res. **59:** 3251–3256.
4. MULLER, H.M. *et al.* 2003. DNA methylation in serum of breast cancer patients: an independent prognostic marker. Cancer Res. **63:** 7641–7645.
5. DULAIMI, E. *et al.* 2004. Tumour suppressor gene Promotor hypermethylation in serum of breast cancer patients. Clin. Cancer Res. **10:** 6189–6193.
6. KOPRESKI, M.S. *et al.* 1999. Detection of tumour messenger RNA in the serum of patients with malignant melanoma. Clin. Cancer Res. **5:** 1961–1965.
7. CHEN, X. *et al.* 2000. Telomerase RNA as a detection marker in the serum of breast cancer patients. Clin. Cancer Res. **6:** 3823–3826.
8. SILVA, J.M. *et al.* 2002. Detection of epithelial tumour RNA in the plasma of colon cancer patients is associated with advanced stages and circulating tumour cells. Gut **50:** 530–534.
9. TSUI, N.B.Y. *et al.* 2002. Stability of endogenous and added RNA in blood specimens; serum; and plasma. Clin. Chem. **48:** 1647–1653.
10. LO, Y.M.D. *et al.* 1998. Quantitative analysis of fetal DNA in maternal plasma and serum. Implications for non-invasive prenatal diagnosis. Am. J. Hum. Genet. **62:** 768–775.
11. GOESSL, C. *et al.* 2001. DNA-based detection of prostate cancer in blood; urine; and ejaculates. Ann. N. Y. Acad. Sci. **945:** 51–58.
12. EADS, C.A. *et al.* 2000. MethyLight: a high-throughput assay to measure DNA methylation. Nucleic Acids Res. **28:** E32.
13. FUNAKI, N.O. *et al.* 1996. Identification of carcinoembryonic antigen mRNA in circulating peripheral blood of pancreatic carcinoma and gastric carcinoma patients. Life Sci. **59:** 2187–2199.
14. LINTULA, S. & U.H. STENMAN. 1997. The expression of prostate specific membrane antigen in peripheral blood leukocytes. J. Urol. **157:** 1969–1972.
15. HANLEY, J.A. & B.J. MCNEIL. 1982. The meaning and use of the area under a receiver operating characteristic (ROC) curve. Radiology **143:** 29–36.
16. PAPADOPOULOU, E. *et al.* 2004. Cell-free DNA and RNA in plasma as a new molecular marker for prostate cancer. Oncol. Res. **14:** 439–445.
17. LEON, S.A. *et al.* 1977. Free DNA in the serum of cancer patients and the effect on therapy. Cancer Res. **37:** 646–650.
18. QUYNH, N. 2004. The ATM gene is a target for epigenetic silencing in locally advanced breast cancer. Oncogene **16:** 9432–9437.
19. SILVA, J.M. *et al.* 2002. Detection of epithelial tumour RNA in the plasma of colon cancer patients is associated with advanced stages and circulating tumour cells. Gut **50:** 530–534.

Early and Specific Prediction of the Therapeutic Efficacy in Non–Small Cell Lung Cancer Patients by Nucleosomal DNA and Cytokeratin-19 Fragments

STEFAN HOLDENRIEDER,[a] PETRA STIEBER,[a] JOACHIM VON PAWEL,[b] HANNELORE RAITH,[c] DOROTHEA NAGEL,[a] KNUT FELDMANN,[c] AND DIETRICH SEIDEL[a]

[a]*Institute of Clinical Chemistry, University Hospital of Munich-Grosshadern, Munich, Germany*

[b]*Department of Oncology, Asklepios Hospital Gauting, Munich, Germany*

[c]*Institute of Clinical Chemistry, Asklepios Hospital Gauting, Munich, Germany*

ABSTRACT: Facing an era of promising new antitumor therapies, predictors of therapy response are needed for the individual management of treatment. In sera collected prospectively from 311 patients with advanced non-small cell lung cancer receiving first-line chemotherapy, changes in nucleosomal DNA fragments, cytokeratin-19 fragments (CYFRA 21–1), carcinoembryonic antigen (CEA), neuron-specific enolase (NSE), and progastrin-releasing peptide (ProGRP) were investigated and correlated with therapy response. In univariate analysis, high levels, slower and incomplete decline in nucleosomal DNA, CYFRA 21–1, and CEA predicted poor outcome. DNA concentrations at day 8 of the first therapeutic cycle and CYFRA 21–1 before start of the second cycle were identified as best predictive variables. In multivariate analysis, they predicted progression with a specificity of 100% in 29% of the cases earlier than imaging techniques. Thus, nucleosomal DNA and CYFRA 21–1 specifically identify a subgroup of patients with insufficient therapy response at the early treatment phase and showed to be valuable for disease management.

KEYWORDS: DNA; nucleosomes; cytokeratin-19 fragments; CYFRA 21–1; serum; plasma; prediction; chemotherapy; lung cancer

INTRODUCTION

Lung cancer accounts for most deaths caused by cancer in males and has an increasing prevalence in women worldwide.[1,2] Often it is detected only in

Address for correspondence: Dr. Stefan Holdenrieder, Institute of Clinical Chemistry, University Hospital of Munich-Grosshadern, Munich, Germany. Voice: 0049-89-7095 3231; fax: 0049-89-7095 6298.

e-mail: Stefan.Holdenrieder@.med.uni-muenchen.de

Ann. N.Y. Acad. Sci. 1075: 244–257 (2006). © 2006 New York Academy of Sciences.
doi: 10.1196/annals.1368.033

advanced stages when therapeutic options are restricted to systemic chemo- and radiotherapy. These therapies often are associated with insufficient success, but there are efforts to improve the situation by the introduction of new drugs.[3,4] Progress in recent years is mirrored by the 2003 therapy guidelines of the American Society of Clinical Oncology (ASCO), which include recommendations for the first-, second-, and third-line chemotherapy while in 1997 only one option for first-line therapy was given.[5] Therefore, it is important to detect as early as possible whether patients will benefit from a specific therapy or not in order to save time and costs by changing early the treatment strategy and by avoiding unnecessary side effects. Because the consequences will be most striking in cases of insufficient response to therapy, the predication has to be highly specific and sensitive.

Macroscopic alterations of the tumor mass are often detected only after several cycles of chemotherapy by imaging techniques. Promising candidates for the early estimation of therapeutic efficacy are biochemical parameters in blood that reflect the biochemical response of the tumor during the initial treatment phase. In lung cancer, various oncological biomarkers are used for diagnosis, prognosis, and therapy monitoring. Carcinoembryonic antigen (CEA) and cytokeratin-19 fragments (CYFRA 21–1) are known to be sensitive, however, not specific markers for non-small cell lung cancer.[6,7] Neuron-specific enolase (NSE) and progastrin-releasing peptide (ProGRP) have high specificity for small-cell lung cancer.[6,8] In addition, circulating DNA, which is supposed to be present in serum and plasma mainly in conjunction with histones as nucleosomes,[9] showed high potential for diagnosis, prognosis, and therapy monitoring.[10–15]

Recently, it was shown that in lung cancer nucleosomal DNA and CYFRA 21–1 during the initial phase of chemotherapy are able to discriminate between responders and nonresponders.[16] On the basis of these promising results, we analyzed in an extended sample of lung cancer patients undergoing first-line therapy whether these parameters and other relevant oncological biomarkers could predict early the response to chemotherapy with high specificity. If a subgroup of patients who do not respond to the treatment can be identified early by these blood markers prior to established imaging techniques, they might be useful in clinical practice as an indicator for the early adjustment of therapy. Here, a model with the best predictive clinical and biochemical parameters was developed that was tested under various therapeutic conditions.

MATERIALS AND METHODS

Patients

Consecutive patients ($n = 311$) with inoperable non-small cell lung cancer (stages III and IV) under the care of the Asklepios Clinics Gauting were included in our study. All patients were investigated initially by whole body computed tomography, bone scan, and bronchoscopy. All patients received first-line chemotherapy regimens containing alternatively carboplatin, mitomycin c, and vinblastin (CMV) or mitomycin c and vinorelbin (MV) or gemcitabine and cisplatin (GC), which were given in three weekly cycles.

In all patients, staging investigations consisting of clinical examination, whole body computed tomography, and laboratory examinations were performed before start of the third cycle of chemotherapy. The response to therapy was classified according to the World Health Organization classifications as follows: "partial remission" as tumor reduction $\geq$50%, "progression" as tumor increase $\geq$25% or appearance of new tumor manifestations, and "no change" as tumor reduction <50% or increase <25%. Of the 311 patients, 126 patients had partial remission (40.5%), 92 showed progression (29.6%), and 93 had no change of disease (29.9%). Patients with no change of disease were followed up until staging before the fifth treatment cycle. Those who presented with partial remission or no change at that time (n = 52) were added to the responsive group (in total n = 178; 57.2%), whereas those with progression at that time (n = 19) joined the nonresponsive group (in total n = 111; 35.7%); 22 patients (7.1%) with no change terminated the therapy before cycle 5 or were lost to follow-up and could not be considered for the evaluation (TABLE 1).

Materials and Methods

Blood samples were collected prospectively before the first and second cycle of therapy for determination of the baseline values (BV1 and BV2), and during the first week of the first cycle, at days 1 (before start of the therapy), 3, 5, and 8.

The samples for nucleosomal DNA determination were centrifugated at 3000 g for 15 min and treated with 10 mM EDTA (pH 8) immediately after centrifugation. Samples were stored at −70°C and analyzed in batches. All samples from a single patient were analyzed in one batch. The details of the preanalytic handling of the samples are described in Holdenrieder *et al.*[17] Nucleosomal DNA fragments were determined by the Cell Death Detection-ELISA^{plus} (Roche Diagnostics, Mannheim, Germany) which was modified for its use in serum matrix as specified in Holdenrieder *et al.*[17] Nucleosomal DNA fragments were quantified using the calibration curve generated from known amounts of DNA.

The baseline values of the oncological biomarkers CEA, CYFRA 21–1, NSE (by Elecsys 2010; Roche Diagnostics), and ProGRP (by ELISA; [ALSI, Japan/IBL, Germany]) were determined before each therapeutic cycle (BV1 and BV2) at the day of sample collection. In addition, CEA and CYFRA 21–1 were determined more frequently during the first week of therapy.

TABLE 1. Characteristics of the patients investigated

	Median	Range
Age		
Years	63.0	25–86
	Number	Percentage
Gender		
Female	99	(31.8%)
Male	212	(68.2%)
Stage		
III A	13	(4.2%)
III B	100	(32.2%)
IV	198	(63.7%)
Performance score		
ECOG 1	110	(35.4%)
ECOG 2	160	(51.5%)
ECOG 3	38	(12.2%)
ECOG 4	3	(0.9%)
Histology		
Squamous cell CA	89	(28.6%)
Adeno cell CA	129	(41.5%)
Large cell CA	8	(2.6%)
Not classified NSCLC	85	(27.3%)
Mode of therapy		
CMV	119	(38.3%)
MV	37	(11.9%)
GC	105	(33.8%)
Others	50	(16.0%)
Therapy response before cycle 3		
PR	126	(40.5%)
NC	93	(29.9%)
PD	92	(29.6%)
Therapy response of NC patients before cycle 5		
PR and NC	52	(16.7%)
PD	19	(6.1%)
Lost to follow-up	22	(7.1%)

Statistics

For all parameters, the baseline values before the first and second cycle (BV1 and BV2), and the percent changes (BV1–2) were considered for statistical analysis. In addition, nucleosomal DNA, CYFRA 21–1, and CEA values at day 8 (d8) of the first therapeutic cycle, the changes in the values between days 1 and 8 (d1–8), and the area under the curve of the values from days 1 to 8 (AUC1–8/d) were evaluated for their predictive power. To calculate the AUC1–8/d, values on days 1 and 8 and at least one of the days 3 or 5 were required.

In the first instance, biochemical parameters were analyzed by Wilcoxon test on their power to discriminate between patients with remission and progression

TABLE 2. Summary of all markers and variables investigated on their discriminating power between patients with remission, no change, and progression of disease

Biochemical parameters		Units	Remission		No change		Progression		
			Median	Range	Median	Range	Median	Range	*P*-value
Nucleosomal DNA	BV1	ng/mL	196.8	9.2–3450	176.2	13.7–6471	256.3	13.7–1936	0.0222
	BV2	ng/mL	89.2	9.2–638.4	140.1	9.2–1439	164.7	11.4–1758	<0.0001
	BV1–2	DEC %	54.7	(−2700)–97.9	27.5	(−825)–98.7	31.5	(−1186)–96.0	0.0265
	d8	ng/mL	64.1	9.2–1135	121.3	22.9–1721	169.3	11.4–1588	<0.0001
	d1–8	DEC %	63.5	(−1650)–99.7	44.6	(−1571)–97.3	40.8	(−334)–94.8	0.0015
	AUC1–8/d	ng/mL	139.6	18.0–1058	205.3	34.3–1910	243.1	35.5–1066	0.0001
CYFRA 21–1	BV1	ng/mL	3.7	0.2–67.0	4.5	0.1–266.2	7.3	0.4–1018	0.0001
	BV2	ng/mL	1.9	0.1–23.0	3.5	0.4–73.5	6.2	0.4–1721	<0.0001
	BV1–2	DEC %	47.1	(−570)–93.0	22.7	(−300)–91.2	22.0	(−13779)–79.3	<0.0001
	d8	ng/mL	4.3	1.1–79.71	6.4	2.0–87.6	10.2	0.7–1114	0.0019
	d1–8	DEC %	2.4	(−446)–83.0	−16.2	(−317)–50.8	−5.5	(−336)–78.9	0.3812
	AUC1–8/d	ng/mL	4.7	0.8–51.9	5.6	1.7–68.7	7.9	0.7–632.0	0.0244
CEA	BV1	ng/mL	6.4	0.2–438.2	5.7	0.5–1437	9.0	0.2–77695	0.4001
	BV2	ng/mL	6.4	0.9–353.6	5.8	0.5–1426	12.5	0.9–1171	0.0665
	BV1–2	DEC %	10.6	(−1050)–73.3	−6.7	(−920)–58.7	−8.9	(−3054)–74.0	<0.0001
	d8	ng/mL	3.9	0.7–52.4	6.7	1.9–220.2	6.1	0.7–174.8	0.1129
	d1–8	DEC %	12.8	(−15,4)–58.2	−5.7	(−77.2)–44.1	−1.1	(−53,5)–38.1	0.0228
	AUC1–8/d	ng/mL	3.6	0.7–42.5	5.5	1.8–179.1	4.0	0.7–121.2	0.4177
NSE	BV1	ng/mL	14.8	5.4–355.0	14.2	3.6–63.0	15.6	8.1–275.3	0.1592
	BV2	ng/mL	12.5	7.0–35.6	11.6	4.2–38.0	16.2	7.8–52.9	0.0118
	BV1–2	DEC %	10.6	(−198)–91.8	7.4	(−144)–90.1	11.5	(−154)–62.8	0.9187

(*continued*)

TABLE 2. (*Continued*)

Biochemical parameters		Units	Remission		No change		Progression		*P*-value
			Median	Range	Median	Range	Median	Range	
ProGRP	BV1	pg/mL	15	2.0–4292	15	3.0–66.0	13	2.0–822	0.2560
	BV2	pg/mL	14	3.0–89.0	15	3.0–789	17	3.0–174	0.4740
	BV1–2	DEC %	13.3	(−1033)–99.6	14.2	(−8667)–93.3	−8.3	(−650)–88.0	0.2851
Clinical parameters									
Age			62	39–78	65	39–86	63	25–83	0.3352
			%	N	%	N	%	N	
Gender	Male		63.5	80	72.0	67	70.6	65	0.2686
	Female		36.5	46	28.0	26	29.4	27	
Performance status	**ECOG 1**		50.8	64	33.3	31	16.3	15	1 vs. 2: **0.0004**
	ECOG 2		46.0	58	61.3	57	48.9	45	2 vs. 3: **<0.0001**
	ECOG 3		3.2	4	5.4	5	31.5	29	3 vs. 4: 0.5224
	ECOG 4		0	0	0	0	3.3	3	
Stage	**M0**		46.0	58	38.7	36	20.6	19	**0.0001**
	M1		54.0	68	61.3	57	79.4	73	
Histology	**SC**		38.1	48	26.9	25	17.4	16	SC vs. (AC + NCC) **0.0009**
	AC		37.3	47	46.2	43	42.4	39	
	NCC		24.6	31	26.9	25	40.2	37	
Therapy	**GC**		52.8	56	35.2	25	23.4	18	GC vs. (CMV + MV) **<0.0001**
	CMV		38.7	41	50.7	36	54.5	42	
	MV		8.5	9	14.1	10	22.1	17	

Parameters which discriminated significantly ($p < 0.05$) according response to therapy were indicated by bold letters. BV = baseline values; d = day; AUCI-8/d = area under the curve of days 1 to 8; DEC = decrease (negative values indicating increases).

of diseases. Clinical variables that were available in defined categories were tested by the chi-square test. In order to identify the best predictive markers, cutoffs for each variable were defined at the 90% specificity for patients with remission, and sensitivities and positive predictive values for having progression of disease were calculated. Among all parameters, those with the best profile of sensitivity and positive predictive value were included in a multivariate analysis. Mantel–Haenszel statistics was used to test whether the predictive power of the markers was independent of relevant clinical parameters. Within the group of patients with clinically good performance status (ECOG1 + 2), the additive effect of the best predictive markers was shown by receiver operating characteristic (ROC) curves. A P value of $P < 0.05$ was considered statistically significant. All statistical analyses were performed with software of SAS (version 8.2, SAS Institute Inc., Cary, NC).

RESULTS

In lung cancer patients undergoing first-line chemotherapy, those with remission could be distinguished from those with progression by the pretherapeutic concentration of nucleosomal DNA fragments ($P = 0.022$), the baseline value before cycle 2 ($P < 0.0001$), values during the first therapeutic week at day 8 ($P < 0.0001$), and the area under the curve of the values from days 1 to 8 ($P = 0.0001$). In addition, the time course of nucleosomal DNA fragments showed a faster decline in patients with remission than in those with progressive disease, between the baseline values of cycles 1 and 2 ($P = 0.027$), and also during the first cycle between the pretherapeutic BV1 and the value at day 8 ($P = 0.002$). Similarly, the following CYFRA 21–1 values discriminated clearly between the groups: the pretherapeutic BV1 ($P = 0.0001$), the baseline value before cycle 2 ($P < 0.0001$), the value at day 8 ($P = 0.002$), the area under the curve of the values from days 1 to 8 ($P = 0.024$), and the kinetics from cycle 1 to 2 ($P < 0.0001$). However, changes in CYFRA 21–1 from days 1 to 8 during the first week of therapy were not significant ($P = 0.381$). Concerning other oncological biomarkers, CEA discriminated patients according to therapy response for the courses from cycle 1 to 2 ($P < 0.0001$) and from day 1 to 8 during the first cycle ($P = 0.023$); NSE for the baseline value before cycle 2 ($P = 0.012$). ProGRP was not capable to distinguish between the groups. Regarding clinical factors, stage (M0 vs. M1: $P = 0.0001$), performance status (ECOG 1 vs. 2: $P = 0.0004$; 2 vs. 3: <0.0001), histology (SC vs. AC + NCC: $P = 0.0009$), and mode of therapy (GC vs. CMV + MV: <0.0001) showed predictive potential, however, not age ($P = 0.3352$) or gender ($P = 0.2686$) (TABLE 2).

In order to identify the best predictive markers for insufficient response to therapy, cutoffs for all parameters were calculated at the 90% specificity for patients with remission, and profiles of sensitivity and positive predictive

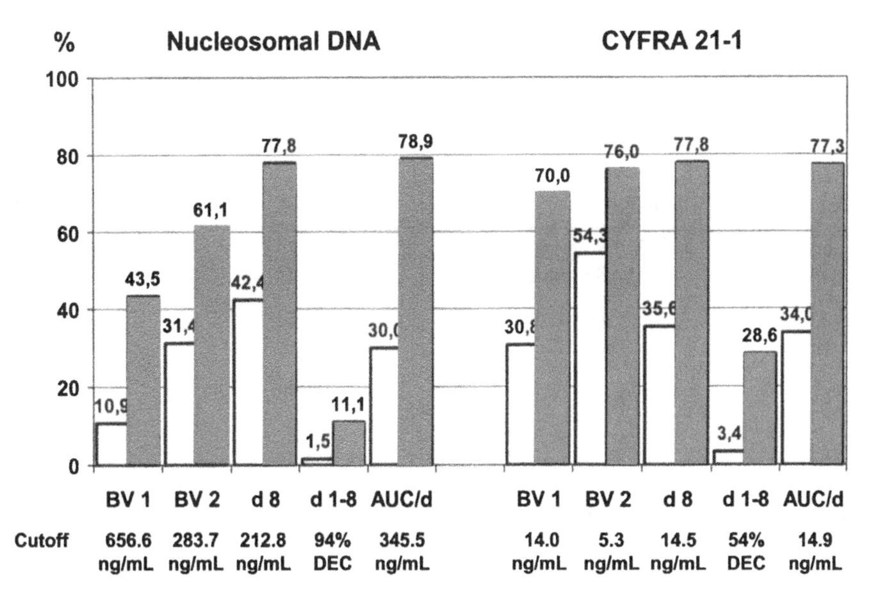

FIGURE 1. Profiles of sensitivity □ and positive predictive values ■ (PPV) for prediction of insufficient response to therapy in patients with newly diagnosed NSCLC undergoing first-line chemotherapy for nucleosomal DNA and CYFRA 21–1 using cutoffs defined at the 90% specificity for patients with remission. Nucleosomal DNA at day 8 and the BV2 of CYFRA 21–1 exhibited the best profiles.

value (PPV) were established. Among all univariately predictive parameters, nucleosomal DNA at day 8 (sensitivity: 42.4%; confidence interval 30.6–55.2%; PPV: 77.8%; confidence interval 60.4–89.3%) and the BV2 of CYFRA 21–1 (sensitivity: 54.3%; confidence interval 42.0–66.1%; PPV: 76.0%; confidence interval 61.5–86.5%) exhibited the best profiles (FIG. 1).

Nucleosomal DNA fragments were found to be independent predictive markers with respect to CYFRA 21–1 ($P = 0.0012$), stage ($P = 0.0003$), and performance status ($P < 0.0001$) using Mantel–Haenszel statistics; and also CYFRA 21–1 predicted response to therapy independently from nucleosomal DNA ($P < 0.0001$), stage ($P < 0.0001$), and performance status ($P < 0.0001$) (TABLE 3).

In a multivariate model including nucleosomal DNA fragments and CYFRA 21–1, both biochemical parameters showed additive information for prediction of insufficient therapy response, particularly in a subgroup of patients with good clinical status (ECOG 1 + 2). As most of the patients with poorer performance status (ECOG 3 + 4) suffered from rapid progression of disease, the application of additional biochemical markers appeared to be superfluous in this setting. However, if, in patients with initially good clinical status, nucleosomal DNA fragments, and CYFRA 21–1 were combined, 100% specificity for prediction of progression was achieved with a sensitivity of 29% (FIG. 2). In a further subgroup of patients with pretherapeutic CYFRA

TABLE 3. Mantel–Haenszel statistics showing independency of nucleosomal DNA values of day 8 (d8) and baseline value 2 of CYFRA 21–1 (BV2) on each other, on stage, and performance status for the prediction of progressive disease

		Nucleosomal DNA (d8)		CYFRA 21–1 (BV2)	
Parameters	Subgroups	Relative risk	*P* value	Relative risk	*P* value
Nucleosomal DNA (d8)	<212.8 ng/mL			3.81	<0.0001
	≥212.8 ng/mL			2.37	
CYFRA 21–1 (BV2)	<5.3 ng/mL	2.51	0.0012		
	≥5.3 ng/mL	1.53			
Stage	UICC III	5.48	0.0003	2.04	<0.0001
	UICC IV	1.84		3.88	
Performance status	ECOG 1	3.00	<0.0001	6.33	<0.0001
	ECOG 2	3.08		3.35	
	ECOG 3	1.10		1.25	

21–1 values >3.3 ng/mL, the sensitivity for prediction of progression could be enhanced to 39.1% at 100% specificity. Although the various treatment protocols showed differences concerning therapeutic success, the power of prediction by nucleosomal DNA and CYFRA 21–1 was comparable in the groups receiving gemcitabine + cisplatin (GC) and carboplatin + mitomycin + vinblastin (CMV) with a sensitivity of 33.3% and 31.1%, respectively, at 100% specificity.

If patients with no change before cycle 3 were added to the "responsive" patient group if remission or stable disease was achieved before cycle 5 and added to the "non-responsive" group if progression occurred before cycle 5, the sensitivity was still 27% at a slightly lower specificity of 98%. The drop in specificity was on account of two patients with high concentrations of nucleosomal DNA and CYFRA 21–1 but nominal "no change" before cycle 5. However, both patients suffered from progressive disease 2 and 4 weeks, respectively, after this staging investigation was done.

DISCUSSION

Along with the development of new therapeutics in oncology, there is a growing need for diagnostic tools for estimating prognosis, treatment monitoring, and early prediction of response to therapy in order to optimize disease management on an individual basis. In patients with non-small cell lung cancer, a panel of clinical and biochemical parameters showed prognostic relevance.[6,7,10,14,18–20] Among them, CYFRA 21–1 was shown to have the strongest evidence as prognostic marker in the early, operable stages as well

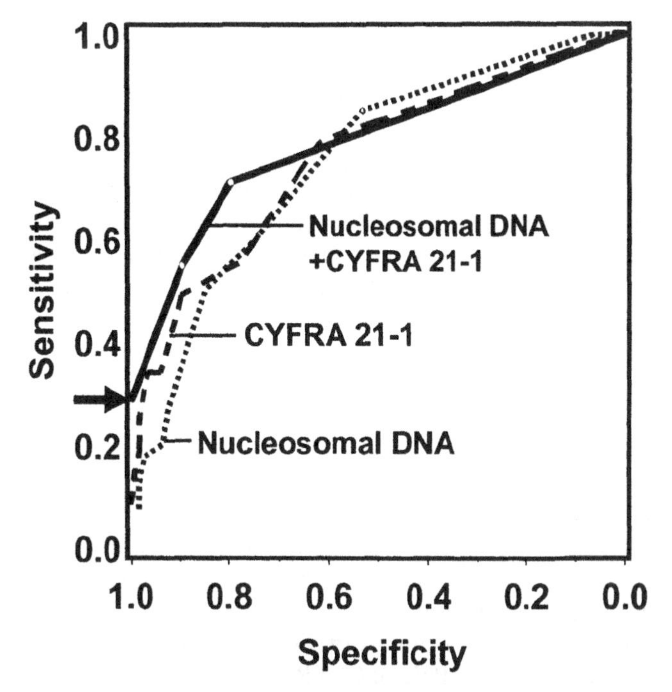

FIGURE 2. Prediction of therapy response by nucleosomal DNA and CYFRA 21–1 in patients with ECOG 1 + 2. ROC curves indicate the predictive power of nucleosomal DNA, CYFRA 21–1, and both markers revealing a clear additive effect. The combination curve meets the 100% specificity axis at a sensitivity level of 29%.

as in the late stages of non-small cell lung cancer.[19] In addition, independent prognostic value of lactate dehydrogenase, albumin, calcium, NSE, CEA, and DNA have been observed in various studies.[7,14,18–20] For monitoring systemic therapy and the early detection of recurrent disease in non-small cell lung cancer patients, CYFRA 21–1 and circulating DNA were frequently reported to be useful.[6,12–15,21–23] However, little is known about the relevance of biochemical markers for the *early* prediction of therapy response prior to imaging techniques.

Recently, we demonstrated that circulating nucleosomal DNA fragments and CYFRA 21–1 at the initial phase of therapy were able to distinguish between responders and nonresponders to chemotherapy.[16] On the basis of these results, we investigated in the present study the predictive power of nucleosomal DNA and CYFRA 21–1, and other relevant lung cancer biomarkers CEA, NSE, and ProGRP in a large sample of patients. As early prediction of insufficient therapy response would result in an early change of disease management in clinical practice, the predication had to be highly specific and sensitive. Therefore, we analyzed the predictive power of these parameters for poor therapy efficacy, especially during the initial treatment phase.

In univariate analysis, we confirmed our previous results that nucleosomal DNA and CYFRA 21–1 were able to distinguish patients according to their response to therapy: Patients with remission had lower levels for the pretherapeutic value and the baseline value before cycle 2, and greater decrease of the baseline values in the kinetic investigations than patients with progression of disease. Already during the first treatment cycle, nucleosomal DNA levels discriminated highly significantly between both groups. Patients with remission exhibited lower values at day 8 and smaller (AUC1–8/d) than patients with progressive disease. This might be on account of less aggressive tumors in responsive patients being related to lower rates of cellular turnover and cell death and more effective elimination of cell death products from circulation. Similar differences between patient groups were also observed concerning early CYFRA 21–1 courses, however, not as pronounced as for nucleosomal DNA during the first week. As the half-life of CEA in serum is considerably longer than the half-life of nucleosomal DNA and CYFRA 21–1, CEA did not show characteristic changes during the initial phase of therapy and discriminated according to response to chemotherapy only before start of the second cycle. As expected, small-cell cancer markers NSE and ProGRP were associated with weak or no predictive potential.

The best predictive markers for insufficient response to therapy were nucleosomal DNA at day 8 and the BV2 of CYFRA 21–1. This finding underlines the importance of the early changes in nonspecific circulating nucleosomal DNA fragments during therapy and the strong impact of somewhat later alterations in more specific CYFRA 21–1. Both variables were shown to provide additive information for therapy prediction independently from each other and from the most relevant clinical factors—stage and performance score. As shown by ROC curves, the combination of both markers reached 100% specificity for prediction of insufficient treatment response at a sensitivity level of 29% in the group of patients with pretherapeutic good clinical status (ECOG 1 + 2). This means that in about one-third of these patients, therapeutic response was predicted with a specificity of 100% after the first application of chemotherapy. In consequence, this information could have enabled an early change of the therapeutic regimen to avoid unnecessary side effects and enable more effective treatments in time. These results are not restricted to one specific chemotherapy but showed similar values in various protocols.

The correct classification of patients with stable disease in advanced non-small cell lung cancer is controversial. On the one hand it is important to stop the progression of the often already metastasized tumor disease. On the other hand, one would wish to achieve at least a partial remission during first-line therapy to prolong the survival of the patients. To take both aspects into account in our setting, patients with "no change" at cycle 3 were followed until staging before cycle 5. Those who showed no progression at that time were added to the responsive group while those with tumor progression joined the nonresponsive group. Following this procedure, the sensitivity for early prediction of progressive disease by combination of DNA and CYFRA 21–1 was still 27%; however, the specificity dropped to 98% because two patients with high levels of DNA and CYFRA 21–1 were in the "no change" group before cycle 5 but later found to show progressive disease, respectively.

It is worth noting that the levels of DNA and CYFRA 21–1 after the first application of chemotherapy were the strongest predictive markers. Many chemotherapeutic drugs are known to induce apoptotic cell death, which may result in a considerable release of these intracellular markers.[24,25] Thus, the increase in serum levels of DNA and CYFRA 21–1 reflect the spontaneous and the induced cell death, which might be the highest in less differentiated, very aggressive, and well-perfused tumors. The slower and incomplete decrease in nonresponsive patients could be influenced in addition by impaired elimination mechanisms and/or newly proliferating cell clones. An additional mechanism for DNA release, the active secretion by lymphocytes is still debated.[26] During apoptosis, most of cellular DNA is cleaved by endonucleases into mono- and oligonucleosomal fragments.[24] In this form, serum and plasma DNA seems to be better conserved from further digestion.[27] Methods that measure all cell-free DNA and those quantifying nucleosomal DNA showed a quite good correlation,[28] confirming earlier observations that most of the circulating DNA is bound to histones in nucleosomal complexes.[9] At least a substantial part of circulating DNA is of cancerous origin. Qualitative changes, such as specific mutations, microsatellite alterations, loss of heterozygosity, and epigenetic modifications were found in cell-free DNA as well as in tumor DNA.[15,29–31] In addition, tumor cells were found to be more susceptible to moderate radiation doses than normal epithelial cells resulting in higher release of nucleosomal DNA *in vitro*.[32] Thus, quantitative and qualitative aspects of circulating DNA have been shown to be helpful for diagnosis, prognosis, and therapy monitoring of various cancers.[10–15,24,29–31,33–36]

CYFRA 21–1 and CEA are oncological biomarkers that are more tumor specific and are currently used in the diagnosis and monitoring of non-small cell lung cancer.[6–8,21–24] However, neither their *early* predictive value for therapy response nor their additive effect to circulating DNA in diagnosis, prognosis, and therapy prediction have yet been shown.

The present study is to our knowledge the most comprehensive one in a welldefined patient population showing the high relevance of the combination of circulating nucleosomal DNA fragments and CYFRA 21–1 in predicting response to chemotherapy during the initial treatment phase. If these findings are confirmed by other prospective trials, the defined use of these parameters could contribute to improve the management of cancer patients.

ACKNOWLEDGMENTS

We appreciate the valuable assistance of E. Dankelmann, B. Faderl, B. Kordes, M.A. Lercher, A. Markus, B. Poell, M. Siakavara, H. Wagner, and the laboratory assistants of the Hospital Gauting in the logistic process of the study.

We are also grateful to A. Mitlewski from the University Hospital Grosshadern for the measurement of nucleosomes, and P. Bialk, H. Bodenmueller, B. Eckert, and P. Heiss from Roche Diagnostics, Germany, for providing the nucleosome assays.

REFERENCES

1. GREENLEE, R.T. *et al.* 2000. Cancer Statistics 2000. CA Cancer J. Clin. **50:** 7–33.
2. PARKIN, D.M. 2004. International variation. Oncogene **23:** 6329–6340.
3. GRIDELLI, C. *et al.* 2003. Treatment of non-small-cell lung cancer: state of the art and development of new biologic agents. Oncogene **22:** 6629–6638.
4. SPIRA, A. & D.S. ETTINGER. 2004. Multidisciplinary management of lung cancer. N. Engl. J. Med. **350:** 379–392.
5. PFISTER, D.G. *et al.* 2004. American Society of Clinical Oncology treatment of unresectable non-small-cell lung cancer guideline: update 2003. J. Clin. Oncol. **22:** 330–353.
6. SCHALHORN, A. *et al.* 2001. Tumor markers in lung cancer. J. Lab. Med. **25:** 353–361.
7. MOLINA, R. *et al.* 2003. Tumor markers (CEA, CA 125, CYFRA 21–1, SCC and NSE) in patients with non-small cell lung cancer as an aid in histological diagnosis and prognosis. Comparison with the main clinical and pathological prognostic factors. Tumour Biol. **24:** 209–218.
8. SCHNEIDER, J. *et al.* 2003. Pro-gastrin-releasing peptide (ProGRP), neuron specific enolase (NSE), carcinoembryonic antigen (CEA) and cytokeratin 19-fragments (CYFRA 21–1) in patients with lung cancer in comparison to other lung diseases. Anticancer Res. **23:** 885–893.
9. RUMORE, P. *et al.* 1992. Hemodialysis as a model for studying endogenous plasma DNA: oligonucleosome-like structure and clearance. Clin. Exp. Immunol. **90:** 56–62.
10. FOURNIE, G.J. *et al.* 1995. Plasma DNA as a marker of cancerous cell death. Investigations in patients suffering from lung cancer and in nude mice bearing human tumours. Cancer Lett. **91:** 221–227.
11. HOLDENRIEDER, S. *et al.* 2001. Nucleosomes in serum of patients with benign and malignant diseases. Int. J. Cancer **95:** 114–120.
12. SOZZI, G. *et al.* 2001. Analysis of circulating tumor DNA in plasma at diagnosis and during follow-up of lung cancer patients. Cancer Res. **61:** 4675–4678.
13. HOLDENRIEDER, S. & P. STIEBER. 2004. Therapy control in oncology by circulating nucleosomes. Ann. N. Y. Acad. Sci. **1022:** 211–216.
14. GAUTSCHI, O. *et al.* 2004. Circulating deoxyribonucleic acid as prognostic marker in non-small-cell lung cancer patients undergoing chemotherapy. J. Clin. Oncol. **22:** 4157–4164.
15. BREMNES, R.M. *et al.* 2005. Circulating tumour-derived DNA and RNA markers in blood: a tool for early detection, diagnostics, and follow-up? Lung Cancer **49:** 1–12.
16. HOLDENRIEDER, S. *et al.* 2004. Circulating nucleosomes predict the response to chemotherapy in patients with advanced non-small cell lung cancer. Clin. Cancer Res. **10:** 5981–5987.
17. HOLDENRIEDER, S. *et al.* 2001. Nucleosomes in serum as a marker for cell death. Clin. Chem. Lab. Med. **39:** 596–605.

18. WATINE, J. *et al.* 2002. Biological variables and stratification of patients with inoperable non-small-cell bronchial cancer: recommendations for future trials. Cancer Radiother. **6:** 209–216.
19. PUJOL, J.L. *et al.* 2004. CYFRA 21–1 is a prognostic determinant in non-small-cell lung cancer: results of a meta-analysis in 2063 patients. Br. J. Cancer **90:** 2097–2105.
20. BARLESI, F. *et al.* 2004. Prognostic value of combination of Cyfra 21–1, CEA and NSE in patients with advanced non-small cell lung cancer. Respir. Med. **98:** 357–362.
21. VOLLMER, R.T. *et al.* 2003. Serum CYFRA 21–1 in advanced stage non-small cell lung cancer: an early measure of response. Clin. Cancer Res. **9:** 1728–1733.
22. EBERT, W. & T. MULEY. 1999. CYFRA 21–1 in the follow-up of inoperable non-small cell lung cancer patients treated with chemotherapy. Anticancer Res. **19:** 2669–2672.
23. YEH, J.J. *et al.* 2002. Monitoring cytokeratin fragment 19 (CYFRA 21–1) serum levels for early prediction of recurrence of adenocarcinoma and squamous cell carcinoma in the lung after surgical resection. Lung **180:** 273–279.
24. HOLDENRIEDER, S. & P. STIEBER. 2004. Apoptotic markers in cancer. Clin. Biochem. **37:** 605–617.
25. LICHTENSTEIN, A.V. *et al.* 2001. Circulating nucleic acids and apoptosis. Ann. N. Y. Acad. Sci. **945:** 239–249.
26. STROUN, M. *et al.* 2000. The origin and mechanism of circulating DNA. Ann. N. Y. Acad. Sci. **906:** 161–168.
27. NG, E.K. *et al.* 2002. Presence of filterable and nonfilterable mRNA in the plasma of cancer patients and healthy individuals. Clin. Chem. **48:** 1212–1217.
28. HOLDENRIEDER, S. *et al.* 2005. Cell-free DNA in serum and plasma: comparison of ELISA and quantitative PCR. Clin. Chem. **51:** 1544–1546.
29. ZIEGLER, A. *et al.* 2002. Circulating DNA: a new diagnostic gold mine? Cancer Treatment Rev. **28:** 255–271.
30. ANKER, P. *et al.* 2003. Circulating nucleic acids in plasma and serum as a noninvasive investigation for cancer: time for large-scale clinical studies? Int. J. Cancer **103:** 149–152.
31. ANDRIANI, F. *et al.* 2004. Detecting lung cancer in plasma with the use of multiple genetic markers. Int. J. Cancer **108:** 91–96.
32. HOLDENRIEDER, S. *et al.* 2004. Nucleosomes indicate the in vitro radiosensitivity of irradiated broncho-epithelial and lung cancer cells. Tumour Biol. **25:** 321–326.
33. KUROI, K. *et al.* 2001. Clinical significance of plasma nucleosomes levels in cancer patients. Int. J. Oncol. **19:** 143–148.
34. TREJO-BECERRIL, C. *et al.* 2005. Serum nucleosomes during neoadjuvant chemotherapy in cervical cancer patients. Predictive and prognostic significance. BMC Cancer **5:** 65.
35. HOLDENRIEDER, S. *et al.* 2005. Circulating nucleosomes and cytokeratin 19-fragments in patients with colorectal cancer during chemotherapy. Anticancer Res. **25:** 1795–1801.
36. KREMER, A. *et al.* 2005. Nucleosomes in pancreatic cancer patients during radiochemotherapy. Tumour Biol. **26:** 44–49.

Circulating Nucleic Acids and Diabetic Complications

ASIF N. BUTT,[a] ZAID SHALCHI,[a] KARIM HAMAOUI,[a]
ANDJENY SAMADHAN,[a] JAKE POWRIE,[b] SHIRLEY SMITH,[b]
SARAH JANIKOUN,[b] AND R. SWAMINATHAN[a]

[a]*Department of Chemical Pathology, St Thomas' Hospital, London, SE1 7EH, UK*

[b]*Department of Diabetes and Endocrinology, St Thomas' Hospital, London, SE1 7EH, UK*

ABSTRACT: Diabetes mellitus is a major health problem across the world. Diabetic retinopathy (DR) and nephropathy are two of the major complications of diabetes. DR is the leading cause of blindness and diabetic nephropathy is the leading cause of end-stage renal failure. We have examined the potential value of circulating nucleic acids in the detection and monitoring of these two complications of diabetes. mRNA for nephrin was significantly higher in all diabetics compared to healthy controls and it was significantly higher in normoalbuminuric patients compared to healthy controls. This may indicate progression to microalbuminuric stage. Circulating rhodopsin mRNA was detectable in healthy subjects and in diabetic patients. It was significantly raised in diabetic patients with retinopathy. Higher rhodopsin mRNA in diabetic patients without retinopathy suggests that some of them may go on to develop it or already have it subclinically. Circulating nucleic acids have the potential to be noninvasive molecular tests for diabetic complications.

KEYWORDS: rhodopsin; diabetic mellitus; retinopathy; DNA; mRNA

INTRODUCTION

Diabetes mellitus is a major public health problem in both developed and developing countries.[1] Incidence of diabetes is increasing at an alarming rate and the health care cost of diabetes is 155 billion international dollars in 2002 and it is expected to rise to 396 billion by 2025 (an international dollar is a common currency unit that takes into account differences in the related purchasing power of various currencies).[1] The World Health Organization

Address for correspondence: Prof. R. Swaminathan, Department of Chemical Pathology, 5th Floor, North Wing, St Thomas' Hospital, Lambeth Palace Road, London, SE1 7EH, UK. Voice: +44-0-20-7188-1285; fax: +44-0-20-7928-4226.
e-mail: r.swaminathan@kcl.ac.uk

**Ann. N.Y. Acad. Sci. 1075: 258–270 (2006). © 2006 New York Academy of Sciences.
doi: 10.1196/annals.1368.034**

estimates that up to 15% of annual health care budget is spent on management of diabetes and diabetes-related illnesses. The medical costs for a person with diabetes are two-to five-fold higher than that of a person without diabetes. The prevalence of diabetes is expected to increase in all countries, but the largest proportional and absolute increase will occur in developing countries where it will rise from 4.2% to 5.6%. By 2025, the adult diabetic population is expected to double in India to about 73 million.[1] The highest prevalence of over 30% is seen in the Nauru islands. Diabetes can affect almost every organ in the body and it is a major cause of death. It is the leading cause of blindness, renal failure, amputation of lower limbs, and also one of the leading causes of death through its affects on cardiovascular disease. Seventy to 80% of patients with diabetes die of cardiovascular disease. Major diabetic complication includes retinopathy, coronary heart disease, nephropathy, cerebral vascular disease, peripheral vascular disease, neuropathy, and diabetic foot.

Diabetic Retinopathy

Diabetic retinopathy (DR) is probably the most characteristic, easily identifiable, and treatable complication of diabetes, and it remains an important cause of visual loss in the developed world.[2] Since type 2 diabetes remains undiagnosed for several years, a significant number of people, even in developed countries, already have retinopathy by the time their diabetes is diagnosed. Within the course of their lifetime, 2% of individuals with diabetes will become legally blind as a result of DR and larger numbers will develop significantly impaired vision. Such consequential vision loss in diabetics is widely recognized as one of the most debilitating complications of the disease. DR is the leading cause of blindness in the 20-to 74-year-old population of the Western world.[3] Almost 25% of all diabetics have some form of retinopathy. In the WESDR study it was shown that 90% of patients under the age of 30 who had diabetes for 10 to 15 years had some form of retinopathy. In patients over 30 years of age prevalence was 84% and 53% in insulin and noninsulin-treated diabetics, respectively.[4] DR increases the risk of legal blindness by 52-to 80-fold and the cost of health care of DR is over 500 million U.S. dollars. Risk factors for the development of DR include hyperglycemia, hypertension, hypercholesterolemia, and pregnancy.[5–8] Pathogenesis of DR is not fully understood but various mechanisms all attributed to insults from the hyperglycemic state are important. Rheological abnormalities associated with diabetes cause intravascular coagulation and disrupt the perfusion to the retina. Disruption of the blood supply causes ischemia in those nonperfused retinal areas, which, in turn, reduces vision and promotes neovascularization in an effort to reestablish the blood supply, causing further damage and deteriorating vision. Although adequate glycemic control seems to stem the rapid development of retinopathy, a large proportion of diabetic patients are unable to maintain such strict treatment options. Several clinical trials have shown that effective laser photocoagulation could reduce the loss of vision by 90%.[8,9] Thus, it has become an important function of any diabetic clinic to assess the eye status of diabetic patients.[10] Typically, eye status examination includes visual accuracy assessment as well as dilated fundoscopy and involves cooperation between the various health care professionals, such as diabetologists, optometrists, general practitioners, and photographic screening clinics. DR is classified into background, preproliferative, and proliferative retinopathy. Because of the risk of developing DR, it is now recommended that all diabetic patients should be screened for retinopathy at regular intervals. The recommendation is that patients should be screened after 5 years of diagnosis in type 1 and at diagnosis in type 2 diabetes and annually thereafter.[11] However, if patients already have some form of retinopathy, then they should be regularly screened at much shorter intervals. In addition, pregnant subjects should be screened every 2 months. Despite these guidelines 30–50% of patients with diabetes still do not get any eye examination.[11] Classification of the eye status of diabetic patients is done according to a scale for the number of abnormalities observed.[12] This type of approach is costly,[13] subjective, and reliable identification requires experience and training for accurate assessment. As yet there is no reliable, independent, quantifiable, and nonsubjective method for categorizing eye status. Such a test is highly desirable.

Diabetic Nephropathy

Diabetic nephropathy (DN) is the most common cause of end-stage renal disease in many countries. Overall incidence of DN is 20% in diabetic patients.[1] It is now a common practice to measure albumin concentration in the urine as a predictor of DN. Microalbuminuria, presence of small amounts of albumin in the urine, is believed to be a strong predictor of DN.[14] However, many nondiabetic patients have microalbuminuria and not all patients with microalbuminuria go on to develop DN.[15] In spite of these reservations it is current recommendation that all diabetic patients should have an annual measurement of albumin in the urine.[16]

In this regard, measurement of tissue-specific nucleic acids in the circulation offers enormous potential for prognostic and diagnostic purposes. It has been shown previously that plasma nucleic acids (DNA and RNA) appear to reflect the amount of cell death occurring in the organism.[17] We have previously shown that total DNA in circulation is increased in patients in the intensive care unit[18] and others have shown similar findings in trauma,[19] stroke,[20] and myocardial infarction. Plasma DNA was found to be the best prognostic indicator of outcome in intensively critically ill patients.[18] As total DNA is nonspecific and does not localize the tissue, the detection of organ-specific mRNA is thought to be better.[21] We have previously shown that retinal-specific mRNA is increased in patients with DR.[22]

Increased albumin excretion in diabetic patients occurs as a result of damage to the glomerular basement membrane; therefore it was argued that measurement of nephron-specific mRNA may offer some potential in the diagnosis and monitoring of DN. Nephrin is a transmembrane protein with a large extracellular portion including eight immunoglobulin-like domains.[23] It is expressed by visceral epithelial cells (podocytes) in the slit diagram of the glomerulus. Nephrin is a product of the *NPHS 1* gene located in chromosome 19 and this protein is crucial for the integrity of the slit diagram. Abnormalities in this protein can lead to proteinuria and eventually to nephrotic syndrome. We, therefore, examined the concentration of mRNA for nephrin in the circulation in patients with and without different degrees of DN.

We have extended these studies and measured total DNA as well as retinal- and nephron-specific mRNA in patients with or without DR and DN.

MATERIALS AND METHODS

Patients

For the DR study, patients with a confirmed history of diabetes with or without retinopathy were recruited from the Diabetic Clinic at Guy's and St Thomas' Hospital, London. The protocol for the study was approved by the local Research Ethics Committee. Blood samples were obtained from 35 diabetic patients without retinopathy, 101 patients with DR, 40 healthy subjects, and 9 subjects with miscellaneous retinal pathology. Patients with DR were divided into background ($n = 34$), preproliferative ($n = 43$), and proliferative ($n = 34$). The eye status of patients was determined from medical notes and independent examination of fundoscopic photographs. Healthy volunteers had no known disease.

Blood samples were also obtained from 26 healthy control subjects and 89 diabetic subjects for the study of mRNA for nephrin in DN. Diabetic subjects were classified according to the albumin creatinine ratio into normal albuminuric, microalbuminuric, and macroalbuminuric.

After obtaining informed consent, blood samples were obtained from all subjects into PAXgene blood RNA tubes (2.5 mL) especially designed for the collection and stabilization of RNA from whole blood and into EDTA tubes (4 mL).

METHODS

Preparation of Cell-Free Plasma

Blood samples taken into the EDTA tube were centrifuged at 1600 $\times$ g for 10 min and the upper 90% of the plasma transferred to microfuge tubes without disturbing the buffy coat or the red cell pellet. The plasma was subjected to a second centrifugation (Eppendoff centrifuge 5415 D, Cambridge, UK) at 16,000 × g for 10 min and again the upper 90% of plasma removed and stored at –80°C until required for further processing.

Extraction of DNA from Cell-Free Plasma

DNA was extracted from 400 μL of EDTA plasma with the QIAamp blood kit (Qiagen, Crawley, UK) using the blood and fluid protocol recommended by the manufacturer.

Extraction and Reverse Transcription of RNA

Whole blood RNA was extracted using the PAXgene blood RNA kit including treatment with DNase 1 to prevent genomic DNA contamination, strictly following the manufacturer's instructions (Qiagen). Extracted RNA was stored at –80°C until required for cDNA synthesis. Reverse transcription was carried out using Superscript II reverse transcriptase following the manufacturer's instruction (Invitrogen Life Sciences, Scotland, UK). The cDNA generated was stored at –80°C until required for quantitation. Separately, samples were also subjected to the foregoing procedure with exception that the reverse transcriptase was replaced with water (negative control).

Real-Time Quantitative PCR

Plasma DNA was measured by a real-time quantitative PCR Taqman assay for the housekeeping gene β-globin using primers and probe sequences described previously.[18] A reaction volume of 50 μL comprising a 10 μL sample or standard, 25 μL X2 Universal Mastermix (Applied Biosystems, Warrington, UK), 300 nM each primer, 100 nM probe, and water was used for amplification conducted in an ABI 7000 sequence detection system (Applied Biosystems). The calibration curve comprising a 10-fold serial dilution of DNA extracted from whole blood was constructed and each data point analyzed in duplicate. Samples and water blanks were also analyzed in duplicate and in parallel with standard curve. The results were expressed as genome equivalents/milliliter where 6.6 pg DNA was equivalent to one genome. One genome equivalent was defined as the amount of a particular target sequence contained in a single-diploid human cell.

The ABI 7000 sequence detection system was used to amplify cDNA and detect PCR product using sequence-specific oligonucleotide probes and intron-spanning-specific primers. cDNA was amplified using the predeveloped assay reagents Taqman assay (PE Applied Biosystems). In the case of the rhodopsin, Taqman assay, 900 nM forward and reverse primers, 250 nM probe, X2 Taqman Universal Mastermix (25 μL), and cDNA sample (5 μL) were present in each reaction. For both assays, standards and samples were analyzed in triplicate in the final reaction volume of 50 μL. Standard curves for rhodopsin were prepared from serial dilutions of cDNA (Clontech Laboratories Inc., Mountain View, CA) obtained from normal healthy human retina and for nephrin from serial dilutions of cDNA (Ambion Corporation, Huntington, UK) obtained from healthy human kidney. A water blank was incorporated in each of the respective assays. Both assays (rhodopsin or nephrin and β-actin) were run simultaneously on 96-well optical reaction plates. PCR amplification included an initial phase of 2 min at 50°C, followed by 10 min at 95°C then 40 cycles of 15 s at 95°C and 1 min at 60°C.

Statistical analysis was performed using SPSS 11. Differences between groups were analyzed by nonparametric statistics. A "*P*" value of <0.05 was considered statistically significant.

RESULTS

TABLE 1 shows the patient characteristics of the four groups in the DR study. Diabetic patients were significantly older than the healthy controls and as expected hemoglobin A1C was higher in patients with complications than in diabetic controls. There was no significant difference in the body mass index between groups. Plasma DNA concentration in healthy subjects and diabetic subjects was significantly different ($P = 0.020$; FIG. 1). When the diabetic subjects were grouped according to the eye status, it was observed that the diabetic control subjects had significantly lower plasma DNA concentration than healthy controls ($P = 0.0098$; FIG. 1) and patients with preproliferative and proliferative retinopathy had significantly higher DNA concentration than the diabetic controls ($P = 0.0098$ and 0.028, respectively). There was no significant difference in plasma DNA levels between the active and quiescent proliferative retinopathy. Patients with miscellaneous retinopathy had significantly higher DNA concentration than diabetic controls ($P = 0.045$).

Rhodopsin mRNA was detectable in the peripheral blood of all subjects in the DR study. Rhodopsin mRNA in diabetic patients as a whole was nearly two and half times higher than that in control subjects ($P \leq 0.001$) (FIG. 2). Messenger RNA for rhodopsin in diabetic control subjects without retinopathy was about 60% higher than that in healthy controls. FIGURE 2 shows the rhodopsin mRNA levels in healthy subjects and diabetics with varying severity of DR. Rhodopsin mRNA was significantly higher in patients with background retinopathy and preproliferative retinopathy and there was a tendency for mRNA levels to increase with increasing severity of retinopathy except in patients with proliferative retinopathy. Patients with proliferative retinopathy

TABLE 1. Characteristics of patient groups and healthy subjects

		Diabetic patients ($n = 106$)				
	Healthy controls	Diabetic controls	Background DR	Preproliferative DR	Proliferative DR	Miscellaneous retinal pathology
n	24	27	25	21	27	6
Age (years)	45.0 ± 2.3	57.6 ± 2.4	65.4 ± 2.9	54.7 ± 2.4	60.0 ± 2.6	64.7
Duration of diabetes (years)	—	4.9 ± 1.0	10.8 ± 1.2	12.7 ± 1.2	19.1 ± 2.0	4.8 ± 1.7
Random blood glucose (mmol/L)	—	7.9 ± 0.9	9.4 ± 0.8	10.9 ± 1.1	12.1 ± 1.2	6.95 ± 1.0
HbA1c (%)	—	7.1 ± 0.4	7.9 ± 0.5	9.1 ± 0.5	8.3 ± 0.5	7.9 ± 0.5
ACR (mg/mmol)	—	4.5 ± 3.4	4.4 ± 1.8	13.4 ± 5.9	22.3 ± 7.0	185.3 ± 182.2
Plasma creatinine (μmol/L)	—	120.2 ± 40.6	100.7 ± 9.6	87.5 ± 5.5	106.5 ± 7.9	87.0 ± 16.6
BMI (kg/m^2)	25.7 ± 1.7	27.8 ± 1.4	30.0 ± 1.4	33.6 ± 2.0	28.8 ± 1.8	33.2 ± 6.6
Systolic blood pressure (mmHg)	—	131.6 ± 5.5	150.2 ± 8.8	137.7 ± 4.3	136.6 ± 5.4	139.3 ± 8.2
Diastolic blood pressure (mmHg)	—	79.4 ± 2.1	81.7 ± 3.6	83.4 ± 2.5	77.2 ± 2.7	68.3 ± 2.4
WBC count ($\times 10^{-9}$/mL)	—	7.2 ± 0.9	8.2 ± 0.6	8.9 ± 0.8	8.0 ± 0.5	7.3 ± 0.4

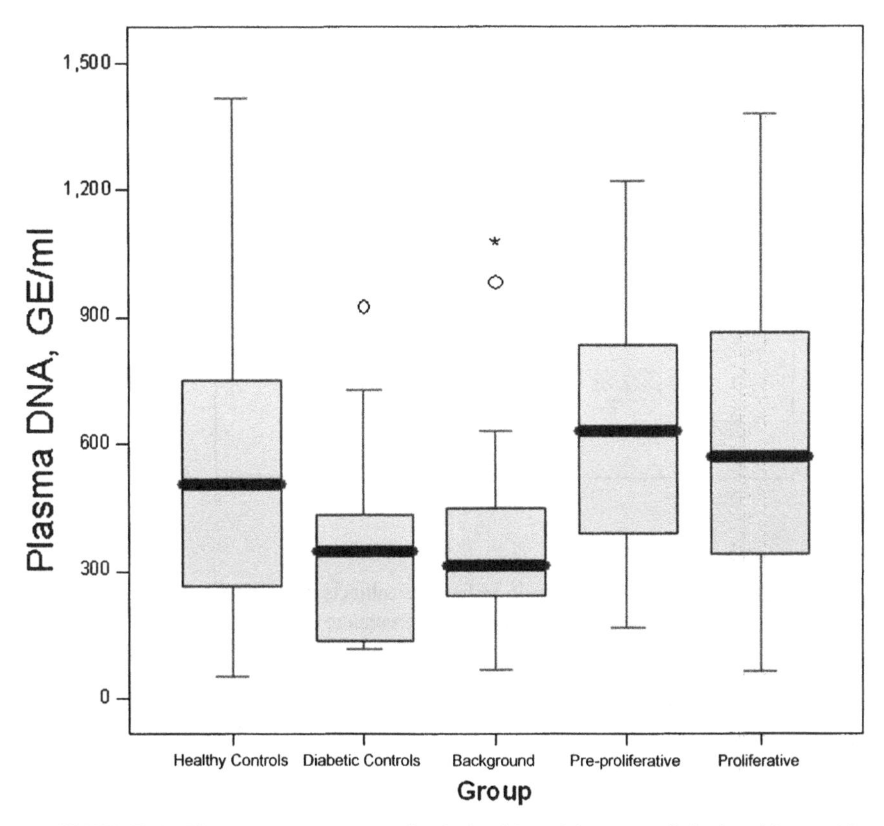

FIGURE 1. Plasma DNA concentration in healthy subjects and diabetic subjects with and without retinopathy.

had higher mRNA compared to healthy subjects and diabetic control subjects, but it was significantly lower than in preproliferative retinopathy. When proliferative retinopathy patients were divided into active and quiescent retinopathy, the quiescent retinopathy subjects had a significantly higher mRNA ($P = 0.002$). Six diabetic patients who had no signs of DR but were diagnosed with miscellaneous retinal pathology (maculopathy, pigment lesions, uveities, and glaucoma) had significantly higher circulating rhodopsin mRNA levels compared to healthy controls ($P = 0.001$) but not compared to diabetic controls ($P = 0.064$). In three patients who gave blood samples before and 20 min after retinal photo coagulation, mRNA levels for rhodopsin decreased.

In the DN study, diabetic patients as a group had significantly lower Ct values compared to control subjects—that is they have significantly higher mRNA for nephrin compared to control subjects (FIG. 3). When diabetic subjects were divided according to the albumin excretion into normal, micro, and

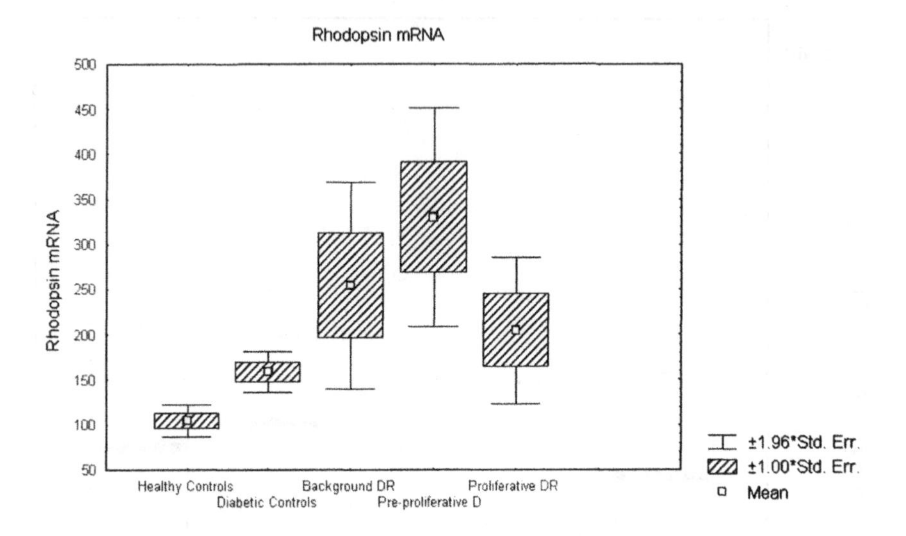

FIGURE 2. mRNA for rhodopsin in healthy subjects and diabetic subjects with and without retinopathy. Results are expressed as a percentage of that in controls.

macroalbumiuric subjects, significant differences were found between the normoalbuminuric and macroalbuminuric subjects compared to control subjects (FIG. 3). Microalbuminuric subjects did not show a significant difference.

DISCUSSION

Previously we have shown for the first time that mRNA for rhodopsin was higher than DR in patients. As far as we are aware there are no other studies examining retinal-specific mRNA in blood to detect or diagnose DR. Recently, Malik *et al.* have reported that plasma concentration of CD105, a glycoprotein strongly expressed in the angiogenic vascular endothelial cells, is increased in patients with diabetes, especially in patients with background or proliferative retinopathy.[24] In this small group of diabetic subjects, they found plasma CD105 was increased nearly twofold in background retinopathy but decreased in advanced retinopathy cases. These results are similar to our findings in that the highest concentration of mRNA for rhodopsin was found in the preproliferative cases and in proliferative cases it was lower than in the preproliferative DR group. Malik *et al.* also measured VEGF levels in vitreous fluid and found it to be significantly higher in patients with advanced retinopathy.[24]

It is known that hyperglycemia can lead to retinal tissue toxicity in a number of ways including increased free radical generation and ischemia.[25] Previous studies have shown that circulating levels of nucleic acids are associated with

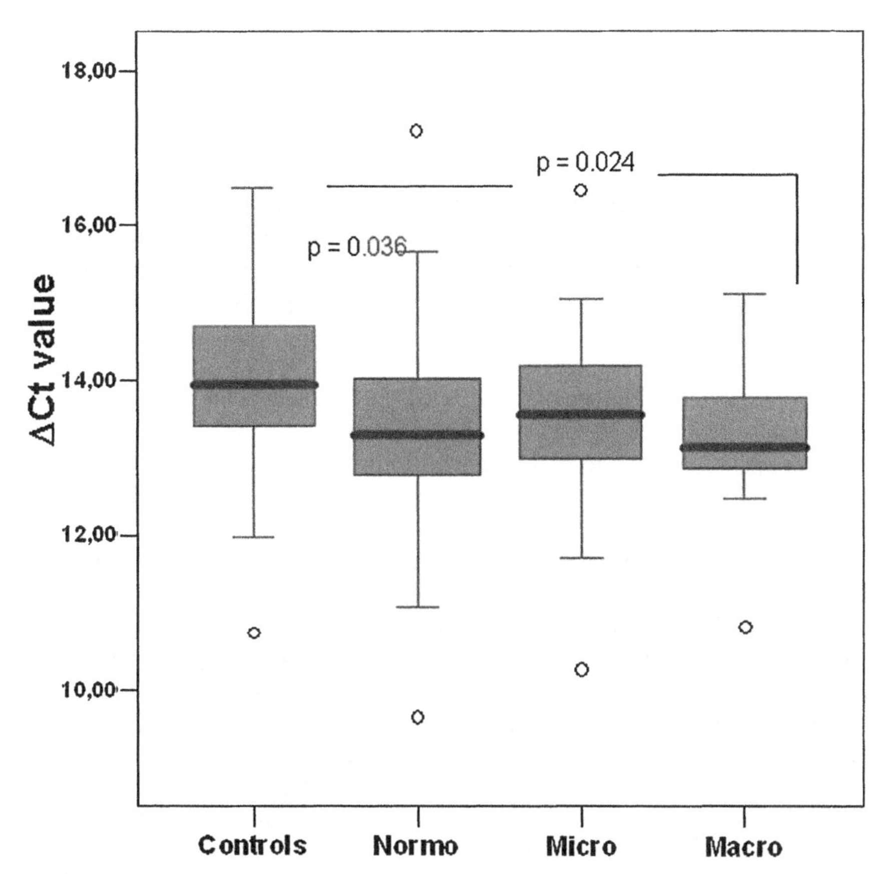

FIGURE 3. Ct values for mRNA for nephrin in healthy subjects and diabetic patients with different degrees of albumin excretion.

cell death.[26] This suggests that hyperglycemic insult in the diabetic subjects leads to increase in death of the rod cells in the outer layer of the retina. The release of rhodopsin mRNA into the blood may result directly from the death of retinal photoreceptors or may be the result of compensatory mechanisms intraretinally that causes increased expression of rhodopsin mRNA in an attempt to maintain rod cell density and hence preserve vision. It is interesting to note that diabetic patients without any clinical features of retinopathy showed significantly higher mRNA levels. This may suggest that the damage to the retina is already present before the detection of retinopathy clinically and it is possible that a significant proportion of these subjects may go on to develop clinical retinopathy. This makes it possible that mRNA for rhodopsin could be used as an early predictor of DR. It is interesting to note that patients with miscellaneous retinopathy had significantly higher mRNA. This finding suggests that mRNA for rhodopsin can be a useful marker not only in the assessment of DR but may also be useful in the detection and evaluation of other retinal lesions: maculopathy, pigment lesions, uveities, and glaucoma. With the exception of the proliferative retinopathy group there was a trend for the relative amounts of rhodopsin mRNA to increase with severity of retinopathy. This upward trend may be related to the more severe damage with widespread retinal pathology associated with the progressive stages of DR. While the precise mechanisms of release of mRNA are not clear, it is possible that release from dead or dying retinal cells due to ischemia, possible upregulation of rhodopsin transcription, or increased active secretion of rhodopsin mRNA may be the underlying mechanisms for the increase in mRNA for rhodopsin observed in this study. Other studies have implicated apoptosis, necrosis, and active secretion as possible mechanisms for the release of nucleic acids from cells into peripheral circulation. In the case of proliferative retinopathy where mRNA for rhodopsin appeared to be lower than in other groups, it may be that the damage has been sustained and that retinal cells are exhausted either in the metabolic capacity or in total number.

To the best of our knowledge, this the first study to quantify nephrin mRNA in peripheral blood as a marker of glomerular function. All healthy and diabetic subjects showed quantifiable levels of nephrin mRNA in the circulation. This study also showed that diabetics have increased levels of nephrin mRNA in blood compared to healthy controls. It was anticipated that there might be a severity-related increase in nephrin mRNA in the circulation. Indeed, normoalbuminuric patients demonstrated a 42% increase in nephrin mRNA compared to healthy controls ($P = 0.036$). However, in the microalbuminuric subjects blood nephrin mRNA levels decreased to 22%. This apparent discrepancy may be explained by downregulation of nephrin mRNA expression. Hence, if all other variables remain equal, less nephrin mRNA might be expected to be released into the circulation in microalbuminuria compared to normoalbuminuria. On the other hand, macroalbuminuric patients showed a 45% increase in blood nephrin mRNA compared to healthy controls. It is possible that in this group, the severity of the renal damage evoked a compensatory synthesis of mRNA for nephrin.

It is well established that ACR, the current test used for assessment of renal status, is influenced by age, ethnicity, and gender. There is some evidence that such factors may also have an influence on nephrin mRNA in blood. For instance, we have demonstrated previously that levels of nephrin mRNA were significantly higher in women, lower in older subjects, and also lower in reduced renal mass.[27] In the present study, the microalbuminuric and macroalbuminuric groups had a male to female ratio of 2:1. Also, the healthy controls were not age matched with the diabetic patients. In addition, the patient and control groups comprised diverse ethnic backgrounds. As the data presented have not been corrected for these factors, it is difficult to draw a firm conclusion about the potential role of nephrin mRNA in the assessment of renal status in diabetic nephropathy. Nevertheless, the abundant presence of nephrin mRNA and the relative ease with which it can be measured in blood shows great promise for its use as a molecular marker of renal function.

CONCLUSION

Circulating mRNA for rhodopsin and nephrin was detectable in healthy subjects and diabetic patients. Rhodposin mRNA was significantly raised in diabetic patients with retinopathy and nephrin mRNA was increased in DN. Circulating organ-specific mRNA may form the basis of a reproducible, quantitative, molecular noninvasive test for DR and DN.

REFERENCES

1. http://www.eatlas.idf.org/Prevalence/(Accessed 28th Dec 2005).
2. KNOTT, R. *et al.* 2003. Diabetic eye disease. *In* Textbook of Diabetes, 3rd ed, Vol. 48. J. Pickup & G. Williams, Eds.: 1–48. Blackwell Publishing. Oxford.
3. CENTERS FOR DISEASE CONTROL AND PREVENTION. National Diabetes Fact Sheet: National Estimates and General Information on Diabetes in the United States. CDC. Atlanta, GA. (accessed Dec28, 2005).
4. KLEIN, R. & B.E. KLEIN. 1998. Relation of glycemic control to diabetic complications and health outcomes. Diabetes Care **21**(Suppl 3): C39–C43.
5. UK PROSPECTIVE DIABETES STUDY GROUP. 1998. Tight blood pressure control and risk of macrovascular and microvascular complications in type 2 diabetes. UKPDS 38. BMJ **317:** 703–713.
6. CHEW, E.Y. *et al.* 1996. Association of elevated serum lipids with retinal hard exudate in diabetic retinopathy. Early Treatment Diabetic Retinopathy Study (ETDRS) Report 22. Arch. Ophthalmol. **114:** 1079–1084.
7. THE DIABETES CONTROL AND COMPLICATIONS TRIAL RESEARCH GROUP. 1993. The effect of intensive treatment of diabetes on the development of long-term complications in insulin-dependent diabetes mellitus. N. Engl. J. Med. **329:** 977–986.
8. 1978. Photocoagulation treatment of proliferative diabetic retinopathy: the second report of diabetic retinopathy study findings. Ophthalmology **85:** 82–106.
9. EARLY TREATMENT DIABETIC RETINOPATHY STUDY RESEARCH GROUP. 1991. Early photocoagulation for diabetic retinopathy. ETDRS Report 9. Ophthalmology **98:** 766–785.
10. AMERICAN DIABETES ASSOCIATION. DIABETIC RETINOPATHY. 1998. Diabetes Care **21:** 157–159.
11. KLEIN, R. 2003. Commentary: screening interval for retinopathy in type 2 diabetes. Lancet **361:** 190–191.
12. BRITISH DIABETIC ASSOCIATION MOBILE RETINAL SCREENING GROUP. 1996. Practical community screening for diabetic retinopathy using mobile retinal camera: report of a 12 centre study. Diabet. Med. **13:** 946–952.
13. VIJAN, S. *et al.* 2000. Cost-utility analysis of screening intervals for diabetic retinopathy in patients with type 2 diabetes mellitus. JAMA **283:** 889–896.

14. CHIARELLI, F. *et al.* 1997. The importance of microalbuminuria as an indicator of incipient diabetic nephropathy: therapeutic implications. Ann. Med. **29:** 439–445.
15. SEGURA, J. *et al.* 2004. Microalbuminuria. Clin. Exp. Hypertens. **26:** 701–707.
16. NEWMAN, D.J. *et al.* 2005. Systemic review on urine albumin testing for early detection of diabetic complications. Health Technol. Assess. **9:** iii–vi.
17. STROUN, M. *et al.* 2001. About the possible origin and mechanism of circulating DNA Apoptosis and active DNA release. Clin. Chim. Acta. **313:** 139–142.
18. WIJERATNE, S. *et al.* 2004. Cell-free plasma DNA as a prognostic marker in intensive treatment unit patients. Ann. N. Y. Acad. Sci. **1022:** 232–238.
19. LAM, N.Y.L. *et al.* 2003. Time course of early and late changes in plasma DNA in trauma patients. Clin. Chem. **49:** 1286–1291.
20. RAINER, T.H. *et al.* 2003. Prognostic use of circulating plasma nucleic acid concentrations in patients with acute stroke. Clin. Chem. **49:** 562–569.
21. CHAN, A.K. *et al.* 2003. Cell-free nucleic acids in plasma, serum and urine: a new tool in molecular diagnosis. Ann. Clin. Biochem. **40:** 122–130.
22. HAMAOUI, K. *et al.* 2004. Concentration of circulating rhodopsin mRNA in diabetic retinopathy. Clin. Chem. **50:** 2152–2155.
23. TRYGGVASON, K. 1999. Unravelling the mechanism of glomerular ultrafiltration: nephrin a key component of the slit diaphragm. J. Am. Soc. Nephrol. **10:** 2440–2445.
24. MALIK, R.A. *et al.* 2005. Elevated plasma CD105 and vitreous VEGF levels in diabetic retinopathy. J. Cell. Mol. Med. **9:** 692–697.
25. CALDWELL, R.B. *et al.* 2005. Vascular endothelial growth factor and diabetic retinopathy: role of oxidative stress. Curr. Drug Targets **6:** 511–524.
26. VASILYEVA, I.N. 2001. Low-molecular weight DNA in blood plasma as an index of influence of ionizing radiation. Ann. N. Y. Acad. Sci. **945:** 221–228.
27. ORLANDI, E. *et al.* 2005. Factors affecting circulating mRNA for Nephrin. Clin. Chem. **51:** 1982–1983.

Circulating Nucleic Acids and Critical Illness

TIMOTHY H. RAINER AND NICOLE Y.L. LAM

Accident and Emergency Medicine Academic Unit, The Chinese University of Hong Kong, Prince of Wales Hospital, Shatin, New Territories, Hong Kong SAR, China

ABSTRACT: This article reviews some of the early work that has been performed to investigate the potential roles of circulating nucleic acids as prediction markers in acute illness and injury. Circulating DNA and RNA concentrations are elevated early in patients with trauma, stroke and ACS, and are generally highest in patients with a high risk of death. Circulating nucleic acids may be useful markers for the evaluation and risk-stratification of such patients.

KEYWORDS: circulating nucleic acids; acute coronary syndrome; cardiovascular accidents; trauma

INTRODUCTION

Acute illness and injury are leading causes of death and morbidity worldwide.[1] In general, the earlier an accurate diagnosis is made and appropriate treatment started, the greater the chance of survival, reduced complications, better quality of life, and reduced health care costs.[2–4] The need to identify markers that aid clinicians in diagnosis, prognosis, and disease monitoring is driving clinical scientific research.

Release of DNA into the circulation may make it a useful, nonspecific marker of tissue injury. This article discusses some of the potential roles of DNA as an early diagnostic and prognostic marker of tissue injury in patients with trauma, stroke, and acute coronary syndrome (ACS).

Most of the studies in this article were conducted at the Prince of Wales Hospital, a tertiary hospital, and a primary trauma center in the New Territories of Hong Kong.[5] The studies involve adult patients and although there have been modifications in protocol over time, in general blood was taken from peripheral veins, collected into EDTA tubes, centrifuged at 1500 *g* for 10 min, and plasma decanted, filtered, and stored at –80°C until further processing.

Address for correspondence: Dr. T. H. Rainer, Accident and Emergency Medicine Academic Unit, The Chinese University of Hong Kong, Prince of Wales Hospital, Shatin, New Territories, Hong Kong SAR. Voice: +852-2632-1033; fax: +852-2648 -1469.
e-mail: thrainer@cuhk.edu.hk

Ann. N.Y. Acad. Sci. 1075: 271–277 (2006). © 2006 New York Academy of Sciences.
doi: 10.1196/annals.1368.035

In this article, the β-*globin* gene was used for determining quantitatively the presence of DNA in plasma. It was measured by real-time quantitative polymerase chain reaction (PCR) as previously described.[5]

The present studies were not conducted in clinical real time but samples were stored and analyzed in batches. Current protocols allow the provision of plasma DNA results within 3 h of blood sampling. However, the recent development of rapid capillary-based instrumentation for quantitative PCR analysis,[6] may further reduce this time to 90 min. However, as treatment is generally most effective the earlier it is started, the ultimate goal will be to develop rapid methods that produce results within minutes rather than hours.

THE EPIDEMIOLOGY OF TRAUMA, STROKE, AND ACS

The epidemiology of disease differs significantly between trauma, stroke, and ACS patients, which make these three groups interesting to study and compare.

Although trauma may affect all ages it is often described as a disease of the young as it tends to affect young, relatively healthy men with little prior comorbidity. Trauma also often results in massive tissue injury and overwhelming acute systemic inflammatory response.

Both stroke and ACS, on the other hand, tend to affect the elderly who have coexisting disease and underlying chronic pathology, such as atherosclerosis and chronic ongoing inflammation. The tissue mass affected is relatively small considering the catastrophic effect on morbidity with paralysis or heart failure.

Trauma

In 1990, trauma accounted for 5.1 million global deaths and if current trends continue this figure will rise to 8.4 million by 2020 A.D.[1,7] Twenty percent of deaths after trauma occur days to weeks after injury as a result of cerebral injury and organ failure.[8]

Circulating plasma DNA concentrations increase within an hour after injury and may increase by as much as a 100-fold in those patients who go on to develop organ failure.[5] Compared with patients with uncomplicated injury, there is a 10- to 18-fold increase in median plasma DNA concentrations in those patients who develop organ failure, multiple organ dysfunction syndrome, acute lung injury, and who die. Plasma DNA concentrations and other predictors, such as the maximal abbreviated injury score and shock index measured in the resuscitation room and incorporated into a classification and regression tree model to guide clinicians, make an early diagnosis of organ failure and mortality. These models have a sensitivity and specificity of 90% and may be used to guide decisions regarding admission to intensive care or modifications in treatment.[9]

The mechanisms by which the levels of circulating DNA are increased following trauma are unclear at present. Such elevations may be the result of increased liberation following cell death, or due to decreased efficiency of DNA clearance following injury.

The release of DNA into the circulation is likely to be a result either of necrosis or apoptosis.[10] Necrosis is characterized by large DNA genomic sequences of >1000 base pairs, while apoptosis—the process of programmed cell death—is characterized by genomic sequences of 80 base pairs. Measuring the relative proportions of 80 and 1000 base-pair sequences may give an indication of whether apoptosis or necrosis is the dominant activity governing the release of DNA into plasma after trauma.

Although both 1000 and 80 base-pair sequences are present in plasma within an hour of injury, the 1000:80 base-pair ratio is much less than one, suggesting that apoptosis may be a dominant process in the release of DNA into the circulation after trauma (Allan Chan, personal communication). If confirmed, this would support the hypothesis that active processes are involved in this release.

The clearance mechanisms for circulating DNA are poorly understood at present but it is possible that direct damage or hemodynamic compromise of the organ systems, such as the liver, kidney, or spleen normally responsible for circulating DNA clearance may also result in increased levels of plasma DNA. For example, evidence of a role in clearance has already been demonstrated in animal experiments.[11,12]

Recent data indicate that human plasma DNA possesses a short half-life in the circulation.[13] The rapid kinetics of plasma DNA suggests that circulating DNA analysis may be useful in monitoring the clinical progress of trauma patients. Serial measurements of plasma DNA concentrations show that levels are much higher between the first and third hour of injury in patients with organ failure.[14] In patients admitted to intensive care units, plasma DNA concentrations are persistently elevated in those patients who develop a complicated course characterized by organ failure compared with patients with an uncomplicated course. These data further support a potential role for plasma DNA in disease monitoring and prognosis.

Stroke

Stroke is the second leading cause of all deaths worldwide, and in 1990 accounted for 4.4 million victims.[1] Cerebrovascular accidents are catastrophic central neurological events, which result in central nervous system injury, compromise of the blood–brain barrier, and the release of neurobiochemical markers into the circulation.[15] The liberation of neurobiochemical protein markers into the circulation may allow the pathophysiology, progress, and prognosis of patients with cerebrovascular disease to be further evaluated.[16–19] Although raised levels of several neurobiochemical protein markers have been detected in the peripheral blood of patients with stroke so far none have found a place in routine clinical practice.

Circulating plasma DNA concentrations are elevated early after the onset of stroke, are related to the extent of brain damage, and may be used to predict short- and long-term neurobehavioral morbidity and poststroke mortality.[17]

Plasma DNA is also an independent predictor of stroke outcome in patients with clear clinical stroke but who have no obvious acute cerebral lesion evident on computerized tomography or magnetic resonance imaging.[18] Plasma DNA concentrations are also higher within 6 h of symptom onset in patients with hemorrhagic stroke compared with patients with ischemic stroke.[19]

There have been few studies comparing circulating DNA with neurospecific protein markers of brain injury, such as S100 protein. S100 may be a better early diagnostic marker than circulating plasma DNA of the presence or absence of stroke.[19] S100 protein levels were elevated in 64% patients with stroke compared with plasma DNA that was elevated above the normal range in only 18% patients. However, plasma β-*globin* DNA alone is a better discriminator of hemorrhagic from nonhemorrhagic stroke than serum S100 protein especially within 6 h of symptoms onset. As current evidence suggests that thrombolysis is only effective if given within 3 to 6 h of stroke onset in patients with ischemic stroke, the discovery that plasma DNA may help differentiate the two pathologies early in the course of the acute pathology is an important finding.

As with most acute clinical syndromes it is likely that multiple rather than a single test will give the most optimal diagnosis. At optimal cutoff levels, higher odds ratios are achieved using combined selected cutoff points of plasma β-*globin* DNA and S100 protein than either agent alone.[19]

The mechanisms by which circulating plasma DNA increases after stroke is unclear. It is possible that the acute event results in tissue damage and the release of DNA into the circulation. This may be a factor in patients with hemorrhagic stroke as plasma DNA levels are elevated compared with healthy controls. However, in patients with ischemic stroke there is little difference in median circulating levels compared with controls. Another possibility, therefore, is that the chronic underlying inflammatory and atherosclerotic processes that predispose a person to, and that can lead to, stroke may also produce chronically elevated levels of plasma DNA. These levels may occur before the acute event and persist after. Temporal age, sex, and premorbidity-matched studies are needed to see whether the acute changes in plasma DNA levels after stroke and which predict later outcome, are part of a chronically sustained pattern, the acute event, or a mixture of both. Free DNA may be liberated from cells undergoing apoptosis or necrosis.[20,21] Which of these mechanisms are involved in patients with stroke is not known, but cellular ischemia and tissue infarction undoubtedly occur and are associated with disruption of the blood–brain barrier such that increased local liberation of DNA from cells might produce increased systemic plasma concentrations. Since both S100 and DNA are markers found in many non-neurological cells in the body, we cannot rule out the possibility that other noncerebral tissue pathologies associated with cell death may contribute to the increased concentrations after stroke. As most patients present to hospital within hours of the onset of stroke, it is unlikely that recumbent posture, or lying on a floor, accounts for elevated DNA concentrations.

Future studies are required to investigate whether neuro-specific markers, for example, tissue-specific RNA may be released into the circulation after stroke, and may function as better predictors than nonspecific tissue markers.

Acute Coronary Syndrome

ACS is a spectrum of cardiac illness, which includes both angina and acute myocardial infarction (AMI). It is the leading cause of death worldwide.[1] Plasma DNA concentrations are elevated in patients with AMI compared with healthy controls.[22] The relationship between plasma DNA concentrations and the spectrum of ACS, which includes stable and unstable angina, non-ST elevation myocardial infarction (NSTEMI), and ST elevation myocardial infarction (STEMI) was unclear until recently. Plasma DNA concentrations measured early after the onset of symptoms, are elevated in patients with minor cardiac injury, elevated further in patients with ST elevation angina (STEA) and STEMI, and highest in those patients who die within 2 years.[23]

The twofold degree of elevation in plasma DNA concentration in patients with ACS is modest compared to that of trauma, which has elevations of up to 100-fold or more but the degree of tissue injury, although at a critical site, is much less in patients with AMI.

Plasma DNA concentrations measured early after the onset of chest pain may also predict post-ACS complications, such as cardiac failure, cardiac reinfarction, cardiac arrest, and readmission to hospital after discharge within 6 months.

As with stroke and trauma, the mechanisms by which the levels of circulating plasma DNA are increased during ACS are also unclear at present. It is also unclear whether elevated plasma DNA levels are the result of an acute coronary event or of pre-event atherosclerosis, chronic inflammation, and associated apoptosis or necrosis. However, circulating plasma DNA levels increase with age and chronic illness.[24]

CONCLUSIONS

These studies show that circulating DNA and RNA concentrations are elevated early in patients with trauma, stroke, and ACS, and are generally highest in patients with a high risk of death. Plasma DNA may be a potentially useful marker for the evaluation of such patients and the assessment of adverse outcome. The ability for rapid risk stratification may allow clinicians to make more rational therapeutic decisions.

The mechanisms by which the levels of circulating DNA are increased during ACS are unclear at present but may result from increased liberation after cell death. It is unclear whether elevated plasma DNA levels are the result of an acute coronary event or of pre-event atherosclerosis, chronic inflammation, and associated apoptosis or necrosis. Circulating plasma DNA levels increase with age and chronic illness.[6]

ACKNOWLEDGMENTS

Dr. Rainer was supported by the Hong Kong Research Grants Council, funded by Research Grants Council Direct Grant Allocation, 2040804, Research Grant Direct Grant Allocation, 2040944, and Research Grants Council Earmarked Grant, CUHK 4402/03M.

REFERENCES

1. MURRAY, C.J.L. & A.D. LOPEZ. 1997. Mortality by cause for eight regions of the world: global burden of disease study. Lancet **349:** 1269–1276.
2. SAUAIA, A. *et al.* 1998. Multiple organ failure can be predicted as early as 12 hours after injury. J. Trauma **45:** 291–303.
3. STEG, P.G. *et al.* 2003. Comparison of angioplasty and pre-hospital thrombolysis in acute myocardial infarction (CAPTIM) investigators. Circulation **108:** 2851–2856.
4. ALBERS, G.W. *et al.* 2002. SA. ATLANTIS trial: results for patients treated within 3 hours of stroke onset. Alteplase thrombolysis for acute non-interventional therapy in ischemic stroke. Stroke **33:** 493–495.
5. LO, Y.M.D. *et al.* 2000. Plasma DNA as a prognostic marker in trauma patients. Clin. Chem. **46:** 319–323.
6. WITTWER, C.T. *et al.* 1997. The LightCycler: a microvolume multi sample fluorimeter with rapid temperature control. BioTechniques **22:** 176–181.
7. MURRAY, C.J.L. & A.D. LOPEZ. 1997. Alternative projections of mortality and disability by cause 1990–2020: global burden of diseases study. Lancet **349:** 1498–1505.
8. MOORE, F.A. *et al.* 1996. Post injury multiple organ failure: a bimodal phenomenon. J. Trauma **40:** 501–510.
9. RAINER, T. *et al.* 2001. Derivation of a prediction rule for posttraumatic organ failure using. Ann. N. Y. Acad. Sci. **945:** 211–220.
10. JAHR, S. *et al.* 2001. DNA fragments in the blood plasma of cancer patients: quantitation and evidence for their origin from apoptotic and necrotic cells. Cancer Res. **61:** 1659–1665.
11. TSUMITA, T. & M. IWANAGA. 1963. Fate of injected deoxyribonucleic acid in mice. Nature **198:** 1088–1089.
12. CHUSED, T.M. *et al.* 1972. The clearance and localization of nucleic acids by New Zealand and normal mice. Clin. Exp. Immunol. **12:** 465–476.

13. Lo, Y.M.D. *et al.* 1999. Rapid clearance of fetal DNA from maternal plasma. Am. J. Hum. Genet. **64:** 218–224.
14. Lam, N.Y.L. *et al.* 2003. Time course of early and late changes in plasma DNA in trauma patients. Clin. Chem. **49:** 1286–1291.
15. Broderick, J.P. *et al.* 1992. The risk of subarachnoid and intracerebral hemorrhages in blacks as compared with whites. N. Engl. J Med. **326:** 733–736.
16. Abraha, H.D. *et al.* 1997. Serum S-100 protein: relationship to clinical outcome in acute stroke. Ann. Clin. Biochem. **34:** 366–370.
17. Rainer, T.H. *et al.* 2003. Prognostic use of circulating plasma nucleic acid concentrations in patients with acute stroke. Clin. Chem. **49:** 562–569.
18. Lam, N.Y.L. *et al.* 2006. Plasma DNA as a prognostic marker for stroke patients with negative neuroimaging within the first 24 hours of symptom onset. Resuscitation 71–78.
19. Rainer, T.H. *et al.* 2005. Comparison of circulating plasma *b-globin* DNA and S-100 protein concentrations in patients with acute stroke. Abstract Syllabus of the 3rd Mediterranean Conference on Emergency Medicine 132.
20. Fournie, G.J. *et al.* 1995. Plasma DNA as a marker of cancerous cell death. Investigations in patients suffering from lung cancer and in nude mice bearing human tumours. Cancer Lett. **91:** 221–227.
21. Stroun, M. *et al.* 1989. Neoplastic characteristics of the DNA found in the plasma of cancer patients. Oncology **46:** 318–322.
22. Chang, C.P.Y. *et al.* 2003. Elevated cell-free serum DNA detected in patients with myocardial infarction. Clin. Chim. Acta. **327:** 95–101.
23. Lam, N.Y.L. *et al.* 2005. Plasma cell-free DNA concentration and stroke mortality and morbidity. Abstracts for CNAPS IV: Fourth International Conference on Circulating Nucleic Acids in Plasma/Serum 32.
24. Fournie, G.J. *et al.* 1993. Plasma DNA as cell death marker in elderly patients. Gerontology **39:** 215–221.

Cell-Free DNA Levels as a Prognostic Marker in Acute Myocardial Infarction

DIONISIOS ANTONATOS, SOTIRIOS PATSILINAKOS, STAVROS SPANODIMOS, PANAGIOTIS KORKONIKITAS, AND D. TSIGAS

Cardiology Department, Konstantopoulio Hospital, Athens, Greece

ABSTRACT: Cell-free DNA that originates from cell death, circulates in peripheral blood. There are indications that the infarcted myocardium contributes to an increase of cell-free DNA levels. Our aims were to quantify levels of cell-free DNA in patients with acute myocardial infarction (AMI) and examine their correlation with myocardial markers and with postinfarction (PI) clinical course. Thirteen patients (age 57 ± 16 year) admitted with AMI and who underwent thrombolysis with reteplase within 6 h from the onset of chest pain were studied. PB samples were collected on admission and for 5 consecutive days. Creatine kinase (CK) and troponin I (TnI) were measured on admission and every 8 h for 3 consecutive days. Clinical events were recorded throughout the hospitalization period. Cell-free DNA levels were also measured in 30 healthy controls. Log-transformed mean (±SE) of maximum free DNA values in patients higher than controls (6873 ± 357 g.e./mL verses 4112 ± 234 g.e./mL, $P < 0.0001$). Log-transformed maximum values of CK and TnI were correlated with log-transformed free DNA values of first ($r = 0.62$, $P = 0.02$/$r = 0.68$, $P = 0.01$) and second ($r = 0.57$, $P = 0.04$/$r = 0.72$, $P = 0.0053$) PI day. Nine patients (group A) had an uncomplicated PI clinical course and four patients (group B) had recorded events (three with angina and one death). Free DNA levels on the second PI day were higher in group B than group A (1298.0 ± 796.0 g.e./mL verses 244.6 ± 257.7 g.e./mL, $P = 0.003$). In conclusion, free DNA levels are significantly higher in patients with AMI than in controls and may play a role in the prognosis of these patients.

KEYWORDS: cell-free DNA; acute myocardial infarction; prognostic marker; necrosis; apoptosis

INTRODUCTION

Increased levels of cell-free DNA in circulation are associated with cell death as a result of tissue injury or inflammatory reaction.[1] Several reports have

Address for correspondence: Dionisios Antonatos, Cardiology Department, Konstantopoulio Hospital, Agia's Olgas's Str., 14342, N. Ionia, Athens, Greece. Voice: +3012102719862; fax: +3012102756372.
e-mail: antonat@otenet.gr

Ann. N.Y. Acad. Sci. 1075: 278–281 (2006). © 2006 New York Academy of Sciences.
doi: 10.1196/annals.1368.037

suggested that the cell-free circulating DNA is derived from both cell apoptosis and necrosis with only a small amount coming from T cell.[2,3] Recently, much interest has developed in the use of cell-free circulating DNA for clinical purpose in many conditions, such as severe trauma, acute respiratory distress syndrome (ARDS), autoimmune disease, cancer, and pregnancy.[4–7] It is known that a combination of necrosis and apoptosis characterizes the acute stages of myocardial infarction (AMI).[8] Thus, myocyte cell death in AMI may lead to the release of free DNA into peripheral blood with possible prognostic significance. To our knowledge, there is only one report on measurement of circulating cell-free DNA in patients with AMI.[9] Classic myocardial markers, such as creatine kinase (CK) and troponin I (TnI) have been used for the diagnosis and follow-up of patients with AMI. It would be of interest to see whether cell-free DNA levels are associated with the elevation of these markers.

In the present study, we investigated the levels of cell-free DNA in patients with AMI and examined their correlation with myocardial markers and with clinical outcome.

MATERIALS AND METHODS

Thirteen patients (seven men and six women), aged 41–73 years admitted to the cardiology unit of our hospital with AMI were studied. All underwent thrombolysis with reteplase within 6 h from onset of pain. Peripheral blood samples were collected on admission and for 5 consecutive days.

Peripheral blood samples were also collected from 30 healthy control subjects matched for sex and age. These individuals were visiting the hospital for their annual health examination, did not have a history of autoimmune disease, did not have tissue injury or trauma at the time of examination, and their hematological–biochemical profile was normal.

Three milliliters of venous blood was collected into EDTA tube and plasma was isolated by double centrifugation (800 *g* and 16,000 *g*). DNA was extracted from 400 μL of serum sample using a QIAamp Blood Kit (Qiagen, Hilden, Germany) according to the manufacturer's protocol. DNA was measured by quantitative RT-PCR with LightCycler TM (Roche Molecular Biochemicals, Lewes, UK).

CK and TnI were measured on admission and every 8 h for 3 consecutive days.

Major clinical events during hospitalization (death, angina, and need for revascularization) were recorded for all patients.

RESULTS

Log-transformed mean (±SE) of maximum cell-free DNA value in patients was higher than controls (6873 ± 357 g.e./mL vs. 4112 ± 234 g.e./mL, $P < 0.0001$).

Log-transformed maximum values for CK and TnI correlated with log-transformed free DNA values of first ($r = 0.62, P = 0.02/r = 0.68, P = 0.01$) and second ($r = 0.57, P = 0.04/r = 0.72, P = 0.0053$) postinfarction days.

Nine patients had uncomplicated postinfarction clinical course and four patients had recorded events (three had angina and one died).

Free DNA levels of second PI day were higher in patients with complications than those without complications (1298.0 ± 796.0 g.e./mL verses 244.6 ± 257.7 g.e./mL, $P = 0.003$).

DISCUSSION

Our study demonstrated that cell-free plasma DNA levels were significantly higher in patients with AMI than in healthy controls. In serial samples, we found that cell-free DNA was higher in the first specimen.[9]

The prolonged ischemia that happens during AMI, leads to necrosis of myocytes causing the release of cell-free DNA into the blood circulation.

A positive correlation of plasma-free DNA levels with CK and TnI suggests that the amount of free DNA released depends on the extent of the myocardial injury. This could explain the significantly higher free-DNA values that were found in the group of patients who suffered major clinical events. Factors other than the extent of necrosis, such as the accelerated apoptosis in the reperfused myocardium,[9] may contribute to the presence of higher free-DNA levels in these patients.

Although the number of the patients studied is small, these preliminary results warrant further study.

REFERENCES

1. Lo, Y.M.D. *et al.* 2000. Plasma DNA as a prognostic marker in trauma patients. Clin. Chem. **46:** 319–323.
2. Lam, N.Y. *et al.* 2003. Time course of early and late changes in plasma DNA in trauma patients. Clin. Chem. **49:** 1286–1291.
3. Jahr, S. *et al.* 2001. DNA fragments in the blood plasma of cancer patients: quantitations and evidence for their origin from apoptotic and necrotic cells. Cancer Res. **61:** 1659–1665.
4. Sauaia, A. *et al.* 1998. Multiple organ failure can be predicted as early as 12 h after injury. J. Trauma **45:** 291–301.
5. Raptis, L. & H.A. Menrad. 1980. Quantitation and characterization of plasma DNA in normals and patients with systemic lupus erythematosus. J. Clin. Invest. **60:** 1391–1399.
6. Shapiro, B. *et al.* 1983. Determination of circulating DNA levels in patients with benign and malignant gastrointestinal disease. Cancer **15:** 2116–2120.
7. Lo, Y.M.D. *et al.* 1999. Increased fetal DNA concentrations in the plasma of pregnant women carrying fetuses with trisomy 21. Clin. Chem. **45:** 1747–1751.

8. Olivetti, G. *et al.* 1994. Acute myocardial infarction in human is associated with activation of programmed myocyte cell death in the surviving portion of the heart. J. Mol. Cell Cardiol. **28:** 2005–2016.
9. Christine, P.-Y. *et al.* 2003. Elevated cell-free DNA detected in patients with myocardial infarction. Clin. Chim. Acta. **327:** 95–101.

MALDI-TOF Mass Spectrometry for Quantitative, Specific, and Sensitive Analysis of DNA and RNA

CHUNMING DING[a] AND YUK MING DENNIS LO[b]

[a]*Centre for Emerging Infectious Diseases, The Chinese University of Hong Kong, Prince of Wales Hospital, Shatin, New Territories, Hong Kong Special Administrative Region, China*

[b]*Department of Chemical Pathology, The Chinese University of Hong Kong, Prince of Wales Hospital, Shatin, New Territories, Hong Kong Special Administrative Region, China*

ABSTRACT: Cell-free fetal DNA and RNA released into the maternal circulation offer new opportunities to study fetal and pregnancy-associated abnormalities. Similarly, tumor cells can release cell-free DNA and RNA into the peripheral circulation, and these cell-free DNA and RNA can be used for cancer diagnosis, monitoring, and prognosis. However, these DNA and RNA often exist at very low concentrations (for fetal DNA, ~20 genome-equivalents (G.E.)/mL of plasma in the first trimester). The analysis is further complicated by the predominant amount of blood cell-derived DNA and RNA. MALDI-TOF mass spectrometry can provide quantitative, specific, and sensitive analysis of DNA and RNA, and thus may be a useful technology for the field.

KEYWORDS: plasma DNA; plasma RNA; prenatal diagnosis; MALDI-TOF mass spectrometry

INTRODUCTION

Cell-free fetal and tumoral DNA and RNA have been detected in the circulating blood of pregnant women and cancer patients, respectively.[1–6] It is thus possible to use serum and plasma as convenient sources for relatively noninvasive prenatal diagnosis and cancer monitoring. However, these DNA and RNA often exist at very low concentrations (for fetal DNA, ~20 G.E./mL of plasma in the first trimester).[7] The analysis is further complicated by the predominant amount of blood cell-derived DNA and RNA. Very recently, MALDI-TOF mass spectrometry (MS) was adopted for high-throughput qualitative

Address for correspondence: Chunming Ding, Centre for Emerging Infectious Diseases, The Chinese University of Hong Kong, Prince of Wales Hospital, Shatin, New Territories, Hong Kong Special Administrative Region, China. Voice: +852-2252-8842; fax: +852-2635-4977.
e-mail: cmding@cuhk.edu.hk

**Ann. N.Y. Acad. Sci. 1075: 282–287 (2006). © 2006 New York Academy of Sciences.
doi: 10.1196/annals.1368.038**

and quantitative analysis of nucleic acids.[8–12] MALDI-TOF MS offers an accurate and direct measurement of the molecular weights of the nucleic acid products, at a resolution about 100 times higher than capillary sequencing. Automated MALDI-TOF MS may be a useful technology platform for analyzing circulating cell-free DNA and RNA. Cell-free fetal DNA and RNA are used in the subsequent sections to illustrate the utilities of MALDI-TOF MS. It is likely that the analyses should also be applicable to tumoral DNA and RNA.

Qualitative Analysis

Clinically, it is useful to detect fetal-specific DNA and RNA. For example, the fetal RhD status determination is useful in the treatment of RhD-sensitized pregnant women.[13] High accuracy has been achieved in clinical settings for *RHD* genotyping.[14] It is relatively easier to detect the fetal-specific DNA sequences like *SRY* and *RHD* genes when pregnant women do not have homologous sequences. In most other situations, such as β-thalassemia and cystic fibrosis, the fetal sequences of interest are overwhelmed by virtually identical maternal sequences that can be more than 99% of the total cell-free DNA in the plasma.[7] Limited successes have been achieved in these situations.[11,15,16]

Sensitivity

Cell-free fetal DNA has been shown to exist at low concentrations in the plasma and serum of pregnant women. In the first trimester, a mean of 25.4 GE (1 GE equals to 6.6 pg of human genomic DNA) per mL of plasma was detected.[7] In some cases, the concentration can be as low as 3 GE/mL of plasma. Real-time quantitative PCR[17] has been used extensively for its simplicity and high sensitivity. In addition, the uracil *N*-glycosylase can be used to destroy the contaminants from previous PCR products. Reliable detection has been achieved for the detection of *SRY* and *RHD* genes.[18–23]

Specificity

Specificity is not a serious issue for fetal *SRY* and *RHD* detection since there is no maternal homologous sequence present. However, many genetic diseases are caused by subtle mutations (point mutations, small insertion/deletions). Thus, the maternal background DNA only differs slightly from the fetal-specific DNA of interest. It is thus technically challenging to achieve the specificity for the detection of fetal DNA in the maternal plasma, without sacrificing the sensitivity. In limited cases, allele-specific PCR has been used successfully.[15,24] But the design of allele-specific PCR is tricky for most single-point mutations. Moreover, some sensitivity is sacrificed to achieve sufficient specificity.

Interestingly, fetal DNA was found to be shorter (<200 bp) than maternal DNA in the maternal circulation.[25] Thus, it is feasible to use size fractionation to enrich fetal DNA.[26,27] However, since a substantial proportion of the maternal DNA is also short,[25] size fractionation can only achieve at most two- to threefold enrichment. The labor-intensive process of size fractionation may also cause loss of fetal DNA, and increase the risk for DNA contamination.

Recently, MALDI-TOF MS has been used widely for genotyping single nucleotide polymorphisms (SNPs).[8] SNPs are indeed point variations. DNA sequences are first amplified by PCR. Subsequently, extra deoxynucleotides are removed by shrimp alkaline phosphatase. The following primer extension reaction is used to distinguish between the mutant and normal allele. In the primer extension reaction, an extension primer binds to the DNA sequence immediately next to the mutation site. A mixture of dideoxynucleotides and deoxynucleotides is added such that one or two bases are added to the mutant allele and the normal allele, generating two products of different molecular weights. MALDI-TOF MS is extremely accurate in determining the molecular weights of the DNA molecules, at a resolution of a few Daltons. Thus, MALDI-TOF MS can detect the two alleles differing by as few as one base robustly with high automation.

We have initially tested the established MassEXTEND technology detailed above in detecting paternally inherited β-thalassemia mutants with mixed results. When fetal DNA is present at a very low percentage compared with the maternal DNA, sometimes the fetal mutant signal is dwarfed by the maternal signal, resulting in false negative results. We subsequently developed the single allele base extension reaction (SABER).[11] In SABER, only one dideoxynucleotide is added for primer extension. This dideoxynucleotide can only be used by the fetal mutant allele for primer extension reaction. The maternal allele thus would not be able to produce any signal in the MALDI-TOF mass spectra. SABER greatly reduced the maternal background signal and allowed for highly sensitive and specific detection of paternally inherited β-thalassemia mutants in our proof of principle study.[11] The extra step of primer extension is a linear amplification step. So it is likely that the sensitivity can also be improved. We have previously shown that single DNA copy sensitivity could be achieved, even when multiplex PCR was carried out.[10] The high sensitivity of MALDI-TOF MS platform has also been demonstrated by several other groups, in a variety of applications.[28–30]

We also demonstrated the ability to detect paternally inherited SNPs.[11] This capability allowed us to use a nearby tagging SNP to distinguish the paternally inherited β-thalassemia mutant from the same maternal β-thalassemia mutant. Thus, the prenatal exclusion of β-thalassemia major can be applied to couples carrying the same β-thalassemia mutations.

Quantitative Analysis

The concentrations of cell-free fetal DNA in the maternal plasma are also of clinical use. For example, fetal DNA level is increased by several fold in pre-eclampsia, which affects 6–8% of pregnant women.[31–33] Increased cell-free fetal DNA level has also been observed in certain aneuploidies[34,35] and other pregnancy-associated complications.[36–40] The clinical utility of these findings has not yet been widely used, possibly due to two reasons: (*a*) biologically, the cell-free fetal DNA levels in normal and abnormal pregnancies overlap significantly; (*b*) quantification inaccuracy causes the overlap of cell-free fetal DNA levels in different pregnancies. It is quite common that only 10 to 20 copies of fetal template are present prior to PCR. The measurement error of real-time quantitative PCR may increase significantly on account of the low concentrations of the cell-free fetal DNA.

To address this deficiency, the real competitive PCR (rcPCR) technology, on the basis of competitive PCR, primer extension, and MALDI-TOF MS, has recently been developed.[9] In rcPCR, a competitor with known concentration and one artificial point mutation compared with the sequence of interest is co-amplified with the sequence of interest in the same PCR mutation. PCR artifacts are thus minimized and good quantification was achieved even at concentrations as low as five copies of DNA.[9] Other groups have subsequently validated the rcPCR technology, in a number of other applications.[28–30] It is possible that rcPCR may offer some advantage over real-time quantitative PCR in analyzing low-level DNA and RNA.[28] The rcPCR technology thus may be a useful technology for circulating cell-free DNA analysis.

CONCLUSIONS

Cell-free fetal DNA and RNA from fetuses or tumors offer a convenient and noninvasive source for clinical diagnosis, monitoring, and prognosis. The low concentrations of these nucleic acids and the overwhelming background DNA and RNA released from blood cells present a challenging analytical problem. Further technological improvements are needed for sensitive, specific, and quantitative analysis so that the full potential of these nucleic acids can be realized.

ACKNOWLEDGMENT

YMDL is supported by a Central Allocation Grant (CUHK 01/03C) from the Research Grants Council of the Hong Kong Special Administrative Region (China).

REFERENCES

1. CHEN, X.Q. *et al.* 1996. Microsatellite alterations in plasma DNA of small cell lung cancer patients. Nat. Med. **2:** 1033–1035.
2. AWROZ, H. *et al.* 1996. Microsatellite alterations in serum DNA of head and neck cancer patients. Nat. Med. **2:** 1035–1037.
3. LO, Y.M. *et al.* 1997. Presence of fetal DNA in maternal plasma and serum. Lancet **350:** 485–487.
4. KOPRESKI, M.S. *et al.* 1999. Detection of tumor messenger RNA in the serum of patients with malignant melanoma. Clin. Cancer Res. **5:** 1961–1965.
5. CHEN, X.Q. *et al.* 2000. Telomerase RNA as a detection marker in the serum of breast cancer patients. Clin. Cancer Res. **6:** 3823–3826.
6. NG, E.K. *et al.* 2003. mRNA of placental origin is readily detectable in maternal plasma. Proc. Natl. Acad. Sci. USA **100:** 4748–4753.
7. LO, Y.M. *et al.* 1998. Quantitative analysis of fetal DNA in maternal plasma and serum: implications for noninvasive prenatal diagnosis. Am. J. Hum. Genet. **62:** 768–775.
8. TANG, K. *et al.* 1999. Chip-based genotyping by mass spectrometry. Proc. Natl. Acad. Sci. USA **96:** 10016–10020.
9. DING, C. & C.R. CANTOR. 2003. A high-throughput gene expression analysis technique using competitive PCR and matrix-assisted laser desorption ionization time-of-flight MS. Proc. Natl. Acad. Sci. USA **100:** 3059–3064.
10. DING, C. & C.R. CANTOR. 2003. Direct molecular haplotyping of long-range genomic DNA with M1-PCR. Proc. Natl. Acad. Sci. USA **100:** 7449–7453.
11. DING, C. *et al.* 2004. MS analysis of single-nucleotide differences in circulating nucleic acids: application to noninvasive prenatal diagnosis. Proc. Natl. Acad. Sci. USA **101:** 10762–10767.
12. DING, C. *et al.* 2004. Simultaneous quantitative and allele-specific expression analysis with real competitive PCR. BMC Genet. **5:** 8.
13. LO, Y.M. *et al.* 1998. Prenatal diagnosis of fetal RhD status by molecular analysis of maternal plasma. N. Engl. J. Med. **339:** 1734–1738.
14. GAUTIER, E. *et al.* 2005. Fetal RhD genotyping by maternal serum analysis: a two-year experience. Am. J. Obstet. Gynecol. **192:** 666–669.
15. CHIU, R.W. *et al.* 2002. Prenatal exclusion of beta thalassaemia major by examination of maternal plasma. Lancet **360:** 998–1000.
16. LI, Y. *et al.* 2004. Size separation of circulatory DNA in maternal plasma permits ready detection of fetal DNA polymorphisms. Clin. Chem. **50:** 1002–1011.
17. GIBSON, U.E. & C.A. HEID *et al.* 1996. A novel method for real time quantitative RT-PCR. Genome Res. **6:** 995–1001.
18. JOHNSON, K.L. *et al.* 2004. Interlaboratory comparison of fetal male DNA detection from common maternal plasma samples by real-time PCR. Clin. Chem. **50:** 516–521.
19. BROJER, E. *et al.* 2005. Noninvasive determination of fetal RHD status by examination of cell-free DNA in maternal plasma. Transfusion **45:** 1473–1480.
20. SEKIZAWA, A. *et al.* 2001. Accuracy of fetal gender determination by analysis of DNA in maternal plasma. Clin. Chem. **47:** 1856–1858.
21. FINNING, K.M. *et al.* 2002. Prediction of fetal D status from maternal plasma: introduction of a new noninvasive fetal RHD genotyping service. Transfusion **42:** 1079–1085.

22. Costa, J.M. *et al.* 2001. First-trimester fetal sex determination in maternal serum using real-time PCR. Prenat. Diagn. **21:** 1070–1074.
23. Rijnders, R.J. *et al.* 2004. Clinical applications of cell-free fetal DNA from maternal plasma. Obstet. Gynecol. **103:** 157–164.
24. Nasis, O. *et al.* 2004. Improvement in sensitivity of allele-specific PCR facilitates reliable noninvasive prenatal detection of cystic fibrosis. Clin. Chem. **50:** 694–701.
25. Chan, K.C. *et al.* 2004. Size distributions of maternal and fetal DNA in maternal plasma. Clin. Chem. **50:** 88–92.
26. Li, Y. *et al.* 2004. Improved prenatal detection of a fetal point mutation for achondroplasia by the use of size-fractionated circulatory DNA in maternal plasma–case report. Prenat. Diagn. **24:** 896–898.
27. Li, Y. *et al.* 2005. Detection of paternally inherited fetal point mutations for beta-thalassemia using size-fractionated cell-free DNA in maternal plasma. JAMA **293:** 843–849.
28. Elvidge, G.P. *et al.* 2005. Development and evaluation of real competitive PCR for high-throughput quantitative applications. Anal. Biochem. **339:** 231–241.
29. Yang, H. *et al.* 2005. Sensitive detection of human papillomavirus in cervical, head/neck, and schistosomiasis-associated bladder malignancies. Proc. Natl. Acad. Sci. USA **102:** 7683–7688.
30. Liu, J. *et al.* 2005. SARS transmission pattern in Singapore reassessed by viral sequence variation analysis. PLoS Med. **2:** e43.
31. Leung, T.N. *et al.* 2001. Increased maternal plasma fetal DNA concentrations in women who eventually develop preeclampsia. Clin. Chem. **47:** 137–139.
32. Lo, Y.M. *et al.* 1999. Quantitative abnormalities of fetal DNA in maternal serum in preeclampsia. Clin. Chem. **45:** 184–188.
33. Levine, R.J. *et al.* 2004. Two-stage elevation of cell-free fetal DNA in maternal sera before onset of preeclampsia. Am. J. Obstet. Gynecol. **190:** 707–713.
34. Lo, Y.M. *et al.* 1999. Increased fetal DNA concentrations in the plasma of pregnant women carrying fetuses with trisomy 21. Clin. Chem. **45:** 1747–1751.
35. Zhong, X.Y. *et al.* 2000. Fetal DNA in maternal plasma is elevated in pregnancies with aneuploid fetuses. Prenat. Diagn. **20:** 795–798.
36. Zhong, X.Y. *et al.* 2000. High levels of fetal erythroblasts and fetal extracellular DNA in the peripheral blood of a pregnant woman with idiopathic polyhydramnios: case report. Prenat. Diagn. **20:** 838–841.
37. Leung, T.N. *et al.* 1998. Maternal plasma fetal DNA as a marker for preterm labour. Lancet **352:** 1904–1905.
38. Lau, T.K. *et al.* 2000. Cell-free fetal deoxyribonucleic acid in maternal circulation as a marker of fetal-maternal hemorrhage in patients undergoing external cephalic version near term. Am. J. Obstet. Gynecol. **183:** 712–716.
39. Sekizawa, A. *et al.* 2002. Increased cell-free fetal DNA in plasma of two women with invasive placenta. Clin. Chem. **48:** 353–354.
40. Sekizawa, A. *et al.* 2001. Cell-free fetal DNA is increased in plasma of women with hyperemesis gravidarum. Clin.Chem. **47:** 2164–2165.

Rapid Prenatal Diagnosis by QF-PCR: Evaluation of 30,000 Consecutive Clinical Samples and Future Applications

VINCENZO CIRIGLIANO,[a,b] GIANFRANCO VOGLINO,[c] ANTONELLA MARONGIU,[c] PAZ CAÑADAS,[a] ELENA ORDOÑEZ,[a] ELISABET LLOVERAS,[d] ALBERTO PLAJA,[d] CARME FUSTER,[a,b] AND MATTEO ADINOLFI[e]

[a]*Departament de Genética Molecular, General Lab, 08029 Barcelona, Spain*

[b]*Unitat de Biologia, Departament de Biologia Cellular, Fisiologia, Immunologia, Universitat Autònoma de Barcelona, E-08193, Bellaterra, Barcelona, Spain*

[c]*Molecular Genetics and Cytogenetics Lab, Promea-Day Surgery, 1026 Turin, Italy*

[d]*Departament de Citogenètica, General Lab, 08029 Barcelona, Spain*

[e]*The Galton Laboratory, University College London, NW1 2HE, London, UK*

ABSTRACT: Rapid prenatal diagnoses of major chromosome abnormalities can be performed on a large scale using highly polymorphic short tandem repeats (STRs) amplified by the quantitative fluorescent polymerase chain reaction (QF-PCR). The assay was introduced as a preliminary investigation to remove the anxiety of the parents waiting for the results by conventional cytogenetic analysis using amniotic fluid or chorionic cells. However, recent studies, on the basis of the analyses of several thousand samples, have shown that this rapid approach has a very high rate of success and could reduce the need for cytogenetic investigations. Its high efficiency, for example, allows early interruption of affected fetuses without the need of waiting for completion of fetal karyotype. The main advantages of the QF-PCR are its accuracy, speed, automation, and low cost that allows very large number of samples to be analyzed by few operators. Here, we report the results of using QF-PCR in a large series of consecutive clinical cases and discuss the possibility that, in a near future, it may even replace conventional cytogenetic analyses on selected samples.

KEYWORDS: prenatal diagnosis; QF-PCR; microsatellite; Down's syndrome

Address for correspondence: Vincenzo Cirigliano, Genética Molecular, General Lab, c/Vila Domat 288, 08029 Barcelona, Spain. Voice: +34-93-2022426; fax: +34-93-4140222.
e-mail: vc@general-lab.com

**Ann. N.Y. Acad. Sci. 1075: 288–298 (2006). © 2006 New York Academy of Sciences.
doi: 10.1196/annals.1368.039**

INTRODUCTION

Prenatal diagnoses of chromosome disorders are carried out on fetal samples (amniotic fluid or chorionic villus sample [CVS]) that need to be cultured for several days to obtain cells in metaphase suitable for cytogenetic analysis. The interval between the collection of the samples and the final report (around 10 to 14 days) is a time of great anxiety for the parents. In recent years, several rapid methods have been devised to allow rapid prenatal diagnoses of the most frequent aneuploidies, including fluorescent *in situ* hybridization (FISH) on cells in interphase[1–4] and the quantitative fluorescent polymerase chain reaction (QF-PCR) test.[5–11]

The QF-PCR is based on the amplification of highly polymorphic, chromosome-specific short tandem repeats (STR) using fluorescent primers; the final products can then be analyzed using an automated DNA sequencer. STRs are chromosome-specific tri-, tetra-, or penta-nucleotide motifs that are reiterated several times to form DNA sequences of different lengths. Because of their variable lengths, STRs are highly polymorphic (up to 15–20 alleles); they are also very stable during the life of an individual.[10,12] The principle of the QF-PCR test is on the basis of the assumption that, within the early phase of amplification, the amounts of DNA generated by two alleles at a given STR locus is proportional to the amounts of the sequences present in the initial target template. In normal heterozygous subjects, the QF-PCR products of a chromosome-specific STR should show two peaks with similar fluorescent activities and thus a ratio close to 1:1; if the employed STR is highly polymorphic, few normal samples should be homozygous and produce a single fluorescent peak (FIG. 1). In samples retrieved from trisomic fetuses (FIG. 1) the presence of three chromosomes may be detected with the corresponding STRs either as three fluorescent peaks with a ratio of 1:1:1 (trisomic triallelic) or as two peaks with a ratio 2:1 (trisomic diallelic). On account of the high polymorphism of the selected STRs, only a few trisomic samples should produce uninformative homozygous patterns. Triploidies are characterized by the presence of three doses of each chromosome-specific STRs.

Sexing can be performed using the Y chromosome-specific SRY sequence (present in males but absent in females) and the modified nonpolymorphic Amelogenin (AMXY) sequences, which produce one product for the X and one for the Y chromosomes (FIG. 1).[13]

However, until a few years ago, the detection of X and Y chromosome abnormalities was hampered by the lack of suitable highly polymorphic markers, but in recent times several DNA sequences have been identified that also allow to test samples for the presence of all types of sex chromosome disorders.[10,14–16]

MATERIALS AND METHODS

Samples were collected between December 1999 and October 2005 in two genetic centers (General Lab, Barcelona, Spain and Sant' Anna Hospital, Turin,

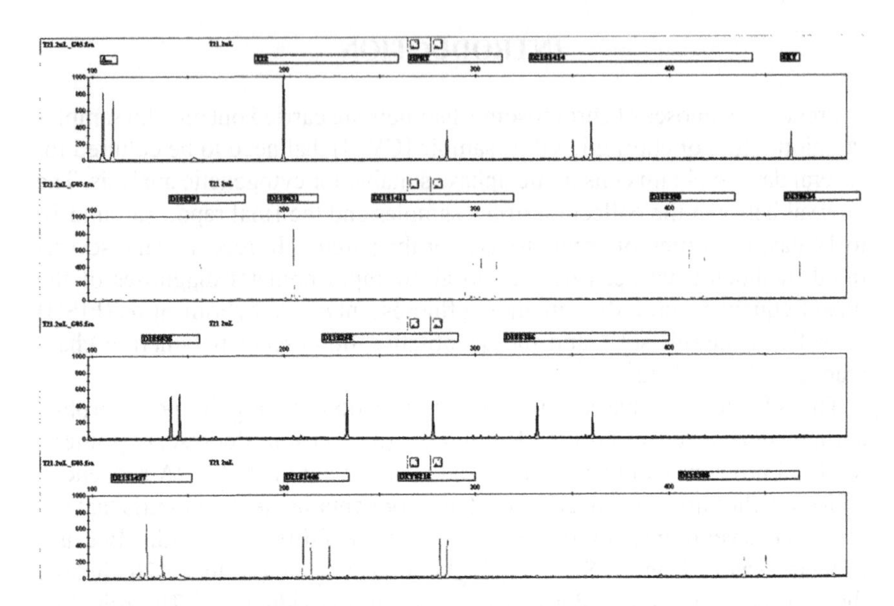

FIGURE 1. Prenatal diagnosis of 47,XY + 21 by QF-PCR. Fetal sex was determined by detecting the X- and Y-specific products of the AMXY. The normal male sex chromosome complement was confirmed by the heterozygous pattern (ratio 1:1) of both pseudoautosomal markers (X22 and DXYS218). Normal chromosome 13 and 18 copy number was detected by the heterozygous patterns (ratio 1:1) for all markers on these chromosomes. Trisomy 21 was detected as trisomic diallelic pattern (2:1) for the D21S1414 and D21S1411. D21S1446 show a trisomic triallelic pattern (1:1:1).

Italy). Both institutions were also offering rapid QF-PCR service to other cytogenetic laboratories, private clinics, and public hospitals, so that several high-risk pregnancy samples were received nationwide on a daily basis.

A large cohort of amniotic samples (28,040) and chorionic villous samples (1120) were collected between 11 and 31 weeks of gestation. The remaining samples were 196 fetal blood samples and 694 tissues retrieved from aborted fetuses.

The most common indications for an invasive procedure were: raised risk of chromosome disorders for advanced maternal age (30%) and biochemical screening performed on maternal serum (32%); 6% of these cases were also associated with an increased nuchal translucency; abnormal ultrasounds were detected in 7% of cases and parental anxiety generated 22% of samples.

All women received genetic counseling, including detailed information on the advantages and limitations of the rapid QF-PCR assay; routine informed consent was obtained in all cases.

Small aliquots (0.5–1 mL) of amniotic fluid were collected for the molecular assay; they were carefully inspected for the presence of contaminating maternal cells and a full record kept. CVS tissues were microdissected under an inverted microscope; two small villi were isolated for DNA extraction and tested separately by QF-PCR. All remaining cells or tissues were stored at 4°C for up to 1 month or frozen for longer periods of time, thus allowing repeat molecular tests, if required. Genomic DNA was extracted from all samples using a resin-based procedure (Instagene Matrix, Bio-Rad Laboratories Inc., Hercules CA, USA) as previously described.[9] This DNA extraction procedure has also the advantage of adding different Chelex volumes, depending on the amount of cells thereby giving similar final concentration of DNA. After centrifugation for 5 min at 14,000 rpm the supernatant was removed and Chelex added to the pellet; thus, DNA extraction can be performed using the same Eppendorf tube that contained the original sample.[9–11] This procedure has been successfully employed using 0.2–1 mL of uncultured amniotic fluid, 200-μL cell cultures, 5 μL of fetal or peripheral blood, and 0.2 mg of fetal or adult tissues. The suspensions were incubated for 8 min at 70°C followed by 4 min at 95°C. After vortexing and a quick spin in a centrifuge, single-strand DNA was ready for QF-PCR amplification.

QF-PCR was performed in both centers using the large set of STR markers for chromosomes X, Y, 13, 18, and 21 as listed in TABLE 1. Primer sequences, retrieved from The Genome Data Base (http://www.gdb.org) and Cooperative Human Linkage Center (http://www.chlc.org), were used for multiplex QF-PCR without modifications. A total of six STRs for the X and Y chromosomes, six for chromosome 21, five for chromosome 18, and four for chromosome 13 were selected because of their high level of heterozygosity, thus reducing the number of uninformative samples.[8–11,13,15] Nonpolymorphic sequence of the *AMXY* gene and SRY were also included to detect fetal sex.

All STRs were stable tri-, tetra-, and penta-nucleotides, suitable for multiplex PCR because of their minimal production of artifacts during amplification. The location of STRs along the chromosomes was also taken into account in order to increase the possibility of detecting partial trisomies. All forward primers (Roche-TIB Molbiol, Berlin, Germany) were labeled with fluorescent molecules allowing accurate sizing and quantification of QF-PCR products. Primers producing amplicons of similar sizes were labeled with different fluo-

TABLE 1. Multiplex QF-PCR assays

Mix 1	Mix 2	Chr. XY	Chr. 21	Chr. 18	Chr. 13
AMXY	SRY	AMXY	D21S11	D18S51	D13S305
DXYS218	X22	HPRT	D21S1411	D18S386	D13S631
D21S1446	HPRT	DXS6803	D21S1412	D18S390	D13S634
D21S1414	D21S1411	DXS6809	D21S1435	D18S391	D13S742
D18S535	D18S386	DXS8377	D21S1437		
D18S391	D13S258	SMBA			
D13S631	D13S634				

Mix 1 and 2 are used to screen all prenatal samples with three markers on chromosomes 13, 18, and 21, two pseudoautosomal X and Y, and one X-linked marker. AMXY and SRY are used for sexing. If required, more markers on selected chromosomes are added with the reported chromosome- specific assays.

rochromes in order to be amplified and analyzed in the same multiplex QF-PCR reactions.[17]

In the course of this study the two centers developed different combinations of primers in multiplex QF-PCR reactions in order to simplify sample handling and data analysis.[10,11,13,18]

Fetal sex and chromosomes X, Y copy numbers were assessed for all cases by amplification of the homologous gene AMXY together with the pseudoautosomal X22 and chromosome X-specific markers (TABLE 1).[13,14] The D21S1411 STR was also used as internal control to quantify the X-linked HPRT marker in order to assess X chromosome copy number in all samples. The combination of the two markers was also useful to control the diagnosis of chromosome 21 trisomies.[14]

Aneuploidy screenings were performed using two multiplex QF-PCR assays that included three STRs on chromosomes X, 13, 18, and 21, and the AMXY. Samples showing homozygosity for two chromosome-specific markers, as well as all aneuploid cases, were retested with other chromosome-specific STRs.[10,11]

The fluorescent QF-PCR products and size standards were analyzed by capillary electrophoresis on ABI 3100-Avant and ABI 3130 automated DNA sequencers and GeneScan or GeneMapper 3.7 Software (Applied Biosystems, Foster City, CA, USA) as previously described.[6–9,12,14]

All prenatal samples were processed and reported within 24–48 h; in 98% of cases results were made available within the next working day.

Conventional cytogenetic analyses were performed on all prenatal samples, cultured and harvested according to standard procedures; depending on the specimen, the results were issued between 5 and 28 days (mean reporting time was 2 weeks for amniotic fluid).

RESULTS

The results of the QF-PCR screenings and conventional cytogenetic analyses are given in TABLES 2 and 3.

DNA extraction and QF-PCR amplification were successful in 29,982 cases (99.9%). Only 18 samples failed to amplify; all these cases were failed cytogenetic cultures heavily contaminated by bacterial or fungal cells. These samples had not been analyzed by QF-PCR soon after collection, since these molecular tests were requested by cytogeneticists only after cell culture failures. Cytogenetic analysis was not possible in 42 prenatal cases because of cell culture failures (33 samples) or maternal cell overgrowth (9 cases). In all these cases the QF-PCR tests were successful and gave normal results thus, a second invasive procedure was avoided.

A high proportion of bloodstained amniotic fluids, suspected of being contaminated by maternal cells, could also be analyzed by QF-PCR. The

TABLE 2. Results of testing 30,000 consecutive fetal samples by QF-PCR and conventional cytogenetic analysis

Karyotype	QF-PCR	Cytogenetics
46,XX; 46,XY	28,668	28,761
47,XX + 21; 47,XY + 21	513*	516[1]
47,XX + 18; 47,XY + 18	216	216
47,XX + 13; 47,XY + 13	86	86
69,XXX; 69,XXY	71	71
45,X	76	77
47,XXY	43*	44
47,XYY	37	37
47,XXX	21[2]	16
49,XXXXX	2	2
49,XXXXY	1	1
48,XXY + 21	1	1
48,XXY + 18	2	2
Mosaics	26	47
92,XXXX	–	2
Other aneuploidies	–	27
Structural balanced	–	31
Structural unbalanced	11	21
Maternal contamination[3]	208	9
Failed tests	18	33
Total abnormalities	1,106	1,197

[1]28 cases of unbalanced Robertsonian translocations, [2] 5 cases diagnosed as 47XXX by QF-PCR were mosaics 46,XX/45,X, and [3] uninformative results of QF-PCR tests due to high levels of maternal cell contamination, or maternal cells overgrowth in *in vitro* culture. * Three trisomies 21 and one 47,XXY were undetectable by QF-PCR because of high level maternal cell contamination in the uncultured amniotic fluid.

TABLE 3. QF-PCR efficiency on 30,000 clinical samples

Sensitivity	92.3%
High-risk abnormalities detection	95%
Specificity	100%
Positive predictive value (PPV)	100%
Negative predictive value (NPV)	99.8%

Sensitivity is the percentage of abnormal karyotypes detected by QF-PCR. Specificity is the percentages of unaffected fetuses detected by QF-PCR. Positive predictive value (PPV) is the probability of a fetus with positive QF-PCR result to be confirmed as aneuploid by cytogenetic analysis. Negative predictive value (NPV) is the probability of a fetus with normal QF-PCR result to be confirmed as normal by cytogenetic analysis.

assessment of the fetal origin of the predominant cell population in the samples was made possible by analysis of maternal buccal cells with the same markers. However, in 208 cases of heavily bloodstained amniotic fluids, QF-PCR showed the presence of high level of maternal cell contamination and no result could be obtained other than fetal sex; in four of these cases cytogenetic analyses revealed the presence of affected fetuses (three trisomies 21 and a 47,XXY) that were not detectable by the molecular test.

Out of 28,761 fetuses diagnosed as normal by conventional cytogenetic tests, 28,668 (99%) were correctly identified as normal by QF-PCR. Fetal sexing was correctly performed in all samples using the AMXY and SRY (FIGS. 1 and 2), even in four cases with deletion of the Y-specific product of the AMXY sequence that were correctly identified as retrieved from normal males. The X-specific AMXY sequence was found duplicated in five males with otherwise normal karyotypes and the Y product in four cases. In these samples, multiplex QF-PCR performed with other X-linked and pseudoautosomal markers allowed the correct diagnosis.

Using the set of autosomal and sex markers shown in TABLE 1, 1106 fetuses were diagnosed as chromosomically abnormal out of 1197 disorders detected by conventional cytogenetic (92.3%) (TABLE 3). They included 815 fetuses affected by trisomies 13, 18, and 21 with 100% specificity. The QF-PCR method was also highly successful in the detection of fetuses with triploidies (69,XXX and 69,XYY) and double trisomies (48,XXY + 21 and 48,XXY + 18). Equally successful was the QF-PCR detection of 47,XXY and 47,XYY and most other X and Y chromosome disorders (TABLE 1 and 2). Five out of 21 fetuses diagnosed as 47,XXX by QF-PCR were mosaics 46,XX/45,X when tested by cytogenetic tests. Seventy-six fetuses with Turner's Syndrome, out of 77, were correctly diagnosed; the exception was a single early case with a rare polymorphic duplication of the X22 marker that was not examined with all X and Y markers listed in TABLE 1 and FIGURE 2.

The selection of the STRs distributed over the examined chromosomes also allowed the detection of 11 out of 21 fetuses with unbalanced chromosomal abnormalities involving chromosomes 13, 18, X, and Y. However, as expected all balanced chromosome disorders—of no or little immediate clinical

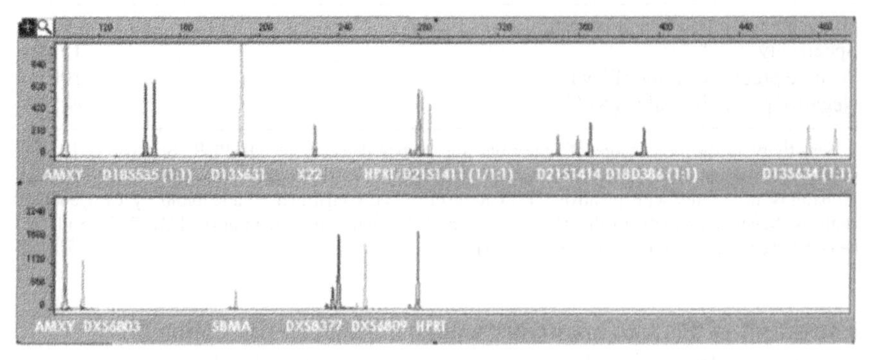

FIGURE 2. Prenatal diagnosis of 45,X by QF-PCR. Female sex was assessed by the lack of Y-specific product of AMXY and SRY sequences. The single HPRT peak has the same dose as D21S1411 alleles; the presence of a single X chromosome is confirmed by detecting single alleles in one pseudoautosomal and five X-linked markers.

consequence—were not detected. Of course, since the QF-PCR was deliberately not aimed at detecting all chromosome disorders, aneuploidies involving chromosomes 5, 15, 16, and 22 were missed. Undetectable by the molecular assay were also two cases of tetrasomy (92,XXXX).

DISCUSSION

The present results confirm that the rapid QF-PCR approach provides correct diagnoses of the majority of prenatal tests. Using the two multiplexes and several STR markers, all normal fetuses were correctly diagnosed, as well all cases of trisomies affecting chromosomes 13, 18, and 21 and triploidies. The QF-PCR showed an overall sensitivity of 95% in detecting chromosome abnormalities with a potential risk for the fetus.

The QF-PCR tests could be carried out even in most samples contaminated with maternal cells, provided maternal buccal cells were also tested, thus allowing to distinguish the fetal from the maternal STR profiles. Contamination with bacteria did not affect the QF-PCR results, while most of these samples could not be analyzed by conventional cytogenetic.[10,11]

The main advantages of the molecular assay are: (*a*) that, within 24–36 h from the collection of the sample it is possible to inform the parents about the results, thus relieving their anxiety—if the fetus is found to be normal—or allowing prompt interruption of pregnancy—if the fetus is affected by a major chromosome abnormality. This policy is now followed in many genetic units, especially if ultrasound examinations also suggest the presence of an affected fetus; (*b*) its low cost per sample, once the DNA sequencer has been purchased; the QF-PCR method is not covered by patent or licensing fees; (*c*) its automation that allows large-scale application. A few operators with a DNA scanner with 16 capillaries can test up to 50 samples per day, but up to 1000 samples per day can be investigated using high performance instruments; and (*d*) in case of technical difficulties, such as failure of PCR amplification, a sample can be retested in a short period of time.

The limitation of the QF-PCR assay is that only major chromosome disorders are detected using the restricted number of selected STRs, missing rare trisomies and duplications or deletions. However, eventually, a third multiplex can be added in selected cases in order to amplify a small number of selected markers, thus detecting these latter disorders.

The overall success of the QF-PCR tests[9–11,19–23] has raised the possibility of replacing conventional cytogenetic analysis with this rapid molecular assay (UK National Screening Committee, 2004), a suggestion recently discussed and rejected by Caine *et al.*[24]

We believe that the QF-PCR could greatly reduce the load of conventional cytogenetic analyses, if pregnancies are, first of all, carefully monitored by combinations of ultrasound, selected biochemical tests, and maternal age-associated risks.[25,26] In young pregnant women, if the noninvasive tests—correctly performed during the first and second trimester of pregnancy—provide normal results, a further invasive procedure (and thus conventional cytogenetic and QF-PCR) need not be performed since the risk of aborting a normal fetus, as a consequence of the invasive procedure, is higher than the risk of having a chromosomal abnormality. However, some pregnant women under 35 years may request an invasive investigation to remove their anxiety; in the majority of these cases the rapid analysis by QF-PCR should reveal that the fetus is normal and further cytogenetic investigations would not be needed.

When some noninvasive tests, particularly ultrasound scan, strongly suggest the presence of a fetal chromosome disorder, but the QF-PCR result is normal, cytogenetic analysis should be performed since the fetus might be affected by a chromosome abnormality not investigated by the molecular assay.

About 6 years ago we suggested that in developing countries, such as China, India, or Pakistan, where prenatal detection of chromosome disorders by conventional cytogenetic analyses are hampered by shortage of funds and specialized technicians, noninvasive procedures should be performed and, when indicated, the QF-PCR assays could be used as the only prenatal test, in view of its very low cost and the high number of analyses that can be carried in a single genetic unit.[27]

The QF-PCR approach can also be used to establish the parental origin of the extra chromosome present in affected fetuses and to confirm paternity, if required. It has also been employed for the detection of RhD genotype of RH-D+ fathers, in order to establish the probability of having Rh-D−children.[28] It is also possible to include primers to detect point mutations causing single gene defects at the same time as testing aneuploidies.[17]

Recently, in a preliminary study, we have successfully investigated the diagnostic value of using a CGH microarray for the detection of chromosome abnormalities by comparing its results with those obtained by QF-PCR and conventional cytogenetics.[29]

At present, many new noninvasive approaches are also under investigations, including the detection of fetal DNA or RNA in maternal blood.[30–35] Large number of samples should be tested before the diagnostic values of these investigations can be assessed, but we believe the we are going through a period of transition, when improved noninvasive screenings, the QF-PCR and selected CGH microarrays could reduce the load of cytogenetic analyses in the prenatal diagnosis of chromosome disorders.

REFERENCES

1. KLINGER, K. *et al.* 1992. Rapid detection of chromosome aneuploidies in uncultured amniocytes by using fluorescence in situ hybridization (FISH). Am. J. Hum. Genet. **51:** 55–65.

2. ADINOLFI, M. & J. CROLLA. 1994. Nonisotopic in situ hybridization. Clinical cytogenetics and gene mapping applications. Adv. Hum. Genet. **22:** 187–255.
3. EIBEN, B. *et al.* 1998. A prospective comparative study on fluorescence in situ hybridization (FISH) of uncultured amniocytes and standard karyotype analysis. Prenat. Diagn. **18:** 901–906.
4. EVANS, M.I. *et al.* 2000. International, collaborative assessment of 146,000 prenatal karyotypes: expected limitations if only chromosome-specific probes and fluorescent in-situ hybridization are used. Hum. Reprod. **14:** 1213–1216.
5. MANSFIELD, E.S. 1993. Diagnosis of Down's Syndrome and other aneuploidies using quantitative polymerase chain reaction and small tandem repeat polymorphisms. Hum. Mol. Genet. **2:** 43–50.
6. PERTL, B. *et al.* 1994. Rapid molecular method for prenatal detection of Down's syndrome. Lancet **343:** 1197–1198.
7. PERTL, B. *et al.* 1996. Rapid detection of trisomies 21 and 18 and sexing by quantitative fluorescent multiplex PCR. Hum. Genet. **98:** 55–59.
8. CIRIGLIANO, V. *et al.* 2001. Assessment of new markers for the rapid prenatal detection of aneuploidies by quantitative fluorescent PCR (QF-PCR). Ann. Hum. Genet. **65:** 421–427.
9. CIRIGLIANO, V. *et al.* 2001. Clinical application of multiplex quantitative fluorescent polymerase chain reaction (QF-PCR) for the rapid prenatal detection of common chromosome aneuploidies. Mol. Hum. Reprod. **7:** 1001–1006.
10. CIRIGLIANO, V. *et al.* 2004. Rapid prenatal diagnosis of common chromosome aneuploidies by QF-PCR. Assessment on 18.000 consecutive clinical samples. Mol. Hum. Reprod. **10:** 839–846.
11. CIRIGLIANO, V. *et al.* 2005. Non invasive screening and rapid QF-PCR assay can greatly reduce the need of cytogenetic analysis in prenatal diagnosis. Repr. Biomed. Online **11:** 671–673.
12. ADINOLFI, M. *et al.* 1997. Rapid detection of aneuploidies by microsatellite and the quantitative fluorescent polymerase chain reaction. Prenat. Diagn. **17:** 1299–1311.
13. SULLIVAN, K.M. *et al.* 1993. A rapid and quantitative DNA sex test: fluorescence-based PCR analysis of X-Y homologous gene amelogenin. Biotechniques **15:** 636–638.
14. CIRIGLIANO, V. *et al.* 1999. Rapid detection of chromosomes X and Y aneuploidies by quantitative fluorescent PCR. Prenat. Diagn. **19:** 1099–1103.
15. CIRIGLIANO, V. *et al.* 2002. X chromosome dosage by quantitative fluorescent PCR and rapid prenatal diagnosis of sex chromosome aneuploidies. Mol. Hum. Reprod. **8:** 1042–1045.
16. CHANG, Y.M. *et al.* 2003. Higher failures of amelogenin sex test in an Indian population group. J. Forensic. Sci. **48:** 1309–1313.
17. SHERLOCK, J. *et al.* 1998. Assessment of diagnostic quantitative fluorescent multiplex polymerase chain reaction assays performed on single cells. Ann. Hum. Genet. **62:** 9–23.
18. VOGLINO, G. *et al.* 2002. Rapid prenatal diagnosis of aneuploidies [letter]. Lancet **359:** 442.
19. VERMA, L. *et al.* 1998. Rapid and simple prenatal DNA diagnosis of Down's syndrome. Lancet **352:** 9–12.
20. ANDONOVA, S. *et al.* 2004. Introduction of the QF-PCR analysis for the purposes of prenatal diagnosis in Bulgaria, estimation of applicability of 6 STR markers on chromosomes 21 and 18. Prenat. Diagn. **24:** 202–208.

21. MANN, K. *et al.* 2001. Development and implementation of a new rapid aneuploidy diagnostic service within the UK National Health Service and implications for the future of prenatal diagnosis. Lancet **358:** 1057–1061.
22. SCHMIDT, W. *et al.* 2000. Detection of aneuploidy in chromosomes X, Y, 13, 18 and 21 by QF-PCR in 662 selected pregnancies at risk. Mol. Hum. Reprod. **6:** 855–860.
23. BILI, C. *et al.* 2002. Prenatal diagnosis of common aneuploidies using quantitative fluorescent PCR. Prenat. Diagn. **22:** 360–365.
24. CAINE, A. *et al.* 2005. UK Association of Clinical Cytogeneticists (ACC). Prenatal detection of Down's syndrome by rapid aneuploidy testing for chromosomes 13, 18, and 21 by FISH or PCR without a full karyotype: a cytogenetic risk assessment. Lancet **366:** 123–128.
25. WALD, N.J. 1995. Antenatal screening for Down's syndrome. Prog. Clin. Biol. Res. **393:** 27–42.
26. WALD, N.J. & A.K. HACKSHAW. 2000. Advances in antenatal screening for Down syndrome. Baillieres Best Pract. Res. Clin Obstet. Gynaecol. **14:** 563–580.
27. ADINOLFI, M. *et al.* 2000. Prenatal screening of aneuploidies by quantitative fluorescent PCR. Community Genet. **3:** 50–60.
28. PERTL, B. *et al.* 2000. RhD genotyping by quantitative fluorescent polymerase chain reaction: a new approach. Br. J. Obstet. Gynecol. **107:** 1498–1502.
29. RICKMAN, L. *et al.* 2005. Prenatal detection of unbalanced chromosomal rearrangements by array-CGH. J. Med. Genet. **43(4):** 353–361.
30. LO, Y.M. 2005. Recent advances in fetal nucleic acids in maternal plasma. J. Histochem. Cytochem. **53:** 293–296.
31. COSTA, J.M. *et al.* 2002. New strategy for prenatal diagnosis of X-linked disorders. N. Engl. J. Med. **346:** 1502.
32. COTTER, A.M. *et al.* 2004. Increased fetal DNA in the maternal circulation in early pregnancy is associated with an increased risk of preeclampsia. Am. J. Obstet. Gynecol. **191:** 515–520.
33. COTTER, A.M. *et al.* 2005. Increased fetal RhD gene in the maternal circulation in early pregnancy is associated with an increased risk of pre-eclampsia. Br. J. Obstet. Gynecol. **112:** 584–587.
34. POON, L.L. *et al.* 2002. Differential DNA methylation between fetus and mother as a strategy for detecting fetal DNA in maternal plasma. Clin. Chem. **48:** 35–41.
35. CHIM, S.S. *et al.* 2005. Detection of the placental epigenetic signature of the maspin gene in maternal plasma. Proc. Natl. Acad. Sci. USA **102:** 14753–14758.

Higher Amount of Free Circulating DNA in Serum than in Plasma Is Not Mainly Caused by Contaminated Extraneous DNA during Separation

NAOYUKI UMETANI, SUZANNE HIRAMATSU, AND DAVE S.B. HOON

Department of Molecular Oncology, John Wayne Cancer Institute, Santa Monica, California 90404, USA

ABSTRACT: Circulating DNA isolated from serum and plasma has been shown to be a useful biomarker in various diseases including cancer. Serum reportedly contains a higher amount of free circulating DNA than it does in plasma. The underlying reason for this is unclear, but important because it may have clinical implications in interpreting results and using the appropriate resource. Twenty-four pairs of serum and plasma samples were collected from patients with tumors, and free circulating DNA was quantified by real-time quantitative PCR (qPCR) for the ALU repeats, which had a sensitivity of 0.1 pg/μL of DNA in serum/plasma. The possibility of DNA loss was eliminated because ALU-qPCR does not require DNA purification from serum/plasma. The DNA concentrations of serum and plasma samples were 970 ± 730 pg/μL and 180 ± 150 pg/μL (mean ± SD), respectively. The amount of DNA in paired serum and plasma specimens was positively correlated ($R = 0.72$ and $P = 0.0002$). An estimated 8.2% of total DNA in serum was extraneous; the concentration of DNA was 6.1 ± 3.5 (mean ± SD)-fold higher in serum than in paired plasma after subtraction of it. Contribution of extraneous DNA from cells in blood ruptured during the separation step was minor for explaining the difference between serum and plasma. A possible explanation was unequal distribution of DNA during separation from whole blood. We advocate that serum is a better specimen source for circulating cancer-related DNA as a biomarker.

KEYWORDS: serum; plasma; circulating DNA; ALU repeats; cancer biomarker

INTRODUCTION

Free circulating DNA has been intensely investigated recently as a biomarker for malignancy and other diseases.[1–8] Free circulating DNA is usually obtained

Address for correspondence: Dr. Dave S.B. Hoon, Department of Molecular Oncology, John Wayne Cancer Institute, 2200 Santa Monica Blvd., Santa Monica, CA 90404, USA. Voice: 310- 449-5264; fax: 310-449-5282.
e-mail: hoon@jwci.org

Ann. N.Y. Acad. Sci. 1075: 299–307 (2006). © 2006 New York Academy of Sciences.
doi: 10.1196/annals.1368.040

from serum or plasma. The most significant difference between these two sources is the existence of coagulation factors and their related proteins, as well as platelets, in plasma. Several reports indicate that the amount of free circulating DNA is significantly lower in plasma than in serum,[9–11] but the reason for this observation is still under controversy.[10,12] If DNA is lost during purification from plasma but not from serum, using serum DNA as a biomarker should be more efficient. However, if extraneous DNA from leukocytes or other sources is accidentally released into serum during its separation from whole blood, using serum DNA would cause erroneous results derived from contaminated DNA. Another possible explanation is unequal distribution of DNA during separation from whole blood; if this is the case, then using serum DNA would increase sensitivity.

To elucidate why the observed amount of DNA is higher in serum than in plasma, we precisely quantified the amount DNA in serum and plasma concurrently separated from same blood without possibility of a DNA loss. The method we used was quantitative real-time PCR (qPCR) of the ALU repeats, which is the most abundant repeat sequence (1.4×10^6 copies) in the human genome.[13] ALU-qPCR was sensitive enough for using minimally processed serum/plasma as a template without DNA purification.

MATERIALS AND METHODS

Blood Samples and Serum/Plasma Separation

Twenty-four patients with breast cancer ($n = 12$), colorectal cancer ($n = 8$), thyroid cancer ($n = 2$), and thyroid adenoma ($n = 2$) were randomly selected from the clinical database at the John Wayne Cancer Institute (JWCI) in 2005. All patients gave consent for blood sampling according to the guidelines set forth by Saint John's Health Center and JWCI Institutional Review Board (IRB) committee. Ten milliliters of blood was collected in a CORVAC serum separator tube (Sherwood-Davis & Geck, St. Louis, MO, USA) and processed within 6 h for serum as follows: the sample was separated by centrifugation ($1{,}000 \times g$ 15 min), and filtered through a 13-mm serum filter (Fisher Scientific, Pittsburgh, PA, USA) to remove potential contaminating cells. An additional 10 mL of blood was collected at the same blood draw in a sodium citrate tube (Becton Dickinson, Franklin Lakes, NJ) and processed within 6 h for plasma by antrifugation and filtration as for serum processing.

Quantification of Free Circulating DNA

To maximize the sensitivity of DNA quantification, ALU repeats, which are the most abundant repeat sequence in the human genome, were used as a target of qPCR. The primer set was designed to amplify the consensus sequence of ALU and produce an amplicon size of 115 bp. The sequence of the forward primer was 5′-CCTGAGGTCAGGAGTTCGAG-3′; the reverse primer was 5′-CCCGAGTAGCTGGGATTACA-3′.

Human serum and plasma contain many substances that interfere with the PCR reaction, such as proteins that bind to template DNA or DNA polymerase. Therefore, we used minimally preprocessed serum/plasma to eliminate inhibitory factors. We mixed 20 μL of each serum/plasma sample with 20 μL of a preparation buffer that contained 2.5% of Tween-20, 50 mM of Tris, and 1 mM of EDTA. This mixture was digested with 16 μg of proteinase K (Qiagen, Valencia, CA, USA) at 50°C for 40 min, and diluted with 160 μL of Tris-EDTA buffer after 5 min of heat deactivation at 95°C. After centrifugation at 10,000 × *g* for 5 min, 1 μL of supernatant containing 0.1 μL equivalent amount of the serum/plasma was used as a template for ALU-qPCR without purification.

The reaction mixture for each ALU-qPCR consisted of a template, 0.2 μM of forward primer and reverse primer, 1 unit of iTaq DNA polymerase (Bio-Rad Laboratories, Hercules, CA, USA), 0.02 μL of fluorescein calibration dye (Bio-Rad Laboratories), and 1× concentration of SYBR Gold (Molecular Probe, Eugene, OR, USA) in a total reaction volume of 20 μL with 5 mM of Mg^{2+}. Real-time PCR amplification was performed with a precycling heat activation of DNA polymerase at 95°C for 10 min, followed by 35 cycles of denaturation at 95°C for 30 sec, annealing at 64°C for 30 sec, and extension at 72°C for 30 sec using iCycler iQ Real-Time PCR Detection System (Bio-Rad Laboratories). The absolute equivalent amount of DNA in each sample was determined by a standard curve with serial dilutions (10 ng to 0.01 pg) of purified DNA obtained from peripheral blood of a healthy volunteer. A negative control (no template) was performed in each reaction plate. PCR products were electrophoresed on 2% agarose gels to confirm product size and specificity of the PCR (FIG. 1A).

To evaluate the effect of inhibitory substances in preprocessed serum/plasma on ALU-qPCR, a known amount (10 ng) of purified DNA obtained from PBL of a healthy volunteer was mixed into the reaction mixture of serum/plasma ALU-qPCR of two patients and quantified. The amount of DNA in the template serum/plasma itself was subtracted from the qPCR result.

Statistical Analysis

The amount of DNA in serum versus plasma was assessed by paired *t*-test and Deming's regression analysis. The statistical package SAS JMP version 5.0.1 (SAS Institute Inc., Cary, NC, USA) and EP Suite 9A version 2.0.a[14] (MarChem Associates, Inc, Concord, MA, USA) were used, and *P* value $<$ 0.05 (two-tailed) was defined as significant.

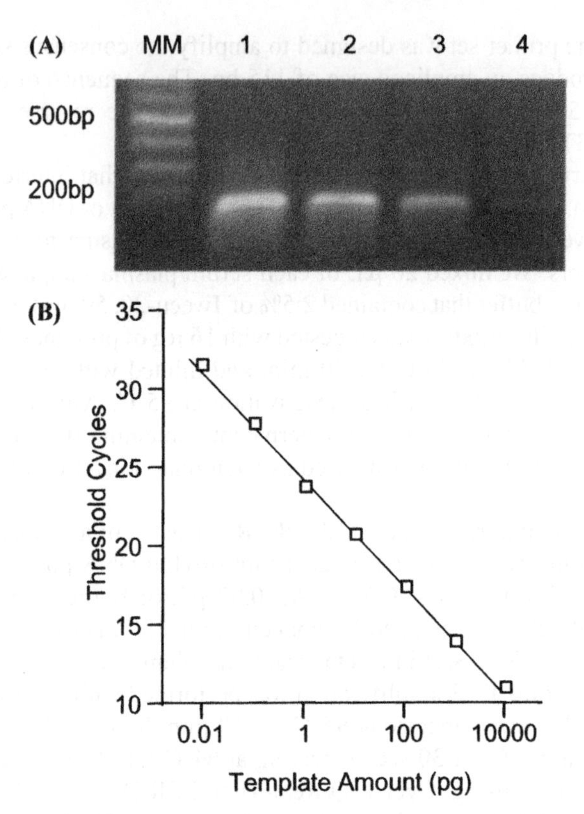

FIGURE 1. (**A**) PCR products of ALU primers on 10, 1, 0.1, and 0.01 pg (*lanes* 1, 2, 3, and 4, respectively) of genomic DNA templates by gel electrophoresis at 28 cycles of thermal cycling of qPCR. MM: molecular marker. (**B**) Serially diluted genomic DNA (10 ng to 0.01 pg) obtained from PBLs of a healthy volunteer was quantified by ALU-qPCR. Linearity was maintained in the 10^6 range, and sensitivity was as low as 0.01 pg, equivalent to about 1/300 copy of genome in a single cell.

RESULTS

Sensitivity and Linearity of ALU-qPCR

ALU-qPCR had linearity ranging from 10 ng to 0.01 pg with a PCR efficiency of 97% and regression coefficient of 0.999 of the standard curve taken from the mean of two sets of serially diluted genomic DNA (FIG. 1B). Thus the lower quantification limit of serum/plasma was 0.1 pg/μL.

Evaluation of Inhibitory Effect on ALU-qPCR by Serum/Plasma

Triplicate ALU-qPCR results for two sets of 10 ng purified DNA with serum was 10.2 ± 1.8 ng and those with plasma was 9.8 ± 1.2 ng (mean ± SD). No significant inhibitory effect by serum or plasma was observed.

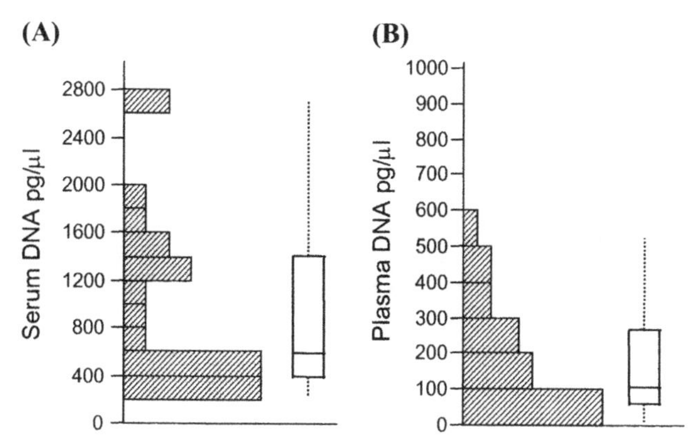

FIGURE 2. ALU-qPCR quantification of DNA in serum (**A**) and plasma (**B**). The DNA concentration in serum and plasma samples was 970 ± 730 pg/µL and 180 ± 150 pg/µL (mean ± SD), respectively.

Amount of DNA in Serum and Plasma by ALU-qPCR

The DNA concentrations of serum and plasma samples by ALU-qPCR were 970 ± 730 pg/µL and 180 ± 150 pg/µL (mean ± SD), respectively (FIG. 2). The serum: plasma ratio of DNA in each pair of specimens followed normal distribution (mean: 7.1, SD: 4.2), with the exception of one pair that had a ratio of 32 (FIG. 3). This outlier was therefore deleted from subsequent statistical

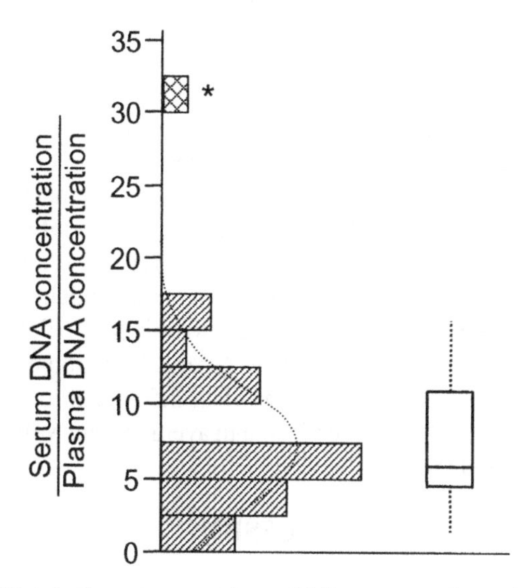

FIGURE 3. Distribution of serum: plasma DNA concentration ratios in pairs of specimens. A fitted normal distribution curve for all but one outlier pair, which had a ratio of 32 (*) and was overlapped.

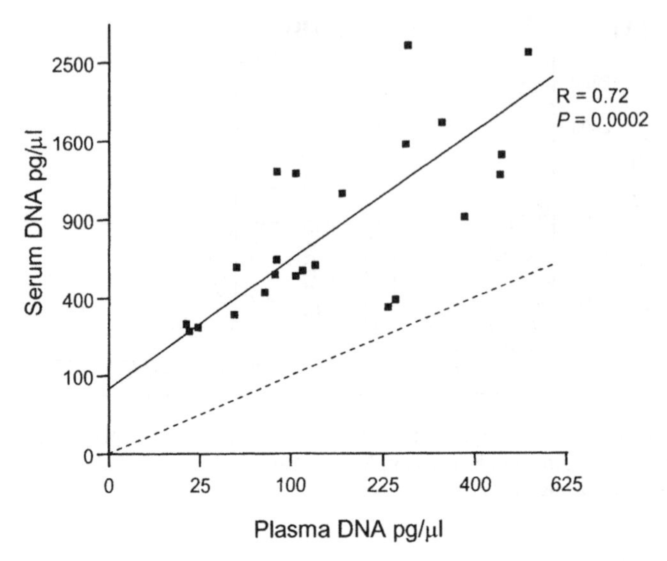

FIGURE 4. Scattergram of serum and plasma DNA concentrations and Deming's regression after square-root variable transformation to normalize the distribution of DNA concentrations. Deming regression model shown with a *solid line* was $(\text{serum DNA})^{0.5} = 1.6 \times (\text{plasma DNA})^{0.5} + 8.7$. The intercept value of serum DNA axis at (plasma DNA) = 0 was 76 (95% CI: 7.8–210) pg/μL. The *dotted line* indicates the line of identity between serum and plasma.

analysis to avoid overvaluation of serum DNA amount. Serum contained a significantly higher amount of DNA than did plasma ($P < 0.0001$, paired *t*-test).

Deming's regression analysis was performed after square-root variable transformation to normalize the distribution of concentrations in serum and plasma (logarithmic transformation was inapplicable because an intercept value of serum DNA at a plasma DNA of zero was required). Deming's regression analysis showed positive correlation between DNA amount in serum and that in plasma ($R = 0.72$ and $P = 0.0002$). The intercept value of regression line at a plasma DNA of zero was 8.7 (95% CI: 2.8–14.5) (FIG. 4). Thus the estimated amount of extraneous DNA in serum that was independent of plasma DNA was 76 pg/μL (95% CI: 7.8–210), equivalent to only 8.2% (95% CI: 0.8–23%) of total amount of serum DNA in average. After subtraction of 76 pg/μL from each qPCR serum value, the serum had 6.1 ± 3.5 (mean ± SD)-fold greater DNA concentration than the paired plasma in average. This value is concordant with results of previous reports.[9,11]

DISCUSSION

Several reports have demonstrated that free circulating DNA is lower in plasma than in serum,[9–11] but none has determined the reason behind this observation. A specious assumption is that leukocytes ruptured during serum separation might release DNA into serum based on the finding that serum DNA concentrations correlated with leukocyte counts.[10] This explanation implies that most of the DNA in serum is extraneous because serum has about 4–6 times more abundant DNA than plasma,[9,11] and thus serum cannot be a good biomarker. However, clinical use of serum has been definitely shown in previous multiple studies,[1,3,6–8] and we considered that such an assumption was incorrect and hypothesized that unequal distribution of DNA during separation from whole blood might be the cause.

Any attempts to investigate the relation between DNA concentrations of serum and plasma have been hindered by the difficulty in handling the low amounts of DNA and the complexity of purifying DNA from serum or plasma. Accurate quantification of DNA in small sample volumes of plasma has been very difficult to date. Our preliminary assessment of the sensitivity of PicoGreen (Molecular Probe) assay by a microplate reader, which is a commonly used method for quantification of serum/plasma DNA, showed that the practical lower limit of linear range was around 20 pg/μL in serum/plasma. It was mainly due to the background noise from the photodetector (data not shown). This assay is not sensitive enough for the accurate assessment of DNA in plasma, especially in samples from healthy individuals. In addition, because the DNA in plasma or serum is highly truncated,[15] recovery of DNA during the purification process is usually not complete and depends on the efficacy of the extraction method. Therefore, we first established a highly sensitive method using ALU repeats for accurate quantification of DNA in serum and plasma without DNA purification. The ALU, which is primate-specific, is the most abundant repeated sequence in the human genome, with a copy number of about 1.4 million per genome.[13,16,17] Because most DNA released from apoptotic cells is truncated into a length of 185–200 bp by a cleavage process during apoptosis,[15] we designed a primer set for 115-bp amplicon in ALU repeats. This method achieved sufficient sensitivity to accurately quantify with high linearity as little as 0.01 pg of DNA, equivalent to about 1/300 copy of genome in a single cell. As a result, it enabled accurate quantification of free circulating DNA without its purification from serum/plasma because of the low requirement of template DNA. This quantification method has potential clinical applications for measurement of DNA in other body fluids, such as the urine of patients with urinary tract cancer or the saliva of patients with salivary gland cancer.

Because our ALU-qPCR technique eliminated the need for DNA purification, it was proven that the lower level of DNA in plasma was not the result of DNA loss during purification of DNA from plasma. In addition, our results showed a significant positive regression between DNA in serum and plasma; the estimated amount of extraneous DNA in serum was only 8.2% of the total serum DNA. Therefore, it was also proven that the excess amount of DNA in serum is not mainly from extraneous DNA released from leukocytes or other sources during the separation of serum. As a result, the most likely explanation for the difference in serum and plasma DNA levels is unequal distribution of DNA during separation from whole blood. Based on the estimated scale factor of serum DNA in relation to plasma DNA, serum has 6.1-fold more DNA than plasma. This might be the result of electrostatic forces between DNA that is anionic and platelets that are also potently anionic existing only in plasma.

In conclusion, serum contains around six times as much amounts of free circulating DNA as plasma. Extraneous DNA, such as DNA from cells in blood ruptured during the separation step was minor for explaining the difference between serum and plasma. A possible explanation was unequal distribution of DNA during separation from whole blood. We advocate that serum is a better specimen source for circulating cancer-related DNA as a biomarker.

ACKNOWLEDGMENT

This study was supported in part by funding from NIH NCI grants P01 CA 29605 and P01 CA 12582, and Martin H. Weil Research Laboratories.

REFERENCES

1. Nawroz, H. *et al.* 1996. Microsatellite alterations in serum DNA of head and neck cancer patients. Nat. Med. **2:** 1035–1037.
2. Chen, X.Q. *et al.* 1996. Microsatellite alterations in plasma DNA of small cell lung cancer patients. Nat. Med. **2:** 1033–1035.
3. Hibi, K. *et al.* 1998. Molecular detection of genetic alterations in the serum of colorectal cancer patients. Cancer Res. **58:** 1405–1407.
4. Mulcahy, H. *et al.* 1998. A prospective study of K-ras mutations in the plasma of pancreatic cancer patients. Clin. Cancer Res. **4:** 271–275.
5. Anker, P. *et al.* 1999. Detection of circulating tumour DNA in the blood (plasma/serum) of cancer patients. Cancer Metastasis Rev. **18:** 65–73.
6. Taback, B. & D.S. Hoon. 2004. Circulating nucleic acids and proteomics of plasma/serum: clinical utility. Ann. N. Y. Acad. Sci. **1022:** 1–8.
7. Taback, B. & D.S. Hoon. 2004. Circulating nucleic acids in plasma and serum: past, present and future. Curr. Opin. Mol. Ther. **6:** 273–278.
8. Fujimoto, A. *et al.* 2004. Allelic imbalance on 12q22-23 in serum circulating DNA of melanoma patients predicts disease outcome. Cancer Res. **64:** 4085–4088.
9. Taback, B. *et al.* 2004. Quantification of circulating DNA in the plasma and serum of cancer patients. Ann. N. Y. Acad. Sci. **1022:** 17–24.
10. Gautschi, O. *et al.* 2004. Circulating deoxyribonucleic acid as prognostic marker in non-small-cell lung cancer patients undergoing chemotherapy. J. Clin. Oncol. **22:** 4157–4164.
11. Holdenrieder, S. *et al.* 2005. Cell-free DNA in serum and plasma: comparison of ELISA and quantitative PCR. Clin. Chem. **51:** 1544–1546.
12. Chan, K. *et al.* 2005. Effects of preanalytical factors on the molecular size of cell-free DNA in blood. Clin. Chem. **51:** 781–784.

13. GU, Z. *et al.* 2000. Densities, length proportions, and other distributional features of repetitive sequences in the human genome estimated from 430 megabases of genomic sequence. Gene **259:** 81–88.
14. MARTIN, R.F. 2000. General Deming regression for estimating systematic bias and its confidence interval in method-comparison studies. Clin. Chem. **46:** 100–104.
15. GIACONA, M.B. *et al.* 1998. Cell-free DNA in human blood plasma: length measurements in patients with pancreatic cancer and healthy controls. Pancreas **17:** 89–97.
16. HWU, H. *et al.* 1986. Insertion and/or deletion of many repeated DNA sequences in human and higher ape evolution. Proc. Natl. Acad. Sci. USA **83:** 3875–3879.
17. SMIT, A.F. 1996. The origin of interspersed repeats in the human genome. Curr. Opin. Genet. Dev. **6:** 743–748.

Improvement of Methods for the Isolation of Cell-Free Fetal DNA from Maternal Plasma

Comparison of a Manual and an Automated Method

DOROTHY J. HUANG,[a] BERNHARD G. ZIMMERMANN,[a,b] WOLFGANG HOLZGREVE,[a] AND SINUHE HAHN[a]

[a]*University Women's Hospital/Department of Research, Spitalstrasse 21, CH 4031 Basel, Switzerland*

[b]*Centre for Research in Biomedicine, UWE Bristol Genomics Research Institute University of the West of England, Coldharbour Lane, Frenchay, BS16 1QY, Bristol, UK*

ABSTRACT: The low amount of cell-free fetal DNA present in the maternal circulation poses significant challenges to its use in future diagnostic applications, and ways of increasing the yield of this potential marker extracted from maternal plasma are constantly being explored. In this study, we compared two methods of DNA extraction, a manual and an automated method. Our analysis revealed that although the manual method yielded overall more total cell-free DNA, the automated system yielded higher quantities of cell-free DNA of fetal origin. Furthermore, the DNA isolated using the automated system appeared to be of greater purity than that isolated by the manual method, with fewer inhibitors to downstream real-time PCR reactions.

KEYWORDS: cell-free fetal DNA; maternal plasma; DNA extraction

INTRODUCTION

Cell-free fetal DNA present in the maternal circulation has great potential for noninvasive prenatal diagnosis and analysis of fetal genetic traits.[1–3] However, only a small fraction of total DNA in the maternal plasma, in the range of 3–6%, is of fetal origin.[4] This poses challenges to its potential applications

Address for correspondence: Sinuhe Hahn, Ph.D., University Women's Hospital/Department of Research, Spitalstrasse 21, CH 4031 Basel, Switzerland. Voice: +41-61-265-9224; fax: +41-61-265-9399.
e-mail: shahn@uhbs.ch

Ann. N.Y. Acad. Sci. 1075: 308–312 (2006). © 2006 New York Academy of Sciences.
doi: 10.1196/annals.1368.041

for clinical diagnosis unless specialized strategies are adopted to increase the overall quantity and quality of the DNA extracted.[4,5] Furthermore, its future applicability to routine use in clinical situations requires the ability to increase throughput and minimize the amount of handling of the samples.

The manual isolation of cell-free DNA from plasma can be a time-consuming process. Techniques involving spin-columns often require multiple reloadings of the columns with sample material, especially when larger volumes of plasma are processed. Use of an automated system for DNA isolation can reduce the time involved, increase the throughput, and help reduce inconsistencies due to human error between processing of different plasma samples. Several companies currently offer a variety of automated systems for the isolation of DNA from different biological sources.

In this study, we compared two techniques of extracting cell-free DNA from maternal plasma, one of which involved an automated DNA isolation system, and the other a manual method frequently used in our laboratory.[5,6] The manual technique involves the use of spin-column technology (High Pure PCR Template Preparation Kit™ from Roche Diagnostics Basel, Switzerland) to isolate DNA. The automated method, using the MagNA Pure™ LC Instrument (Roche Applied Science), employs magnetic glass particles to separate out the DNA from the other plasma components.

MATERIALS AND METHODS

Following approval from the Cantonal Institutional Review Board of Basel, nine plasma samples obtained from women carrying male fetuses were used to test the methods of DNA extraction. Identical samples were used for both methods, and the gestational ages during which the samples had been drawn ranged from the first to the third trimester. The plasma was separated from other blood components using our standard protocol (centrifugation at 1,600 $\times$ g for 10 min, followed by a second centrifugation of the plasma fraction at 16,000 $\times$ g for 10 min) and stored at $-70°C$.[5] For the manual extraction method, cell-free DNA was extracted from 400 μL plasma according to the manufacturer's instructions and eluted into 100 μL elution buffer. For the automated method, DNA was extracted from 1,000 μL plasma per protocol using the Roche MagNA Pure™ LC DNA Isolation Kit–Large Volume and eluted into 200 μL buffer. On the following day, a second identical set of samples was processed using the same procedures as described above, yielding a total of 18 samples tested per method. Real-time PCR was then performed with 5 μL of DNA by TaqMan® real-time PCR assay (Applied Biosystems) for GAPDH (glyceraldehydes-3-phosphate dehydrogenase), representing total cell-free DNA. For cell-free fetal DNA we used 5 μL of DNA quantified with the multi-copy DYS14 sequence on the Y chromosome, as we have recently determined that this provides a more accurate assessment of low amounts of this material[7]. Real-time PCR reactions were performed in reaction volumes of 25 μL. Each sample was analyzed in triplicate, and multiple water blanks were included with each run.

RESULTS

Our analysis revealed that extraction of DNA overall manually yielded more total cell-free DNA than the automated method (TABLE 1). On average, the plasma concentration of total DNA as calculated from the real-time PCR results for GAPDH was 2890 GE/mL (range: 797–10059 GE/mL), while that for the automated method was 2215 GE/mL (range: 620–9010 GE/mL), equaling a 23% greater yield for the manual method. However, when the quantity of cell-free DNA of fetal origin obtained was examined, it was found that the automated system resulted in a greater yield of free fetal DNA. The average plasma concentration of manually obtained fetal DNA as determined from the real-time PCR results for DYS14 was 51 GE/mL (range: 9–160 GE/mL), while that obtained by use of the MagNA Pure LC Instrument was 86 GE/mL (range: 40-204 GE/mL), equivalent to a 41% greater yield for the automated system.

Interestingly, the amplification efficiencies of the late cycles of the real-time PCR reactions appeared to be affected by the method of DNA isolation used. This was evidenced by the overall decreased steepness of the amplification curve for the manually prepared samples than the automated samples (FIG. 1). The signal amplification efficiencies were calculated from the slope of the amplification plots using the equation:

$$\text{amplification efficiency} = 10^{(-1/\text{slope})} - 1$$

An efficiency value of 100% corresponds to a maximally efficient, ideal amplification reaction. For the samples prepared by the automated method, the average efficiency of amplification reactions for the GAPDH locus was 85%, while that for manually extracted samples was 73%. The average amplification reaction efficiencies for the DYS14 locus for the automated and manual

TABLE 1. Average total and fetal plasma DNA concentrations for manual and automated sample preparations as determined by real-time PCR for GAPDH and DYS loci

	Manual		Automated	
	1st set	2nd set	1st set	2nd set
Mean total DNA[a] (range)	3009 (797–10059)	2774 (814–7175)	2042 (664–9010)	2389 (620–8208)
Mean fetal DNA[a] (range)	48 (15–75)	54 (9–160)	89 (41–204)	83 (40–180)

[a]Cell-free DNA concentrations represented in genome equivalents (GE)/mL maternal plasma.

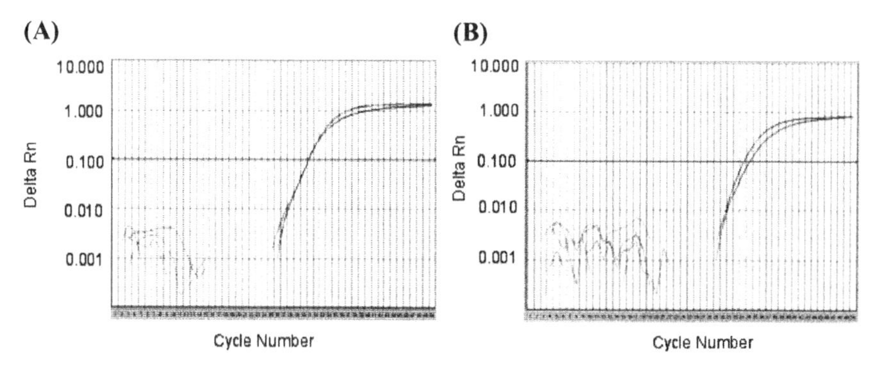

FIGURE 1. (**A**) Real-time PCR amplification curves for one representative sample probed for the GAPDH locus. Cell-free DNA was extracted from maternal plasma samples by the automated system (*upper steeper curve*) and manual method (*lower curve*). (**B**) Real-time PCR amplification curves for one representative sample probed for the DYS14 locus. Cell-free DNA was extracted from maternal plasma samples by the automated system (*upper steeper curve*) and manual method (*lower curve*).

preparations were 66% and 46%, respectively. The amplification efficiencies of the diluted genomic DNA standards were equal and greater than 90% for both assays.

DISCUSSION

The optimization of cell-free fetal DNA extraction from maternal plasma is of major importance in the movement away from prenatal diagnostic methods involving invasive techniques that carry a definite risk of fetal loss.[1,2] As we demonstrated in this small preliminary study, the use of an automated DNA extraction system, such as the MagNA Pure LC® instrument can assist in improving the yield of cell-free fetal DNA from maternal plasma, thereby contributing to the fulfillment of this goal. The mechanism behind this finding is unclear, but may possibly be a reflection of the size differences between maternal and fetal cell-free DNA. With free fetal DNA being generally about 200–300 bp in size and considerably smaller than maternal DNA,[5,8] it is possible that these smaller fragments are better extracted by the automated system than the spin-column method, or that perhaps the smaller-size fragments are more easily lost through the columns. In addition to the findings that higher amounts of cell-free DNA are obtained through use of the automated system, our data also suggest that this DNA is of a greater purity, which in turn improves the efficiency of subsequent real-time PCR reactions. This aspect may have considerable implications, particularly for quantitative analyses of cell-free DNA. Future studies will include exploration of the mechanism behind the differences observed in this exploratory work and also involve a significantly larger sample size to confirm the findings presented here.

REFERENCES

1. Hahn, S. & W. Holzgreve. 2002. Prenatal diagnosis using fetal cells and cell-free fetal DNA in maternal blood: what is currently feasible? Clin. Obstet. Gynecol. **45:** 649–656.
2. Chiu, R.W. & Y.M. Lo. 2004. The biology and diagnostic applications of fetal DNA and RNA in maternal plasma. Curr. Top. Dev. Biol. **61:** 81–111.
3. Wataganara, T. & D.W. Bianchi. 2002. Fetal cell-free nucleic acids in the maternal circulation: new clinical applications. Ann. N.Y. Acad. Sci. **1022:** 90–99.
4. Lo, Y.M. *et al.* 1998. Quantitative analysis of fetal DNA in maternal plasma and serum: implications for noninvasive prenatal diagnosis. Am. J. Hum. Genet. **62:** 768–775.
5. Li, Y. *et al.* 2004. Size separation of circulatory DNA in maternal plasma permits ready detection of fetal DNA polymorphisms. Clin. Chem. **50:** 1002–1011.
6. Zhong, X.Y. *et al.* 2000. Detection of fetal Rhesus D and sex using fetal DNA from maternal plasma by multiplex polymerase chain reaction. Br. J. Obstet. Gynaecol. **107:** 766–769.
7. Zimmermann, B. *et al.* 2005. Optimized real-time quantitative PCR measurement of male fetal DNA in maternal plasma. Clin. Chem. **51:** 1598–1604.
8. Chan, K.C. *et al.* 2004. Size distributions of maternal and fetal DNA in maternal plasma. Clin. Chem. **50:** 88–92.

The Importance of Careful Blood Processing in Isolation of Cell-Free DNA

KAREN PAGE,[a] TOM POWLES,[b] MARTIN J. SLADE,[b] MANUELA TAMBURO DE BELLA,[b] ROSEMARY A. WALKER,[a] R. CHARLES COOMBES,[b] AND JACQUELINE A. SHAW[a]

[a]*Department of Cancer Studies and Molecular Medicine, University of Leicester, Robert Kilpatrick Clinical Sciences Building, Leicester Royal Infirmary, Leicester LE2 7LX, UK*

[b]*Department of Cancer Medicine, Hammersmith Campus, Imperial College School of Medicine, MRC Cyclotron Building, Du Cane Road, London W12 0NN, UK*

ABSTRACT: In healthy individuals, the source of cell-free plasma DNA is predominantly apoptotic, whereas, increased plasma DNA integrity is seen in cancer patients. Therefore, it is important to carefully isolate absolutely "cell-free" plasma DNA. Plasma DNA from 30 healthy females was analyzed using 4 PCR amplicons of increasing size, comparing standard blood processing with additional centrifugation steps prior to DNA extraction. Cellular DNA contamination, indicated by positive amplicons >300 bp was eliminated only after the extra centrifugation step. This highlights the importance of careful processing in preparation of cell-free plasma DNA as a tool for cancer detection and we recommend the use of a microcentrifuge spin, prior to DNA extraction.

KEYWORDS: cell-free plasma DNA; PCR; apoptosis

INTRODUCTION

The presence of free-circulating nucleic acid in the blood has been recognized since the 1970s,[1] with increasing quantities found in cancer patients.[2] The identification of DNA exhibiting tumor-specific LOH in plasma from patients with lung[3] and head and neck cancer[4] respectively, prompted a wider investigation of other cancer types. For breast cancer patients, we[5] and others[6] have shown that tumor DNA in plasma at diagnosis may be a valuable predictor

Address for correspondence: Dr. Karen Page, Department of Cancer Studies and Molecular Medicine, Robert Kilpatrick Clinical Sciences Building, University of Leicester, Leicester, LE2 7LX, United Kingdom. Voice: +44-116-252-3239; fax: +44-116-252-3274.
e-mail: kd18@leicester.ac.uk

Ann. N.Y. Acad. Sci. 1075: 313–317 (2006).
doi: 10.1196/annals.1368.042

of disease-free survival. However, in order to develop novel methods of cancer detection using analysis and quantitation of plasma DNA, it is important to control for contamination by cellular DNA, which might interfere with the analysis. In healthy individuals the source of cell-free plasma DNA is predominantly apoptopic[7] and so only small DNA fragments (<300 bp) should be detected. Therefore, we have used PCR analysis of increasing-sized amplicons to analyze plasma DNA isolated from normal, healthy female donors to investigate the utility of standard blood processing for successful isolation of plasma DNA. We have also compared the use of an additional centrifugation step both before and after storage of plasma on the recovery of cell-free DNA. Recently, Chiu *et al.* have recommended an extra centrifugal step, in addition to initial plasma separation, to produce absolutely cell-free DNA.[8]

MATERIALS AND METHODS

Blood Collection, Processing, and DNA Extraction

Twenty milliliters venous blood samples were withdrawn from a peripheral vein into EDTA-containing tubes from three groups of normal healthy female donors ($n = 30$). Following plasma separation by centrifugation at 2000 rpm for 10 min (×2), the first two sets of samples were stored directly at −80°C. The third set was subjected to an extra spin, 13,500 rpm for 5 min in a bench-top microfuge and the supernatant was removed to a clean Eppendorf tube prior to storage. After thawing of plasma, all samples were split into two aliquots. The first was subjected to no further treatment and the second to centrifugation at 13,500 rpm for 5 min in a bench-top microfuge. Plasma DNA was extracted from all samples using the QiAamp Blood Kit (Qiagen, Hilden, Germany) according to the blood and body fluid protocol, modifying the initial plasma sample (500 μL) and final elution (100 μL) volume.

PCR Analysis

DNA recovered from stored plasma was analyzed by PCR using five amplicons of increasing size (100 bp and 350 bp GAPDH and 512 bp, 843 bp, 1271 bp Bcl-2 amplicons). Primer details are given in TABLE 1. Ten microliters of each plasma DNA sample was analyzed by PCR as described previously[5] using the following cycles: 98°C for 3 min, followed by 35 cycles of 94°C for 30 sec, 60°C for 30 sec, and 72°C for 30 sec to 2 min depending on the amplicon length, with a final extension of 7 min at 72°C. All reactions were carried out on a GeneAmp PCR System 9700 (Perkin–Elmer Applied Biosystems, Warrington, UK). PCR products were resolved on 1–4% agarose gels.

TABLE 1. PCR primers for 100 bp and 350 bp GAPDH and 512 bp, 843 bp, 1271 bp Bcl-2 amplicons

Primer/amplicon	Sequence
GAPDHF (100 bp)	GCAAGAGCACAAGAGGAAGA
GAPDHR (100 bp)	ACTGTGAGGAGGGGAGATTC
GAPDH (350 bp)	Given in Hall *et al.*[9]
Bcl-2 RP1(common)	ATAGCAGCACAGGATTGG
Bcl-2 FP2 (512 bp)	GTGCATTTCCACGTCAAC
Bcl-2 FP3 (843 bp)	GATGGAATAACTCTGTGGC
Bcl-2 FP4 (1271 bp)	AGAAGGACATGGTGAAGG

RESULTS

When we compared the effect of different processing steps prior to DNA extraction on the range of amplicons detected following PCR of plasma DNA, we found a wide variation between the three sets of samples. With no extra processing, all samples from sets 1 and 2 showed amplification of the two smallest amplicons, and several samples also showed amplification of the larger Bcl-2 amplicons (FIG. 1). However, for plasma DNA from sets 1

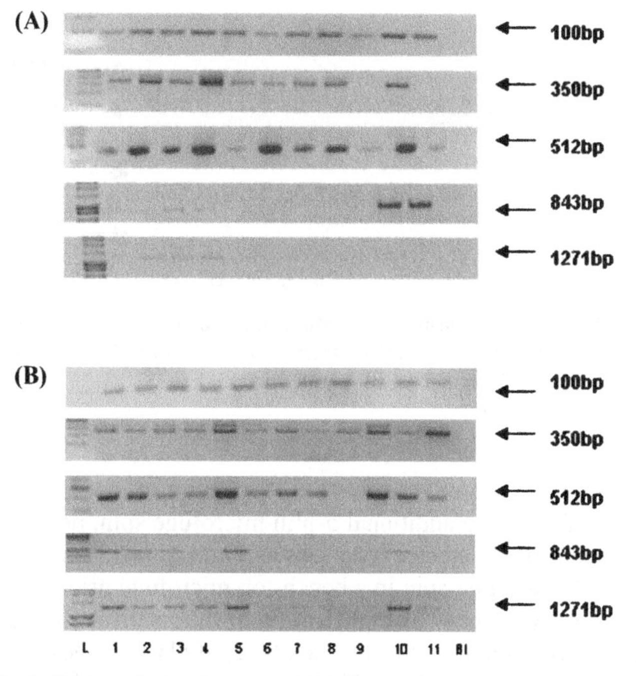

FIGURE 1. PCR analysis of plasma DNA from (A) control set 1 (n = 11) and (B) control set 2 (n = 11) with no additional centrifugation. Analysis of 100 bp and 350 bp GAPDH and 512 bp, 843 bp, 1271 bp Bcl-2 amplicons. Bl = water blank.

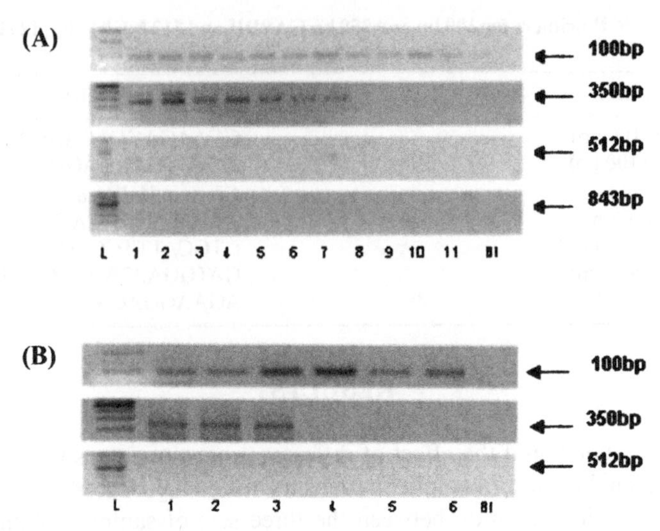

FIGURE 2. Analysis of plasma DNA from (A) control set 1 and (B) control set 2 with additional 5-min microfuge spin prior to DNA extraction. Analysis of 100 bp and 350 bp GAPDH and 512 bp, 843 bp Bcl-2 amplicons. Bl = water blank.

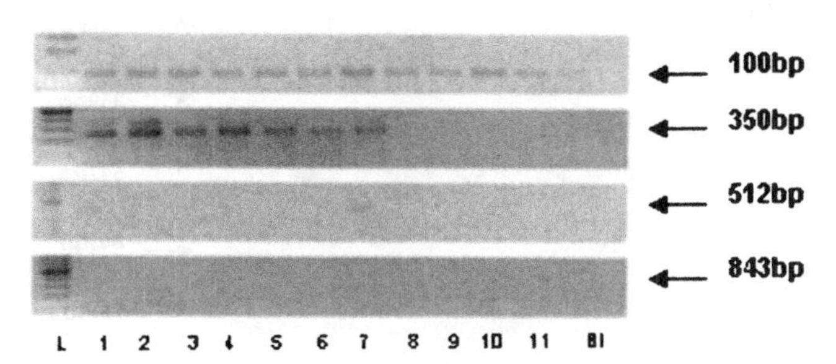

FIGURE 3. Analysis of plasma DNA from control set 3 ($n = 7$) with additional 5-min microfuge spin prior to storage and DNA extraction. Analysis of 100 bp and 350 bp GAPDH and 512 bp, 843 bp Bcl-2 amplicons. Bl = water blank.

and 2 recovered after the additional 5-min microfuge spin, none of the three largest amplicons were detected (control sets 1 and 2, FIG. 2). Set 3, which had been subjected to a 5-min spin in a bench-top microfuge prior to storage and DNA extraction, only showed amplification of the smallest 100-bp amplicon (FIG. 3). In comparison, in our routine analyses of plasma DNA isolated from breast cancer patients, with the addition of a microfuge spin prior to DNA extraction, most plasma DNA samples successfully amplify large amplicons (>500 bp in size) (data not shown).

DISCUSSION

Recently, it has been shown that additional centrifugation may be required to eliminate contaminating cellular DNA from plasma DNA samples.[8] We examined whether additional microfuge centrifugation would eliminate contaminating cellular DNA from "normal" plasma samples, isolated from healthy females (age range 25–55 years), using PCR analysis of a number of increasing-sized amplicons. Cellular DNA contamination was indicated by successful amplification of products >300 bp in size, evident in control sets 1 and 2 (FIG. 1), which then decreased or disappeared after additional microfuge centrifugation (FIG. 2).

Since healthy individuals have very low levels of free-circulating DNA, the amplified DNA observed prior to additional centrifugation was most likely derived from contaminating cellular DNA. The addition of a 5-min spin in a bench-top microfuge prior to storage and or DNA extraction from plasma, was sufficient to eliminate contaminating cellular DNA, as revealed by the absence of amplification of products >300 bp in length (control set 3, FIG. 3).

Since plasma DNA is a promising biomarker for cancer detection and increased tumor DNA integrity in plasma is associated with cancer,[7] it is vital to process blood samples appropriately in order to minimize contamination by cellular DNA. Our data highlight the importance of careful processing in preparation of cell-free plasma DNA for cancer detection and we recommend the use of a high-speed spin in a bench to microcentrifuge, prior to storage and/or DNA extraction.

REFERENCES

1. LEON, S.A. *et al.* 1977. Free DNA in the serum of cancer patients and the effect of therapy. Cancer Res. **37:** 646–650.
2. KOFFLER, D. *et al.* 1973. The occurrence of single stranded DNA in the serum of patients with SLE and other diseases. J. Clin. Invest. **52:** 198–204.
3. CHEN, X.Q. *et al.* 1996. Microsatellite alterations in plasma DNA of small cell lung cancer patients. Nat. Med. **2:** 1033–1035.
4. NAWROZ, H. *et al.* 1996. Microsatellite alterations in serum DNA of head and neck cancer patients. Nat. Med. **2:** 1035–1037.
5. SHAW, J.A. *et al.* 2000. Microsatellite alterations in plasma DNA of primary breast cancer patients. Clin. Cancer Res. **6:** 1119–1124.
6. SILVA, J.M. *et al.* 1999. Presence of tumour DNA in plasma of breast cancer patients: clinicopathological correlations. Cancer Res. **59:** 3251–3256.
7. WANG, B.G. *et al.* 2003. Increased plasma DNA integrity in cancer patients. Cancer Res. **63:** 3966–3968.
8. CHIU, R.W. *et al.* 2001. Effects of blood-processing protocols on foetal and total DNA quantification in maternal plasma. Clin. Chem. **47:** 1607–1613.
9. HALL, L.L. *et al.* 1998. Reproducibility in the quantification of mRNA levels by RT-PCR-ELISA and RT Competitive-PCR-ELISA. Biotechniques **24:** 652–658.

Nucleosomal DNA Fragments in Autoimmune Diseases

STEFAN HOLDENRIEDER,[a] PETER EICHHORN,[a] ULRICH BEUERS,[b] WALTER SAMTLEBEN,[c] ULF SCHOENERMARCK,[c] REINHART ZACHOVAL,[b] DOROTHEA NAGEL,[a] AND PETRA STIEBER[a]

[a]*Institute of Clinical Chemistry, University Hospital Munich-Grosshadern, Munich, Germany*

[b]*Medical Clinic II, University Hospital Munich-Grosshadern, Munich, Germany*

[c]*Medical Clinic I, University Hospital Munich-Grosshadern, Munich, Germany*

ABSTRACT: The inadequate response of immune cells to circulating apoptotic products, such as nucleosomal DNA fragments, is assumed to be a potent stimulus for the production of autoantibodies during the pathogenesis and progression of systemic lupus erythematosus (SLE). Here, we analyzed the levels of circulating nucleosomes, caspases, and C-reactive protein in sera of 244 individuals with various autoimmune diseases (155 with autoimmune hepatic disorders, 25 with ANCA-associated vasculitis, and 64 with various connective tissue diseases), and 32 healthy controls. Nucleosomes and caspase activities were significantly elevated in sera of patients with hepatic autoimmune diseases, connective tissue diseases, and particularly in ANCA-associated vasculitis when compared with healthy individuals. Nucleosomes showed a correlation with caspases, and caspases with C-reactive protein, but nucleosomes did not correlate with C-reactive protein. Serum levels of the apoptotic products, nucleosomes, and caspases are increased in various autoimmune diseases but may not be solely responsible for antinucleosome antibody production in SLE patients. It remains to be clarified whether qualitative changes in nucleosomes are linked with pathogenesis and disease progression in SLE.

KEYWORDS: DNA; nucleosomes; caspases; serum; plasma; autoimmune disease; connective tissue disease; vasculitis; AIH; PSC; PBC; SLE

BACKGROUND

Structural and Functional Characteristics of Nucleosomes

Nucleosomes are supposed to be the major immunogens in systemic lupus erythematosus (SLE).[1,2] They are the basic elements of chromatin and are

Address for correspondence: Dr. Stefan Holdenrieder, Institute of Clinical Chemistry, University Hospital Munich-Grosshadern, Munich, Germany. Voice: 0049-89-7095-3231; fax: 0049-89-7095-6298.

e-mail: Stefan.Holdenrieder@med.uni-muenchen.de

Ann. N.Y. Acad. Sci. 1075: 318–327 (2006). © 2006 New York Academy of Sciences.
doi: 10.1196/annals.1368.043

mainly located in the nucleus of eukaryotic cells. Nucleosomes are formed by a core protein consisting of an octamer of histones and about 146 bp of ds-DNA that is wrapped around it.[3]

The histones H2A, H2B, H3, and H4, which are present in duplicates in the nucleosomal core are linked with each other by hydrogen bonds and hydrophobic interactions.[3,4] Protruding tails of the histones can establish contact with other nucleosomes or can be modified by methyl-, acetyl-, phosphate- or ubiquitin-groups.[5] DNA surrounds the protein core about 1.5 times on the flat outside and is connected with histones at 14 sites.[3] Various nucleosomes are joined by 10 to 100 bp of so-called linker-DNA and form a chain-like structure, which is further stabilized by the H1 histone on the outside.[3] This primary order prevents chromatin from embroilment and allows transcription factors to bind to specific DNA sequences.[3,4] However, nucleosomal arrangement is far from being a rigid construction. The bonds between DNA and the histone core in a single nucleosome can be loosened to enable nucleosomes scrolling along the DNA. Alternatively, chromatin can build up loops and translocate nucleosomes to another DNA region. This mobility of nucleosomes is essential for a high flexibility in granting access to transcription factors according to the changing requirements of the cells.[4,6]

Release of Nucleosomes during Cell Death

Although nucleosomes are parts of the highly organized chromatin and are involved in active intracellular and nuclear processes they can also be found in low amounts in the serum and plasma of healthy individuals.[7] Apoptotic cell death has been identified as an important source of the nucleosomal release from cells.[8–10] During apoptosis, specific endonucleases are activated, which enter the nucleus and cleave the chromatin at the easily accessible linking sites into oligo- and mononucleosomal fragments.[11] Like other cellular degradation products, nucleosomes are physiologically packed into membrane-bound apoptotic bodies, which are shed and phagocytized by macrophages and neighboring cells, where they are digested by lysosomal proteases and can be used as raw material for new cells.[12,13]

Besides this very effective recycling process, a small amount of nucleosomes might also be released from dying cells or apoptotic bodies into the extracellular space and reach the blood circulation.[9,12] In addition, an active physiological secretion of DNA by lymphocytes was suggested.[14] As various endonucleases are present and active in serum and plasma, most forms of pure nucleic acids would be removed very rapidly. However, in the arrangement (or organization) as nucleosomes, they seem to be better protected from digestion in plasma.[15] This hypothesis is consistent with the observation of Rumore *et al.*, who found mainly mono- and oligonucleosomal complexes in the serum of lupus patients and in patients after hemodialysis.[16,17] The good correlation between real-time PCR and a nucleosome ELISA supports these findings.[18]

Circulating Nucleosomes in Various Diseases

Elevated levels of circulating nucleosomal DNA were found in various benign and malignant pathological conditions, for example, after either exhaustive sports, trauma, ischemia, sepsis, graft rejection, autoimmune diseases,[19–24] and numerous neoplastic diseases.[7,12,25–27] In malignancies, the pretherapeutic level of DNA was reported to have predictive and prognostic relevance.[28] During chemo- and radiotherapy, the kinetics of circulating DNA have been shown to correlate with the response to antitumor therapy and with tumor recurrence.[7,12,27–30] The elevated amounts of circulating nucleosomal DNA are mainly due to spontaneous or therapy-induced cell death as in apoptosis, necrosis, or aponecrosis. Confronted with the increased release of DNA, the physiological recycling system, even if it is upregulated, is often overwhelmed. In addition, during some pathological situations, the removal of DNA from serum and plasma is defective or impaired, leading to persistently higher concentrations of circulating DNA.[12]

Role of Nucleosomes in Autoimmune Diseases

Despite many efforts, knowledge about the specific role of circulating nucleosomal DNA in the pathogenesis of many diseases is scarce. Only SLE has been investigated more intensively, probably because of the high frequency of antinucleosome and anti-dsDNA antibodies and their potential association with the progression of SLE.[1,24,31,32]

The autoimmune response in SLE is T cell–dependent and autoantigen-driven. Various factors contribute to the development of autoimmunity. The apoptosis rate of lymphocytes is enhanced both spontaneously and after *in vitro* stimulation[31,33] and the amount of circulating nucleosomes is higher in SLE patients as compared to healthy individuals.[24,34] The activity of plasma DNase, which is mainly responsible for the plasmatic degradation of nucleosomes, was found to be reduced in SLE patients[35] and correlated inversely with higher antibody production.[34] Enhanced lymphocyte apoptosis, insufficient clearance from the circulation and the resulting increased and prolonged persistence of nucleosomes in serum might contribute to the immunogeneity of nucleosomes in SLE.[1,33–36]

Furthermore, nucleosomes are phagocytized, processed, and exposed on the surface of antigen-presenting cells[1,37] and it has been speculated that nucleosomes might be modified to stimulate the autoantibody production in SLE.[1] In SLE, different autoantibodies have been identified, such as antinucleosome-, anti-DNA-, and antihistone antibodies, which are mainly of the IgG subtype.[38–41] Antinucleosome antibodies are highly specific for SLE[1,41,42] and the extent of antinucleosome antibody production correlates with anti-ds-DNA-antibody titers.[41] The appearance of antinucleosome antibodies clearly precedes the formation of anti-DNA- and antihistone antibodies, suggesting that the whole nucleosome represents the driving autoantigen in SLE.[1,24]

Antinucleosome antibodies seem to be particularly important in the pathogenesis of renal manifestations in SLE. Circulating nucleosomes and antinucleosome antibodies can form complexes that aggregate at the glomerular basement membrane with strong interactions between the cationic histones and the anionic heparin sulfate of the glomerular basement membrane. In SLE patients with renal involvement, higher levels of antinucleosome- as well as anti-DNA- and antihistone antibodies are found, although the nucleosome–antinucleosome-antibody complexes were reported to damage the glomerular basement membrane most severely.[1,38,39,43–46]

So far, the correlation of serum nucleosome levels with disease extent and activity in SLE shows conflicting results. Whereas Williams *et al.* found a marked association between elevated nucleosome levels and the SLE disease activity score,[43] Amoura *et al.* could not support these results.[24]

The relevance of circulating nucleosomes for other autoimmune diseases has not been investigated in detail so far. In the present study, the levels of nucleosomes and the activity of caspases as markers of apoptosis were analyzed in sera of patients with various autoimmune disorders and correlated with C-reactive protein as marker of disease activity.

PATIENTS AND METHODS

Circulating nucleosomes, caspases, and C-reactive protein were analyzed in sera of 276 individuals, among them 155 with autoimmune hepatic disorders (34 with autoimmune hepatitis, AIH; 74 with primary biliary cirrhosis, PBC; 42 with primary sclerosing cholangitis, PSC; and 5 with overlap syndromes); 25 with ANCA-associated vasculitis (11 Wegener's granulomatosis, WG; 14 microscopic polyangiitis, MPA); 64 with various connective tissue diseases (40 systemic lupus erythematosus, SLE; 13 Sjegren's syndrome, SS; 11 progressive systemic sclerosis, PSS); and 32 healthy individuals.

Nucleosomes were determined by an immunoassay system using two monoclonal mouse antibodies specific for nucleosomes; caspases were determined by an immunofluorimetric research assay detecting the caspases 2, 3, 6, 7, 8, 9, and 10 (homogeneous caspases assay, both Roche Diagnostics GmbH, Germany); and C-reactive protein was measured by latex-consolidated immunoturbidimetry (AU2700, Olympus, Germany).

Results were expressed as medians of the marker concentrations for each group and as a percentage of values above the 95th percentile of controls.

Figures show dot plots of the marker levels and medians for all subgroups. Differences between various patient groups and healthy individuals were calculated by the Wilcoxon test. Correlations of the various markers were tested by Spearman's rank correlation.

RESULTS

In sera of healthy individuals only low levels of nucleosomes were found (median 5 ng/mL). In contrast, nucleosome concentrations were significantly elevated in sera of patients with hepatic autoimmune diseases (median 101 ng/mL; $P < 0.001$), ANCA-associated vasculitis (456 ng/mL; $P < 0.001$), and connective tissue diseases, including SLE (33 ng/mL; $P = 0.007$). In 77% of patients with hepatic autoimmune diseases, nucleosome levels were higher than the 95th percentile of controls (41.5 ng/mL) as well as in all patients with ANCA-associated vasculitis. Higher values were also observed in 45% of patients with connective tissue diseases. In the subgroup of SLE patients, only 23% had elevated serum nucleosome concentrations (FIG. 1).

With respect to caspase activity, a significantly higher level was observed in sera of patients with hepatic autoimmune diseases (median 0.1 mU/L;

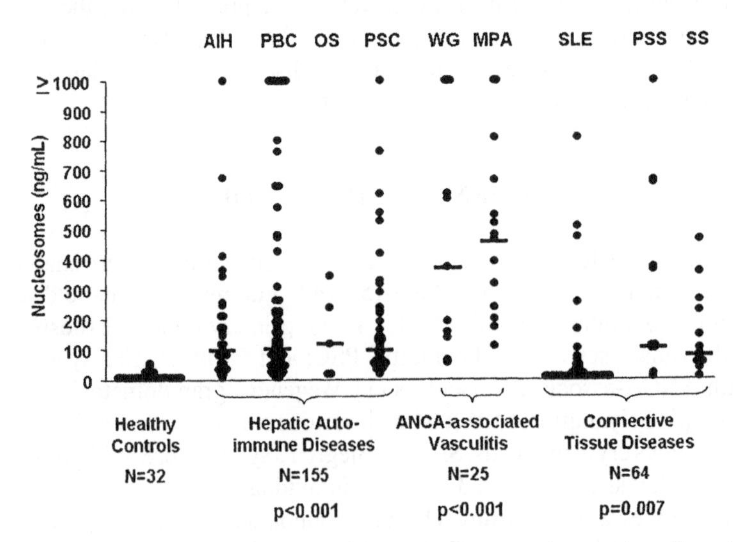

FIGURE 1. Distribution of serum nucleosome concentrations and medians (—) in healthy individuals, patients with autoimmune hepatic diseases (autoimmune hepatitis, AIH; primary biliary cirrhosis, PBC; primary sclerosing cholangitis, PSC; overlap syndromes, OS), with ANCA-associated vasculitis (Wegener's granulomatosis, WG; microscopic polyangiitis, MPA), and with various connective tissue diseases (systemic lupus erythematosus, SLE; Sjögren's syndrome, SS; progressive systemic sclerosis, PSS). *P* values indicate the differences between healthy individuals and the various autoimmune disease groups.

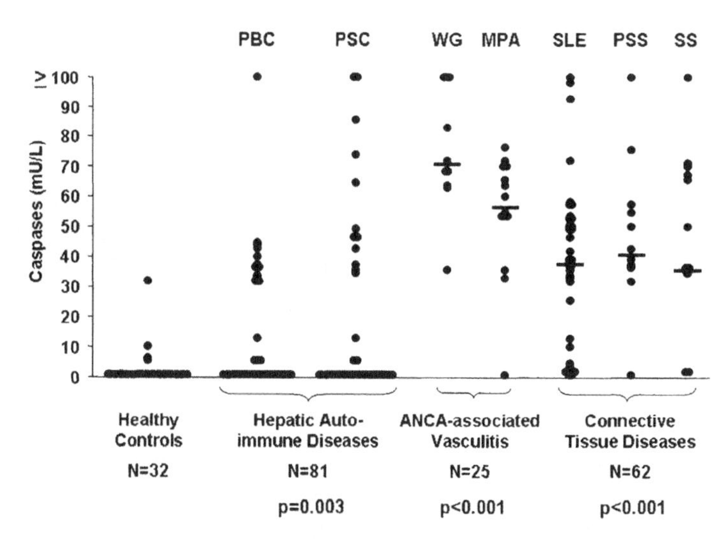

FIGURE 2. Distribution of serum caspase activity and medians (———) in healthy individuals, patients with autoimmune hepatic diseases (autoimmune hepatitis, AIH; primary biliary cirrhosis, PBC; primary sclerosing cholangitis, PSC; overlap syndromes, OS), with ANCA-associated vasculitis (Wegener's granulomatosis, WG; microscopic polyangiitis, MPA) and with various connective tissue diseases (systemic lupus erythematosus, SLE; Sjögren's syndrome, SS; progressive systemic sclerosis, PSS). *P* values indicate the differences between healthy individuals and the various autoimmune disease groups.

$P = 0.003$), ANCA-associated vasculitis (64.7 mU/L; $P < 0.001$), and connective tissue diseases (38.3 mU/L; $P < 0.001$) in comparison with healthy individuals (0.1 mU/L). Serum caspase levels higher than the 95th percentile of controls (17.4 mU/L) were observed in 30% of patients with hepatic autoimmune diseases, in 96% of patients with ANCA-associated vasculitis, and in 77% of patients with connective tissue diseases. In the subgroup of SLE patients, 71% of the patients had elevated serum caspase activity (FIG. 2).

Significant correlations were found between circulating nucleosomes and caspases ($r = 0.164$, $P = 0.017$) as well as between caspases and C-reactive protein ($r = 0.363$, $P < 0.001$), but not between nucleosomes and C-reactive protein ($r = 0.100$, $P = 0.145$).

DISCUSSION

For the diagnosis of SLE, antinucleosome antibodies have shown higher sensitivity than anti-dsDNA antibodies and antihistone antibodies (93% vs. 71% and 40%, respectively) at a high specificity of 98% versus that in patients with other systemic autoimmune diseases.[46] In routine clinical practice, antinucleosome antibodies are particularly valuable in the diagnosis of those SLE patients who are negative for anti-dsDNA antibodies.[46] The high specificity of antinucleosome antibodies for SLE was confirmed by several studies that reported on only low antibody levels in most other autoimmune disorders as well as in hepatitis and HIV infections (reviewed in Ref. 1). Only two studies found higher percentages of antinucleosome antibody positivity in patients with systemic sclerosis and mixed connective tissue diseases, which might, however, be due to different antigen preparation and cut-off definition.[39,47] Within the group of SLE patients, antinucleosome antibodies are strongly associated with disease activity and the development of renal complications.[1,43–46] Especially, IgG3 subclasses are present at high levels in patients with active SLE, but absent from those with inactive SLE or other connective tissue diseases.[39] Thus, IgG3 subclasses of antinucleosome antibodies were suggested as a specific marker for lupus nephritis.[39]

Concerning the levels of circulating nucleosomes, Amoura *et al.* found elevated concentrations in only 25% of 58 SLE patients investigated.[24] While nucleosome levels correlated inversely with antinucleosome antibodies, no correlation was found between nucleosomes and disease activity.[24] In contrast, Williams *et al.* recently reported elevated nucleosome levels in 47% of 140 SLE patients and a strong correlation of nucleosomes and disease activity.[43] These observations raised two questions: (*a*) whether the presence of high amounts of serum and plasma nucleosomes as a result of enhanced lymphocyte apoptosis and delayed removal from blood circulation is necessary and sufficient for the stimulation of the antibody production during the pathogenesis of SLE or (*b*) whether other processes, such as the modification of the nucleosomes or the exposition of nucleosome particles on the surface of apoptotic cells and antigen-presenting cells are more relevant?

In our study, we addressed these questions indirectly by investigating the levels of nucleosomes in sera of patients with various autoimmune diseases. Our findings of elevated nucleosome concentrations not only in SLE, but also in other autoimmune diseases, which are known not to be associated with the formation of antinucleosome antibodies, suggest that the increase in the apoptotic rate and/or the defective removal of apoptotic products is a general phenomenon in autoimmune disorders. The high levels of caspase activity that were found in sera of patients with these disorders as well as the correlation of both apoptosis markers further support this hypothesis. Our observations confirm and extend the results of other studies reporting on elevated rates of apoptosis not only in SLE, but also in other autoimmune diseases, such as Wegener's granulomatosis or Takayasu arteriitis.[1,24,31,33,34,37,39,43]

Correlation with disease activity was performed using the serum concentrations of C-reactive protein. This procedure might constitute only a rough approximation as the specific scores of various autoimmune diseases are more informative. According to our findings, nucleosome levels were not related with disease activity, in line with earlier reports of Amoura *et al.*, who could not find any correlation of nucleosome levels and SLE disease activity score (SLEDAI) in SLE patients.[24] In contrast, caspase activity showed a strong correlation with C-reactive protein levels. Whether this observation is due to an interaction of the two markers or whether some apoptotic indicators are linked differently with disease progression remains to be clarified.

It is remarkable that high levels of circulating nucleosomes were found in most of the autoimmune diseases that are known *not* to be associated with the formation of antinucleosome antibodies. This observation suggests that the pure presence of the nucleosome antigens in serum is not sufficient to stimulate the antinucleosome antibody production. On the other hand, nucleosome concentrations were only elevated in about 23% of the patients with SLE. This percentage seems to be too low to explain the high prevalence of antinucleosome antibodies reported in this patient subset.

According to our results, the formation of these antibodies is less likely to be dependent on the amount of circulating nucleosomes, but is probably caused by qualitative changes in nucleosomal particles circulating in serum and plasma and/or exposed on the surface of apoptotic or antigen-presenting cells. Additionally, costimulatory signals by T cells might be necessary to support the effective stimulation of B cells for the antinucleosome antibody production.[1] Future nucleosome research should focus on the identification of nucleosome subsets or particles that are specific for driving autoimmune processes in SLE and that can be targeted by therapeutic interventions.

REFERENCES

1. KOUTOUZOV, S. *et al.* 2004. Nucleosomes in the pathogenesis of systemic lupus erythematosus. Rheum. Dis. Clin. North Am. **30:** 529–558.
2. BERDEN, J.H. 2003. Lupus nephritis: consequence of disturbed removal of apoptotic cells? Neth. J. Med. **61:** 233–238.
3. LUGER, K. *et al.* 1997. Crystal structure of the nucleosome core particle at 2.8 A resolution. Nature **389:** 251–260.
4. KORNBERG, R.D. & Y. LORCH 1999. Twenty-five years of the nucleosome, fundamental particle of the eukaryote chromosome. Cell **98:** 285–294.
5. STRAHL, B.D. & C.D. ALLIS. 2000. The language of covalent histone modifications. Nature **403:** 41–45.
6. NELSON, S.M. *et al.* 2004. DNA and the chromosome—varied targets for chemotherapy. Cell Chromosome **3:** 2.
7. HOLDENRIEDER, S. *et al.* 2001. Nucleosomes in serum of patients with benign and malignant diseases. Int. J. Cancer **95:** 114–120.
8. VAN NIEUWENHUIJZE, A.E. *et al.* 2003. Time between onset of apoptosis and release of nucleosomes from apoptotic cells: putative implications for systemic lupus erythematosus. Ann. Rheum. Dis. **62:** 10–14.
9. LICHTENSTEIN, A.V. *et al.* 2001. Circulating nucleic acids and apoptosis. Ann. N. Y. Acad. Sci. **945:** 239–249.
10. HOLDENRIEDER, S. *et al.* 2001. Nucleosomes in serum as a marker for cell death. Clin. Chem. Lab. Med. **39:** 596–605.
11. HENGARTNER, M.O. 2000. The biochemistry of apoptosis. Nature **407:** 770–776.

12. HOLDENRIEDER, S. & P. STIEBER. 2004. Apoptotic markers in cancer. Clin. Biochem. **37:** 605–617.
13. HENGARTNER, M.O. 2001. Apoptosis: corralling the corpses. Cell **104:** 325–328.
14. STROUN, M. *et al.* 2000. The origin and mechanism of circulating DNA. Ann. N. Y. Acad. Sci. **906:** 161–168.
15. NG, E.K. *et al.* 2002. Presence of filterable and nonfilterable mRNA in the plasma of cancer patients and healthy individuals. Clin. Chem. **48:** 1212–1217.
16. RUMORE, P.M. & C.R. STEINMAN. 1990. Endogenous circulating DNA in systemic lupus erythematosus. Occurrence as multimeric complexes bound to histone. J. Clin. Invest. **86:** 69–74.
17. RUMORE, P. *et al.* 1992. Haemodialysis as a model for studying endogenous plasma DNA: oligonucleosome-like structure and clearance. Clin. Exp. Immunol. **90:** 56–62.
18. HOLDENRIEDER, S. *et al.* 2005. Cell-free DNA in serum and plasma: comparison of ELISA and quantitative PCR. Clin. Chem. **51:** 1544–1546.
19. ATAMANIUK, J. *et al.* 2004. Increased concentrations of cell-free plasma DNA after exhaustive exercise. Clin. Chem. **50:** 1668–1670.
20. ZEERLEDER, S. *et al.* 2003. Elevated nucleosome levels in systemic inflammation and sepsis. Crit. Care Med. **31:** 1947–1951.
21. LO, Y.M. *et al.* 2000. Plasma DNA as a prognostic marker in trauma patients. Clin. Chem. **46:** 319–323.
22. GEIGER, S. *et al.* 2006. Nucleosomes in serum of patients with early cerebral stroke. Cerebro. Vasc. Dis. **21:** 32–37.
23. FISHBEIN, T.M. *et al.* 2004. Increased apoptosis is specific for acute rejection in rat small bowel transplant. J. Surg. Res. **119:** 51–55.
24. AMOURA, Z. *et al.* 1997. Circulating plasma levels of nucleosomes in patients with systemic lupus erythematosus: correlation with serum antinucleosome antibody titers and absence of clear association with disease activity. Arthritis Rheum. **40:** 2217–2225.
25. KUROI, K. *et al.* 1999. Plasma nucleosome levels in node-negative breast cancer patients. Breast Cancer **6:** 361–364.
26. SOZZI, G. *et al.* 2001. Analysis of circulating tumor DNA in plasma at diagnosis and during follow-up of lung cancer patients. Cancer Res. **61:** 4675–4678.
27. TREJO-BECERRIL, C. *et al.* 2003. Circulating nucleosomes and response to chemotherapy: an *in vitro*, *in vivo* and clinical study on cervical cancer patients. Int. J. Cancer **104:** 663–668.
28. GAUTSCHI, O. *et al.* 2004. Circulating deoxyribonucleic acid as prognostic marker in non-small-cell lung cancer patients undergoing chemotherapy. J. Clin. Oncol. **22:** 4157–4164.
29. HOLDENRIEDER, S. *et al.* 2004. Circulating nucleosomes predict the response to chemotherapy in patients with advanced non-small cell lung cancer. Clin. Cancer Res. **10:** 5981–5987.
30. KREMER, A. *et al.* 2005. Nucleosomes in pancreatic cancer patients during radiochemotherapy. Tumor Biol. **26:** 44–49.
31. BELL, D.A. & B. MORRISON. 1991. The spontaneous apoptotic cell death of normal human lymphocytes *in vitro*: the release of, and immunoproliferative response to, nucleosomes *in vitro*. Clin. Immunol. Immunopathol. **60:** 13–26.
32. KOUTOUZOV, S. *et al.* 1997. Comparison of structural characteristics of antisubnucleosome and anti-DNA monoclonal antibodies derived from lupus mice. Ann. N. Y. Acad. Sci. **815:** 327–330.

33. EMLEN, W. *et al.* 1994. Accelerated *in vitro* apoptosis of lymphocytes from patients with systemic lupus erythematosus. J. Immunol. **152:** 3685–3692.
34. D'AURIA, F. *et al.* 2004. Accumulation of plasma nucleosomes upon treatment with anti-tumour necrosis factor-alpha antibodies. J. Intern. Med. **255:** 409–418.
35. SALLAI, K. *et al.* 2005. Antinucleosome antibodies and decreased deoxyribonuclease activity in sera of patients with systemic lupus erythematosus. Clin. Diagn. Lab. Immunol. **12:** 56–59.
36. LICHT, R. *et al.* 2004. Decreased phagocytosis of apoptotic cells in diseased SLE mice. J. Autoimmun. **22:** 139–145.
37. RADIC, M. *et al.* 2004. Nucleosomes are exposed at the cell surface in apoptosis. J. Immunol. **172:** 6692–6700.
38. GABLER, C. *et al.* 2003. The putative role of apoptosis-modified histones for the induction of autoimmunity in systemic lupus erythematosus. Biochem. Pharmacol. **66:** 1441–1446.
39. AMOURA, Z. *et al.* 2000. Presence of antinucleosome autoantibodies in a restricted set of connective tissue diseases: antinucleosome antibodies of the IgG3 subclass are markers of renal pathogenicity in systemic lupus erythematosus. Arthritis Rheum. **43:** 76–84.
40. RAHMAN, A. & F. HIEPE. 2002. Anti-DNA antibodies—overview of assays and clinical correlations. Lupus **11:** 770–773.
41. GHIRARDELLO, A. *et al.* 2004. Antinucleosome antibodies in SLE: a two-year follow-up study of 101 patients. J. Autoimmun. **22:** 235–240.
42. SUER, W. *et al.* 2004. Autoantibodies in SLE but not in scleroderma react with protein-stripped nucleosomes. J. Autoimmun. **22:** 325–334.
43. WILLIAMS, R.C. JR. *et al.* 2001. Detection of nucleosome particles in serum and plasma from patients with systemic lupus erythematosus using monoclonal antibody 4H7. J. Rheumatol. **28:** 81–94.
44. GHILLANI-DALBIN, P. *et al.* 2003. Testing for anti-nucleosome antibodies in daily practice: a monocentric evaluation in 1696 patients. Lupus **12:** 833–837.
45. CORTES-HERNANDEZ, J. *et al.* 2004. Antihistone and anti-double-stranded deoxyribonucleic acid antibodies are associated with renal disease in systemic lupus erythematosus. Am. J. Med. **116:** 165–173.
46. SIMON, J.A. *et al.* 2004. Anti-nucleosome antibodies in patients with systemic lupus erythematosus of recent onset. Potential utility as a diagnostic tool and disease activity marker. Rheumatology **43:** 220–224.
47. WALLACE, D.J. *et al.* 1994. Antibodies to histone (H2A-H2B)-DNA complexes in the absence of antibodies to double-stranded DNA or to (H2A-H2B) complexes are more sensitive and specific for scleroderma-related disorders than for lupus. Arthritis Rheum. **37:** 1795–1797.

Concentrations of Circulating RNA from Healthy Donors and Cancer Patients Estimated by Different Methods

ELENA Y. RYKOVA,[a] WINFRIED WUNSCHE,[b] OLGA E. BRIZGUNOVA,[a] TATYANA E. SKVORTSOVA,[a] SVETLANA N. TAMKOVICH,[a] ILIJA S. SENIN,[a] PAVEL P. LAKTIONOV,[a] GEORG SCZAKIEL,[b] AND VALENTIN V. VLASSOV[a]

[a]*Institute of Chemical Biology and Fundamental Medicine SB of RAS, Lavrentiev Ave. 8, Novosibirsk, Russia*

[b]*Institut für Molekulare Medizin, Universität zu Lübeck, Ratzeburger Allee 160, Lübeck, Germany*

ABSTRACT: Circulating RNA (cirRNA) was isolated from plasma and cell surface-bound fractions of blood of healthy women and breast cancer patients. RNA samples were DNase treated and quantified by a SYBR Green II assay. Concentrations of RNA sequences of GAPDH, Ki-67 mRNA, and 18S rRNA were measured by real-time quantitative PCR (RT-qPCR) after reverse transcription with random hexamer primers. The obtained data spread over three orders of magnitude for GAPDH and Ki-67 mRNA signals and two orders of magnitude for the copy number of 18S rRNA in blood fractions in both groups. In blood of healthy donors, no correlation was found between the copy number of GAPDH, Ki-67 mRNA, and 18S rRNA and RNA concentrations measured by the SYBR Green II assay. Within the group of breast cancer patients, the concentration GAPDH and Ki-67 mRNA correlated with the concentration of total RNA only in the cell surface-bound fraction; whereas the concentration of 18S rRNA correlated with total RNA in both, the cell surface-bound fraction and blood. The copy number of Ki-67 mRNA correlated with copy numbers of GAPDH mRNA in all fractions of cirRNA of healthy donors and breast cancer patients. A correlation between copy numbers of Ki-67 mRNA and 18S rRNA was found only in cell surface-bound fraction of breast cancer patients. The data described here demonstrate the necessity of searching for more suitable RNA markers in order to estimate total cirRNA concentrations by RT-qPCR, although mRNA of GAPDH could be used for normalization of the level of cancer-specific mRNA among patients.

Address for correspondence: Elena Y. Rykova, Institute of Chemical Biology and Fundamental Medicine SB of RAS, Lavrentiev Ave. 8, Novosibirsk, Russia. Voice: +7-383-3-304654; fax: +7-383-3333677.

e-mail: rykova@niboch.nsc.ru

Ann. N.Y. Acad. Sci. 1075: 328–333 (2006).
doi: 10.1196/annals.1368.044

KEYWORDS: **extracellular RNA; plasma; blood; breast cancer; RT-qPCR**

INTRODUCTION

It has been clearly demonstrated that RNA is not only located inside cells but is also found circulating in the bloodstream in a cell-free form or is cell surface associated. Along with RNA molecules, derived from normal cells, tumor-derived RNA was also found in the circulation, enabling an approach for noninvasive investigation of gene expression of transformed cells and for cancer diagnostic development.[1] The origin of circulating RNA (cirRNA) is not clear to date. Its active secretion by cells seems possible[2] though cell destruction might be the main source for cirRNA.[3] cirRNA in blood is protected from ribonucleases. For example, it was shown that cirRNA can be packaged in apoptotic bodies separately from DNA,[4] it can be contained within nucleoprotein complexes,[5] lipids, proteolipids, phospholipids,[6,7] proteins,[8] and it can be associated with circulating particles.[9] Such complexed forms of cirRNA provide stability of RNA in blood and can reflect the origin of specific RNA that depends on its intracellular localization and function. Messenger RNAs for c-abl and tyrosinase mRNAs were shown to be degraded after a freeze–thaw cycle[10] whereas GAPDH mRNA was stable under these conditions.[11] The concentration of cirRNA estimated by sequence-specific quantitative RT-PCR (RT-qPCR) thus could not be considered to quantitatively monitor the total cirRNA concentration. It has to be kept in mind that total cirRNA are composed of many different RNA species, which are differently expressed, secreted, or affected by other factors in the circulation. The data on the contribution of different RNAs in the total RNA concentration can reflect general processes regulating the circulation of extracellular RNA in health and disease, and are necessary for searching of the accurate RNA marker that could be used for comparison of the levels of cancer-specific RNA between patients.

Here we investigated the potential role of messenger RNA and ribosomal RNA as quantitative markers for total RNA measured by RT-qPCR and compared this with total RNA concentration as measured by a fluorescence assay.

MATERIALS AND METHODS

Blood samples were obtained from Novosibirsk Regional Oncologic Dispensary from previously untreated breast cancer patients. Tumor staging was performed according to the TNM classification. Blood samples of healthy donors were obtained from the Novosibirsk Central Clinical Hospital. Blood samples were stored at 4°C and treated within 4 h of collection. Blood was collected into tubes with sterile PBS with 10 mM EDTA; the plasma fraction,

PBS–EDTA, and trypsin eluates were prepared as described in Laktionov *et al.*[12] omitting procedures of separation of red and white blood cells. PBS–EDTA and trypsin eluates were mixed together as the cell surface-bound fraction. Plasma and cell surface-bound fractions were stored frozen at −40°C in aliquots and were thawed once before investigation. A glass milk-based protocol providing quantitative isolation of nucleic acids was used for the isolation of nucleic acids from plasma and the cell surface-bound fraction.[13] All samples were treated with 2U of RNase-free DNase (EN0531, Fermentas St. Leon-Rot, Germany) for 1 h at 37°C. The concentration of total RNA was estimated using the SYBR Green II assay. The detection limit of the assay calculated to the initial blood volume was 0.75 ng/mL for plasma and 3.2 ng/mL of cell surface-bound fraction.

Reverse transcription of RNA was performed using random hexamer primers and SuperScript™ II Reverse Transcriptase kit (Invitrogen, Karlsruhe, Germany). DNA contamination of the RNA samples was controlled by GAPDH mRNA amplification without previous reverse transcription.

18S rRNA was quantified by real-time RT-PCR using Taqman technology and 18S rRNA Control Kit Yakima Yellow®-Eclipse® D15ark Quencher (Eurogentec, Seraing, Belgium) (amplicon 121 bp). Quantification of other genes was performed using qPCR™ Core Kit (Eurogentec) reagents. GAPDH mRNA was detected by an 86-bp amplicon using a TaqMan probe labeled with Yakima Yellow®-Eclipse® D15ark Quencher. The TaqMan probe was placed on an exon/exon boundary. Ki-67 was detected by a 77-bp amplicon consisting of an exon/exon-boundary primer and a FAM/TAMRA-labeled TaqMan Probe. Real-time RT-PCR data were analyzed using ABI Prism 7000 Sequence Detection System software (Applied Biosystems, Freiburg, Germany).

Correlation of the RNA concentrations estimated by SYBR Green II assay and by real-time PCR were evaluated using Statistica 6.0 software by Spearman correlation test.

RESULTS AND DISCUSSION

The concentration of total cirRNA was measured using SYBR Green II assay. The sensitivity of this method was greater than that of the assay for simultaneous detection of DNA and RNA, developed earlier.[14] The detection limit of the modified assay based on the initial blood volume was 0.75 ng/mL (verses 8 ng/mL) in plasma and 3.2 ng/mL (verses 20 ng/mL in trypsin and 40 ng/mL in PBS–EDTA eluates).[15] Using this assay we found measurable amounts of cirRNA in about 80% (in contrast to 35%, described earlier) of the plasma samples and at the surface of blood cells of all healthy donors and breast cancer patients. No statistically significant differences were found in the concentration and distribution of cirRNA in the blood of healthy donors compared with breast cancer patients. The average concentration of cirRNA in plasma of healthy donors and breast cancer patients was 2.5 ± 2.8 ng/mL and 1.2 ± 0.9 ng/mL, respectively. In healthy women the average concentration of cell surface-associated RNA was 31.2 ± 29.9 ng/mL and in cancer patients 46.4 ± 58.4 ng/mL. Further, we found that major portions of cirRNA in blood were associated with the cell surface of circulating cells. Cell surface-bound RNA comprised 89 ± 12% in healthy women and 95 ± 4% in breast cancer patients of the total cirRNA. The average concentration of total cirRNA in blood of healthy women was estimated to be 33.7 ± 29.7 ng/mL. This concentration is slightly lower when compared with that previously reported,[15] which could be due to DNase treatment of the samples in the modified assay. It should be emphasized that cirRNA were found in amounts higher than the detection level of the previous assay only in a limited number of samples.[15] Thus, the more sensitive assay used in the current investigation provides more precise and accurate statistical evaluation of the data. The data obtained here are in accordance with the previous observation, as far as the concentrations of cell surface-bound RNA in patients with breast cancer are close to the detection limit of the assay used in the previous investigation, which were 20 ng/mL in trypsin and 40 ng/mL in PBS–EDTA eluates.

Concentrations of 18S rRNA and mRNA of GAPDH and Ki-67 were quantified by real-time RT-PCR method using Taqman technology. It was shown that 69–73% of ribosomal RNA and 71–79% of messenger RNA circulate in blood of healthy donors and breast cancer patients being bound with the surface of blood cells. Thus, distribution of 18S rRNA, mRNA of GAPDH, and Ki-67 genes differs from distribution of total RNA between cell surface-bound and cell-free fractions in blood. It is not clear which specific RNA contributes additional 15% to the cellsurface-bound RNA (89–95%) in blood.

The RT-qPCR data spread over three orders of magnitude for the GAPDH and Ki-67 mRNA and about two orders of magnitude for the 18S rRNA copy number in blood fractions in both healthy women and breast cancer patients. In healthy donors, no correlation was found between 18S rRNA, mRNA of GAPDH, and Ki-67 copy numbers and the concentration of RNA measured by SYBR Green II assay. It is noteworthy that we described previously that the main part of cell surface-bound RNA of healthy donors has the electrophoretic mobility similar to that of 500-bp-long DNA.[16] As qPCR amplification regions were < 121 bp, degradation of RNA could not be considered as an explanation for the absence of correlation between copy number and SYBR Green II assay at least when the concentration of cell surface-bound RNAs in healthy donors are compared. The lack of a correlation between concentrations of rRNA, mRNA, and total RNA in blood of healthy donors might indicate that processes, which are involved in generation and circulation of extracellular mRNA and rRNA are not *universal* and are different for specific RNA forms.

In contrast to healthy women, the concentration of GAPDH and Ki-67 mRNA as well as of 18S rRNA copy number in breast cancer patients correlated with concentration of total RNA in cell surface-bound fraction (TABLE 1). These

TABLE 1. Correlation of copy number of circulating Ki-67 mRNA with copy number of GAPDH mRNA and 18S rRNA

Ki-67 mRNA	Healthy donors $N = 21$, P value		Breast cancer patients $N = 22$, P value	
	18S rRNA	GAPDH mRNA	18S rRNA	GAPDH mRNA
Plasma	no	0.001098	no	0.000001
Cell surface-bound RNA	no	0.000002	0.044692	0.000000
Total blood	no	0.000013	no	0.000000

No correlation between 18S rRNA and GAPDH mRNA were found.
Spearman correlation test was used for analyses of the data.

TABLE 2. Correlation between copy number of circulating GAPDH, 18S, Ki-67 RNA and RNA concentration measured by the SYBR Green II assay

	Healthy donors $N = 21$, P value			Breast cancer patients $N = 22$, P value		
Total RNA (SYBR Green II assay)	18S rRNA	GAPDH mRNA	Ki-67 mRNA	18S rRNA	GAPDH mRNA	Ki-67 mRNA
Plasma	no	no	no	no	no	no
Cell surface-bound RNA	no	no	no	***0.001***	***0.001***	***0.046***
Total blood	no	no	no	***0.022***	no	no

Spearman correlation test was used for analyses of the data.

data indicate that in the circulation of breast cancer patients similar mechanisms are involved in appearance and circulation of cell surface-bound RNA. However, different mechanisms might be responsible for the distribution of ribosomal RNA and messenger RNA among cell surface-bound and cell-free fractions as there was correlation between total RNA concentration in blood and 18S rRNA copy number and there was no correlation of total RNA with copy number of both messenger RNAs.

A correlation between copy number of GAPDH and Ki-67 mRNA was found in plasma, cell-bound, and total blood of breast cancer patients and healthy women, and in contrast no correlation was found between copy number of 18S rRNA and either GAPDH mRNA or Ki-67 genes (TABLE 2). These data support the existence of different mechanisms of appearance and/or circulation of 18S rRNA and specific messenger RNA in human blood and demonstrate the possibility of using GAPDH mRNA as a dominator for the comparison of mRNA in the pool of cirRNA. Despite the fact that the mechanisms leading to the appearance and redistribution of RNA are to be clarified, the data obtained demonstrate the necessity of searching for more suitable RNA markers for estimation of cirRNA concentration by RT-qPCR in blood of healthy donors and in plasma of breast cancer patients.

ACKNOWLEDGMENTS

The present work was supported by Russian Academy of Sciences Program "Science to medicine," Russian Foundation for Basic Research Grant 06-04-49732, Interdisciplinary Project from Siberian Division of Russian Academy of Sciences No. 13, Scientific Schools Grant SS-1384.2003.4, and Grant for Young Scientists from Siberian Division of Russian Academy of Sciences 2005.

REFERENCES

1. Lo, K.W. *et al.* 1999. Analysis of cell-free Epstein-Barr virus associated RNA in the plasma of patients with nasopharyngeal carcinoma. Clin. Chem. **45:** 1229–1234.
2. Stroun, M. *et al.* 1978. Presence of RNA in the nucleoprotein complex spontaneously released by human lymphocytes and frog auricles in culture. Cancer Res. **38:** 3546–3554.
3. Boettcher, K. *et al.* 2006. Investigation of the origin of extracellular RNA in human cell culture. Ann. N. Y. Acad. Sci. In press.
4. Hasselmann, D.O. *et al.* 2001. Extracellular tyrosinase mRNA within apoptotic bodies is protected from degradation in human serum. Clin. Chem. **47:** 1488–1489.
5. Sisco, K.L. *et al.* 2001. Is RNA in serum bound to nucleoprotein complexes? Clin. Chem. **47:** 1744–1745.
6. Wieczorek, A.J. *et al.* 1985. Isolation and characterization of a RNA-proteolipid complex associated with the malignant state in humans. Proc. Natl. Acad. Sci. USA **82:** 3455–3459.
7. Rosi, A. *et al.* 1988. RNA-lipid complexes released from the plasma membrane of human colon carcinoma cells. Cancer Lett. **39:** 153–160.
8. Masella, R. *et al.* 1989. Characterization of vesicles, containing an acylated oligopeptide, released by human colon adenocarcinoma cells. FEBS Lett. **246:** 25–29.
9. Ng, E.K. *et al.* 2002. Presence of filterable and nonfilterable mRNA in the plasma of cancer patients and healthy individuals. Clin. Chem. **48:** 1212–1217.
10. Kopreski, M.S. *et al.* 1999. Detection of tumor messenger RNA in the serum of patients with malignant melanoma. Clin. Cancer Res. **5:** 1961–1965.
11. Tsui, B.Y.N. *et al.* 2002. Stability of endogenous and added RNA in blood specimens, serum, and plasma. Clin. Chem. **48:** 1647–1653.
12. Laktionov, P.P. *et al.* 2004. Extracellular circulating nucleic acids in human plasma in health and disease. Nucleosides Nucleotides Nucleic Acids **23:** 879–883.
13. Tamkovich, S.N. *et al.* 2004. Simple and rapid procedure suitable for quantitative isolation of low and high molecular weight extracellular nucleic acids. Nucleosides Nucleotides Nucleic Acids **23:** 873–877.
14. Morozkin, E.S. *et al.* 2003. Fluorometric quantification of RNA and DNA in solutions containing both nucleic acids. Anal. Biochem. **322:** 48–50.
15. Tamkovich, S.N. *et al.* 2005. Circulating nucleic acids in blood of healthy male and female donors. Clin. Chem. **51:** 1317–1319.
16. Laktionov, P.P. *et al.* 2004. Cell-surface-bound nucleic acids: free and cell-surface-bound nucleic acids in blood of healthy donors and breast cancer patients. Ann. N. Y. Acad. Sci. **1022:** 221–227.

Isolation and Comparative Study of Cell-Free Nucleic Acids from Human Urine

OLGA E. BRYZGUNOVA,[a] TATYANA E. SKVORTSOVA,[a]
ELENA V. KOLESNIKOVA,[a] ANDREY V. STARIKOV,[b]
ELENA YU. RYKOVA,[a] VALENTIN V. VLASSOV,[a]
AND PAVEL P. LAKTIONOV[a]

[a]*Institute of Chemical Biology and Fundamental Medicine, Siberian Division of the Russian Academy of Sciences, 8, Lavrentiev Ave., Novosibirsk 630090, Russia*

[b]*National Novosibirsk Regional Oncologic Dispensary, 6, Plahotnogo St., Novosibirsk 630090, Russia*

ABSTRACT: Cell-free nucleic acids (NA) from human urine were investigated. Concentrations of DNA and RNA in the urine of healthy people were independent of gender and were in the range of 6 ng/mL to 50 ng/mL and 24 ng/mL to 140 ng/mL, respectively. DNA fragments of 150–400 bp represent the main part of cell-free DNA, along with DNA fragments up to 1,300 bp, which were found in male urine, and DNA fragments up to 19 kbp, which were found in female urine. Analysis of circulating DNA, isolated from blood of breast cancer patients and cell-free DNA isolated from their urine by methylation-specific PCR, demonstrates that the presence of methylated promoters of RASSF1A and RARβ2 genes in plasma was accompanied by the detection of the same methylated markers in urine. The data obtained demonstrate applicability of cell-free urine DNA in cancer diagnostics.

KEYWORDS: cell-free DNA; cell-free RNA; urine; blood; breast cancer; methylation-specific PCR

INTRODUCTION

Circulating nucleic acids (NA) from human blood have been successfully used for the development of early noninvasive cancer diagnostics. Detection of tumor-, fetus-, or transplanted-organ/cell-specific genetic markers in the bloodstream has been described in many publications.[1–3] It has been shown

Address for correspondence: Pavel P. Laktionov, Institute of Chemical Biology and Fundamental Medicine, Siberian Division of the Russian Academy of Sciences, 8, Lavrentiev Ave., Novosibirsk 630090, Russia. Voice: +7-383-3304654; fax: +7-383-3333677.
e-mail: lakt@niboch.nsc.ru

**Ann. N.Y. Acad. Sci. 1075: 334–340 (2006). © 2006 New York Academy of Sciences.
doi: 10.1196/annals.1368.045**

that small amounts of cell-free circulating NA from the blood can pass through the kidney barrier into the urine. Tumor-specific sequences were detected in DNA, isolated from urine of patients with tumors of different localization.[4–6] The obvious advantage of urine sampling makes urine a useful source of NA for the development of a noninvasive cancer test. However, cell-free NA from human urine are yet poorly investigated and low concentration of free NA in urine limits their use in diagnostics. The current investigation was devoted to the investigation of cell-free NA recovered from urine and a comparison of cell-free NA from urine with NA circulating in the bloodstream.

MATERIALS AND METHODS

Urine samples from healthy donors were obtained from Novosibirsk SBRAS Central Clinical Hospital. Cells were pelleted from urine samples (15 mL) by low-speed centrifugation (600 *g*, 10 min). The upper part of the supernatant (10 mL) was collected and stored frozen at −40°C in aliquots. Aliquots were thawed once just before investigation. Two protocols based on the binding of NA with glass-filters activated as described in Ref. 7 were used for isolation of NA from urine. Isolation of NA in the presence of chaotropic salts was performed in accordance with a protocol described for glass-milk.[8,9] Isolation of NA in the absence of chaotropic salts was developed for the isolation of NA from large volumes of urine. At the first step, urine was filtered through a glass filter. NA bound with the glass filter was washed and eluted as described.[8,9] The efficiency of isolation of NA from urine without chaotropic salts was evaluated by comparison with isolation in the presence of chaotropic salts and with the isolation of DNA markers added into urine just before isolation.

NA was quantified using a Hoechst 33258 assay for DNA and SYBR Green II assay for RNA[10] after treatment of the samples with 2U RNase-free DNase I (Fermentas, EN0523) for 1 h at 37°C. Detection limits for DNA and RNA were 10 ng/mL and 0.5 ng/mL, respectively.

Urine and blood samples of breast cancer patients were obtained from the National Novosibirsk Regional Oncologic dispensary. Urine and blood samples were stored at 4°C and treated within 4 h after sampling. Urine samples were treated and stored as described for healthy donors. Blood samples were collected and treated as described[11] omitting the procedure of the separation of red and white blood cells. Plasma, PBS-EDTA, and trypsin eluates from the surface of blood cells were stored frozen at −40°C in aliquots and were thawed once just before investigation.

Aberrant methylation of RASSF1A and RARβ2 gene promoters was determined by methylation-specific PCR (MSP) according to a modified protocol.[12] DNA isolated from 2 mL of plasma and trypsin samples, 6 mL of PBS-EDTA eluates by guanidine thiocyanate/glass-filter protocol and from 6 mL of urine by the glass-filter protocol in the absence of chaotropic salts was denatured by NaOH, modified by sodium bisulfite, purified using the glass-milk protocol and eluted into 40 μL of water.[13] PCR amplification was performed with 4 μL of the modified DNA as template using specific primer sequences for the methylated and unmethylated forms of the genes. The sequences of primers used in this study have been reported previously for RARβ2[14] and for RASSF1A.[15] PCR products were visualized on 6% PAA gel stained with ethidium bromide.

RESULTS AND DISCUSSION

A large volume of urine provided an opportunity for the isolation of a considerable amount of NA. However, all procedures of NA isolation were limited by the reagent consumption. We compared two isolation protocols for NA from human urine. NA was bound to a glass surface in the presence or absence of a chaotropic salt. It was shown that the guanidine thiocyanate/activated glass-filter protocol provided an isolation efficiency of not less than 85% for 100-bp DNA and not less than 80% for 100-bp RNA[7] when NA is isolated from a buffer solution. The binding of NA with a glass-filter in the absence of chaotropic salts provided 90% efficiency for DNA isolation and 60% efficiency for RNA isolation from cell-free urine as estimated by Hoechst 33258 and SYBR Green II assays. Isolation of DNA markers from human urine by guanidine thiocyanate/glass-filter protocol demonstrated approximately 50% efficiency for isolation of DNA fragments longer than 300 bp and about 25% efficiency for isolation of 200-100-bp DNA fragments (FIG. 1). Apparently,

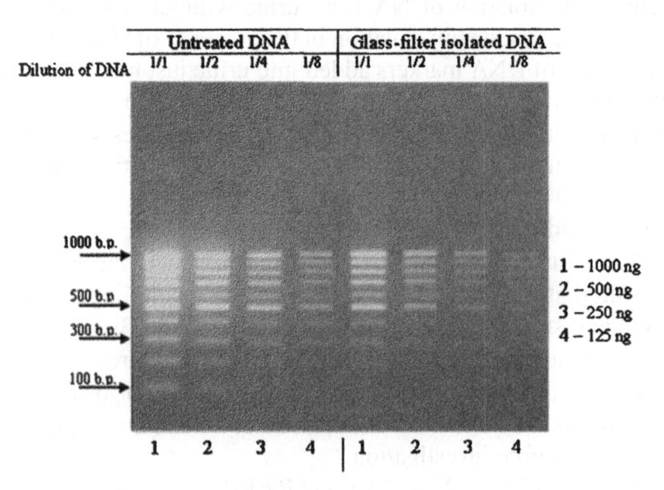

FIGURE 1. Isolation of DNA marker M15 (100–1,000 bp) from human urine. Seven micrograms of M15 (Sibenzyme) DNA in 70 μL of TE buffer was isolated using guanidine thiocyanate/activated glass-filter protocol and eluted into 70 μL. Dilutions of starting DNA solution (7 μg in 70 μL) and DNA after glass-filter isolation were applied on 1.5% agarose gel.

isolation of NA from urine by guanidine thiocyanate/activated glass-filter protocol was less efficient than that from a buffer solution. Thus, the binding of cell-free NA from urine with a glass surface in the absence of chaotropic salts could be used for the isolation of the cell-free NA from large volumes of urine. To quantify cell-free NA in human urine, NA was isolated by guanidine thiocyanate/activated glass-filter protocol from 3 mL of urine. Both cell-free DNA and cell-free RNA were found in the urine from men and women. The concentration of free DNA in urine of healthy subjects was in the range of 6 ng/mL to 50 ng/mL, and the free RNA concentration was in the range of 24 ng/mL to 140 ng/mL (TABLE 1). There was no difference in concentration of cell-free NA in urine between males and females. The main part of the cell-free DNA in urine of both genders was represented by 150- to 400-bp DNA fragments. Urine DNA is less degraded in women than in men: DNA fragments longer than 1,500 bp were not seen in male urine and 19-kbp DNA fragments were present in female urine (FIG. 2). Treatment of cell-free NA from female urine with RNase H led to partial digestion of the NA, suggesting the presence of DNA/RNA hybrids. Ethidium bromide staining of agarose gels revealed a band corresponding to 1300-bp DNA from male urine, which was not degraded by DNase I, RNase A, and RNase H (FIG. 2.). The nature of this substance is not clear, but the high affinity of the compound to a glass surface and staining with ethidium bromide indicates the presence of polynucleotides in its content. Despite a high level of RNase activity in human urine,[16] the

TABLE 1. Extracellular NA in the urine of healthy donors

Donors	Gender of Patient	DNA[a] ng/mL	RNA* ng/mL	Total Extracellular NA[a]
1	F	6	24	30
2	F	7	37	44
3	F	9	43	52
4	F	50	139	189
5	F	25	62	87
6	F	22	71	93
7	F	23	66	89
8	F	24	57	81
9	F	16	38	54
10	F	28	84	112
11	M	25	64	89
12	M	8	37	45
13	M	32	69	101
14	M	7	20	27
15	M	50	94	144
16	M	22	68	90
17	M	46	79	125
18	M	23	137	160
19	M	12	25	37

[a]concentration of NA is calculated for 1 mL of urine.

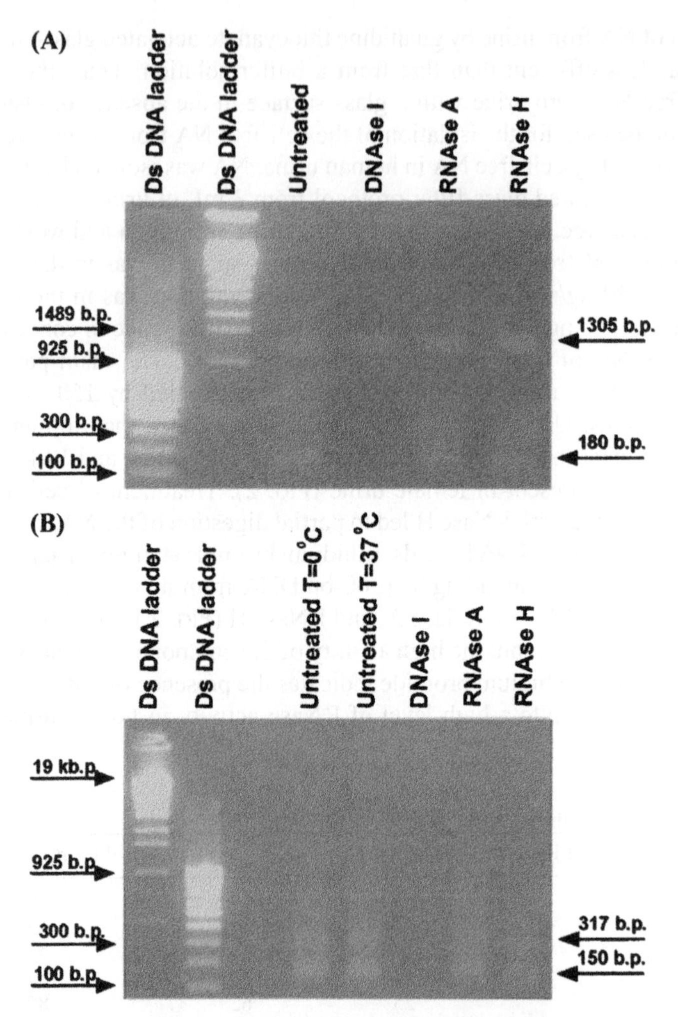

FIGURE 2. Extracellular NA of human urine. Extracellular NA was isolated in male (**A**) and female (**B**) urine and analyzed by electrophoresis in 1.5% agarose gel before or after treatment with nucleases. NA isolated from 340 mL male urine and 380 mL female urine was digested with 2 U DNase I (37°C, 30 min in 10 mM Tris-HCl, 2.5 mM $MgCl_2$, 0.1 mM $CaCl_2$, pH = 7.5), 1 U RNase A (37°C, 30 min) or 2.5 U RNase H (37°C, 30 min in 20 mM HEPES·KOH, 50 mM KCl, 4 mM $MgCl_2$, 1 mM DTT, 50 mg/mL BSA, pH = 8.0).

concentration of cell-free RNA in urine was high and comparable with that in blood plasma. The origin of this RNA is not clear, and RNA in urine is protected from nuclease degradation by forming complexes with biopolymers, as was found in human plasma DNA.[17] Low isolation efficiency of RNA in comparison with DNA without chaotropic salts also confirms this suggestion.

TABLE 2. RASSF1A and RARβ2gene promoter methylation in blood and urine from patients with breast cancer ($n = 5$)

Plasma	PBS-EDTA Fraction	Trypsin Fraction	Urine
+/−	+/−	−/−	+/−
+/−	+/−	+/−	+/−
−/−	−/−	−/−	−/−
−/−	+/−	+/−	+/−
+/−	+/−	+/−	+/−

NOTE: RASSF1A in specimen is shown over slash; RARβ2 is shown under slash.

Earlier it was shown that NA can circulate in blood not only in serum, but also can be absorbed at the surface of blood cells.[11] It was shown that methylated RARβ2 and RASSF1A gene promoters were found in 65% of patients with breast cancer if total circulating DNA (cell-free together with cell-surface-bound) was used for the assay.[18] The presence of methylated RARβ2 and RASSF1A gene promoters was investigated by MSP in DNA extracted from the urine, plasma, and cellular eluates. Samples from five patients with breast cancer obtained 2 weeks after a course of combined anticancer therapy (TABLE 2) were analyzed. Methylation of the RARβ2 gene promoter was not revealed in this group of patients. Detection of a methylated RASSF1A gene promoter in circulating DNA from blood was accompanied by the detection of the same methylated marker in their urine. The data obtained demonstrate the applicability of free urine DNA for the development of noninvasive tests for cancer diagnostics.

ACKNOWLEDGMENTS

The present work was supported by Russian Academy of Sciences program “Science to medicine,” Russian Foundation for Basic Science Research Grant 06-04-49732, Interdisciplinary. Project from Siberian Division of Russian Academy of Sciences No. 13, Scientific Schools Grant SS–1384.2003.4, Grant for Young Scientists from Siberian Division of Russian Academy of Sciences 2006, Russian Science Support Foundation, and in part by Award REC-008 from CRDF.

REFERENCES

1. ANKER, P. *et al.* 2003. Circulating nucleic acids in plasma and serum as a noninvasive investigation for cancer: time for large-scale clinical studies? Int. J. Cancer **103:** 149–152.
2. CHAN, A. *et al.* 2003. Cell-free nucleic acids in plasma, serum and urine: a new tool in molecular diagnosis. Ann. Clin. Biochem. **40:** 122–130.

3. LICHTENSTEIN, A. *et al.* 2001. Circulating nucleic acids and apoptosis. Ann. N. Y. Acad. Sci. **945:** 239–249.
4. YING-HSIU, Su. *et al.* 2004. Human urine contains small, 150 to 250 nucleotide-sized, soluble DNA derived from the circulation and may be useful in the detection of colorectal cancer. J. Mol. Diagn. **6:** 101–107.
5. BOTEZATU, I. *et al.* 2000. Genetic analysis of DNA excreted in urine: a new approach for detecting specific genomic DNA sequences from cells dying in an organism. Clin. Chem. **46:** 1078–1084.
6. YING-HSIU, Su. *et al.* 2004. Transrenal DNA as a diagnostic tool: important technical notes. Ann. N. Y. Acad. Sci. **1022:** 81–89.
7. LAKTIONOV, P. *et al.* RF Patent No. 2004126133/13, priority date 26/8/2004.
8. LAKTIONOV, P. *et al.* Russian patent No. 8470; 20022134341.
9. LAKTIONOV, P. *et al.* Russian patent No. 8470; 2002126328.
10. HAUGHLAND, R. 1996. Handbook of Fluorescent Probes and Research Chemicals, 6th ed.: 165. Molecular Probes, Inc. Eugene, OR.
11. LAKTIONOV, P.P. *et al.* 2004. Cell-surface-bound nucleic acids: free and cell-surface-bound nucleic acids in blood of healthy donors and breast cancer patients. Ann. N. Y. Acad. Sci. **1022:** 221–227.
12. HERMAN, J.G. *et al.* 1996. Methylation-specific PCR: a novel PCR assay for methylation status of CpG islands. Proc. Natl. Acad. Sci. USA **93:** 9821–9826.
13. LAKTIONOV, P. *et al*, invertors; Laktionov, P. & Vlassov V, assignee. 2002. Russian patent 20022134341. Date of application: December 19.
14. SIRCHIA, S.M. *et al.* 2000. Evidence of epigenetic changes affecting the chromatin state of the retinoic acid receptor beta2 promoter in breast cancer cells. Oncogene **19:** 1556–1563.
15. BURBEE, D.G. *et al.* 2001. Epigenetic inactivation of RASSF1A in lung and breast cancers and malignant phenotype suppression. J. Natl. Cancer Inst. **93:** 691–699.
16. NASKALSKI, J.W. *et al.* 1994. Ribonuclease in human serum and urine expressed in arbitrary standard units of activity. Folia. Med. Cracov. **35:** 31–38.
17. TSUI, N. *et al.* 2002. Stability of endogenous and added RNA in blood specimen, serum, and plasma. Clin. Chem. **48:** 1647–1653.
18. DULAIMI, E. *et al.* 2004. Tumor suppressor gene promoter hypermethylation in serum of breast cancer patients. Clin. Cancer Res. **10:** 6189–6193.

Influence of Mycoplasma Contamination on the Concentration and Composition of Extracellular RNA

EVGENIY S. MOROZKIN, ELENA Y. RYKOVA, VALENTIN V. VLASSOV, AND PAVEL P. LAKTIONOV

Institute of Chemical Biology and Fundamental Medicine, Siberian Division of the Russian Academy of Sciences, 8, Lavrentiev Ave., Novosibirsk 630090, Russia

ABSTRACT: Kinetics of extracellular RNA accumulation in culture medium and at the cell surface along with their composition and distribution among cell-free and cell-surface-bound fractions were investigated in mycoplasma-contaminated and mycoplasma-free HeLa cells. It was shown that the mycoplasma infection influenced the concentration and kinetics of accumulation of total extracellular RNA and the distribution of specific RNA fragments among cell-free and cell-surface-bound fractions. Fragments of immature rRNA were found in culture of mycoplasma-infected HeLa cells. The data obtained indicate the existence of selective mechanisms providing binding of RNA with cell surface and their excretion out of cells.

KEYWORDS: extracellular RNA; circulating RNA; mycoplasma contamination

INTRODUCTION

Ribonucleic acids are found outside of cells as free molecules or bound with the cell surface in tissue culture, in blood, and in other biological fluids.[1–3] The biological importance of extracellular RNA is not clear, but the data, which demonstrate inhibition of gene expression by interfering RNAs, provide an opportunity to speculate on the involvement of extracellular RNA in cell-to-cell communication.[4] Cell death (apoptosis and necrosis) seems to be the main process leading to the release of RNA from cells. However, the existence of active RNA transport by living cells has been described.[1,5,6] Moreover, it was shown that some stimuli, such as mitogen, can induce release of RNA by lymphocytes.[1]

Address for correspondence: Pavel P. Laktionov, Institute of Chemical Biology and Fundamental Medicine, Siberian Division of the Russian Academy of Sciences, 8, Lavrentiev Ave., Novosibirsk 630090, Russia. Voice: +7-383-3304654; fax: +7-383-3333677.
e-mail: lakt@niboch.nsc.ru

**Ann. N.Y. Acad. Sci. 1075: 341–346 (2006). © 2006 New York Academy of Sciences.
doi: 10.1196/annals.1368.046**

One of the potential factors influencing the exRNA appearance from cells could be an intracellular obligate infection. Since mycoplasma infection is widespread in the human population, we investigated the concentration and composition of extracellular RNA in mycoplasma-contaminated (myc+) and mycoplasma-free (myc–) HeLa cells.

MATERIALS AND METHODS

Human cervical carcinoma cell line (HeLa) and HeLa cells contaminated with mycoplasma (*M. hyorhinis* and *M. fermentans*) were cultivated in 100 × 20 mm culture dishes at 37°C in 5% CO2 in DMEM medium (Sigma, USA) supplemented with 10% heat-inactivated FBS (Sigma) and antibiotics, penicillin, and streptomycin at 100 u/mL (Gibco, USA). To investigate the time course of RNA accumulation outside cells, HeLa cells were seeded at a density of 5×10^5 cells per culture dish. Twelve hours after seeding, cells were washed with phosphate-buffered saline containing 5 mM EDTA (PBS/EDTA) (2 mL, 3 min, room temperature) and fresh growth medium was added. At different time intervals after the PBS/EDTA treatment, growth medium was collected, cells were rinsed with 2 mL of PBS/EDTA for 3 min and exposed to 1 mL of 0.25% trypsin (Sigma) for 1 min at room temperature. Trypsin was inactivated by the addition of a trypsin inhibitor (Sigma). Total cell number was estimated and the viability of the cells was assessed using the trypan-blue dye exclusion assay and MTT test. To remove the cells growth medium, PBS/EDTA and trypsin eluates were centrifuged and RNAs from the supernatants were analyzed.

Quantitative guanidine thiocyanate/glass-milk methods[7,8] were used for RNA isolation. Isolated RNA was quantified using a fluorescence-based assay as described previously.[9] This procedure allows the accurate determination of RNA concentrations from a range of 10 to 1000 ng/mL. Nucleic acids were 5′-radiolabeled using [γ-^{32}P] ATP and T4 polynucleotide kinase. DNase I or RNase A (Fermentas, Latvia) treatment was used to ascertain the nucleic acid composition followed by analysis using polyacrylamide gel electrophoresis.

Chemical sequencing of ^{32}P-labeled ribo-oligonucleotides was performed as briefly described.[10] Chemical modifications of ribo-oligonucleotide were as follows: G – 0.3% solution of dimethylsulfate (40 sec, 90°C) followed by treatment with 0.5 M solution of $NaBH_4$ (10 min, chill on ice); A – 0.6% solution of diethylpyrocarbonate (10 min, 90°C); U – 2 M solution of NH_2OH, pH 10 (40 sec, 90°C); C – 2 M NH_2OH, pH 10 (40 sec, 90°C); followed by cleavage with 1 M aniline solution, pH 4.5, for 20 min at 60°C. The RNA samples obtained were analyzed on 18% polyacrylamide gel with 8 M urea.

Isolated RNA was treated with DNase I (RNase free) and then reverse-transcribed into cDNA using specific primers (20 pmol) and M-MuLV reverse transcriptase (Fermentas) at 42°C for 1 h, followed by incubation at 70°C for

TABLE 1. Primers, length of PRC products, and regions of rRNA fragments tested

Primers	Size, bp	Regions of rRNA
Reverse: 5′-TACCTCTTAACGGTTTCACGCCCT Forward: 5′-TGGCGACCCGCTGAATTTAAGCAT	405	5′-fragment of 28S rRNA
Reverse: 5′-TCTACGAATGGTTTAGCGCCAGGT Forward: 5′-AGAGGAACCGCAGGTTCAGACATT	370	3′-fragment of 28S rRNA
Reverse: 5′-ACCACACCGCACGCAACA Forward: 5′-AAGACGGAGAGGGAAAGAGAGAGC	208	ITS 2 fragment
Reverse: 5′-TTCACTCGCCGTTACTGAGGGAAT Forward: 5′-CTCTCTCTCCCGTCGCCTCT	219	Fragment containing both 28S rRNA and ITS 2
Reverse: 5′-GGTCCAAGAATTTCACCTCTAGC Forward: 5′-TACCTGGTTGATCCTGCCAGTAG	974	18S rRNA fragment
Reverse: 5′- AAGCGACGCTCAGACAGGCGTAG Forward: 5′-GACTCTTAGCGGTGGATCACTC	158	5.8S rRNA fragment

10 min. The resulting cDNA was used in PCR under the following conditions: DNA was denatured at 95°C for 10 min followed by 35 cycles at 95°C for 30 sec, 60°C for 30 sec, and 72°C for 30 sec. PCR products were analyzed by 6% PAA gel containing ethydium bromide. The sequences of primers and length of PCR products are presented in TABLE 1.

RESULTS AND DISCUSSION

To investigate the kinetics of RNA secretion by mycoplasma-free (myc–) and mycoplasma-infected (myc+) HeLa cells, 12 h after seeding, attached cells, were washed with PBS/EDTA in order to remove dead and weakly attached cells, and fresh culture medium was added. FIGURE 1 shows time course of the accumulation of RNA in the culture medium and at the cell surface of myc– and myc+ cells. In contrast to myc+ cells, myc– cells did not have detectable exRNA level in the trypsin eluate (FIG. 1). The concentration of exRNA in PBS/EDTA eluates of myc– cells was stable during cultivation, whereas exRNA found in the growth medium rapidly increased after the PBS/EDTA treatment and then slowly decreased and approached zero level in 24 h (FIG. 1). In general, the concentration of exRNA in the growth medium and PBS/EDTA eluates of myc+ cells was higher and was found outside cells for a longer time. The amount of RNA found at the cell surface and in the growth medium of myc+ HeLa cells rapidly increased after the PBS/EDTA treatment and then slowly decreased (FIG. 1). The concentration of RNA in the culture medium approached zero level 48 h after the PBS/EDTA treatment.

PAGE analysis of 5′-[^{32}P]-labeled exRNA did not demonstrate any difference between myc+ and myc– cells (FIG. 2). Culture medium and trypsin supernatants from both cell lines contained similar 5′-end-labeled exRNA

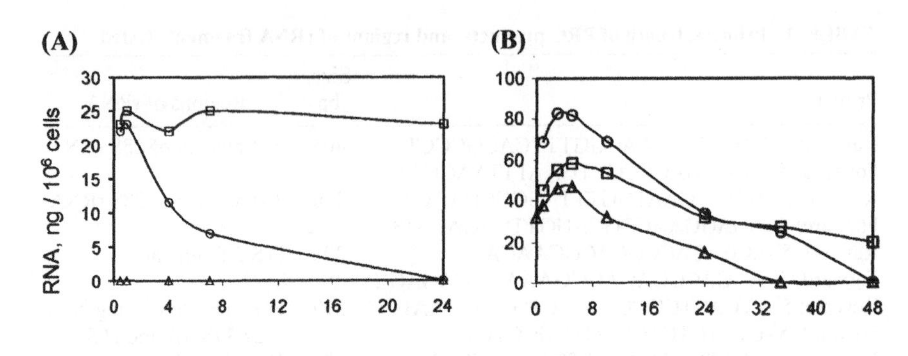

FIGURE 1. Concentration of extracellular RNA in cultures of myc– (*panel A*) and myc+ (*panel B*) HeLa cells. Nucleic acids from growth medium (*circles*), PBS/EDTA eluate (*squares*), trypsin eluate (*triangles*) were isolated and quantified. Average coefficient of variation values for triplicate independent experiments was 5%.

fragments, whereas exRNA from PBS/EDTA eluates contained a characteristic set of the labeled fragments, which were different from those in the culture medium and trypsin eluates.

^{32}[P]-labeled ribo-oligonucleotide (35 b) from the PBS-EDTA fraction of HeLa myc+ was isolated by PAGE and sequenced by a chemical method for RNA sequencing.[10] The nucleotide sequence of ribo-oligonucleotide according to the results of BLAST analysis (5′-AC GGG UGG GGU CCG CGC AGU CCG CCC GGA GG-3′) completely coincided with 426–456 sequence of human ribosomal 28S RNA.

Since ribosomal RNAs are the most abundant cellular RNA, their presence out of cells could be predicted. However, which species of rRNA are present in the extracellular media is still unclear. Thus, we have investigated the secretion and circulation of rRNA by RT-PCR using a set of primers specific for the different regions of rRNA (FIG. 3). All tested fragments of mature rRNA were found at the cell surface of HeLa myc– cells and in the culture medium except for the 18S rRNA fragment, which was only at the cell surface and was not found in the culture medium (TABLE 2). In contrast to myc– cells, all fragments of mature rRNA were found in culture medium of HeLa myc+ cells and in the PBS/EDTA and trypsin eluates from the cell surface (TABLE 2).

Despite the fact that the maturation of rRNA occurs in the nucleus, we attempted to detect the pre-rRNA fragment in the extracellular media using RT-PCR assay. Pre-rRNA fragments containing ITS2 and ITS2/5′-28S regions were not found in the pool of exRNA of HeLa myc– cells. However, these fragments (ITS2, ITS2/5′-28S) were found in the culture of HeLa myc+ cells and only at the cell surface (TABLE 2). It should be noted that exact location of pre-rRNA fragments is unclear. They could be localized at the surface of HeLa cells or inside mycoplasma cells.

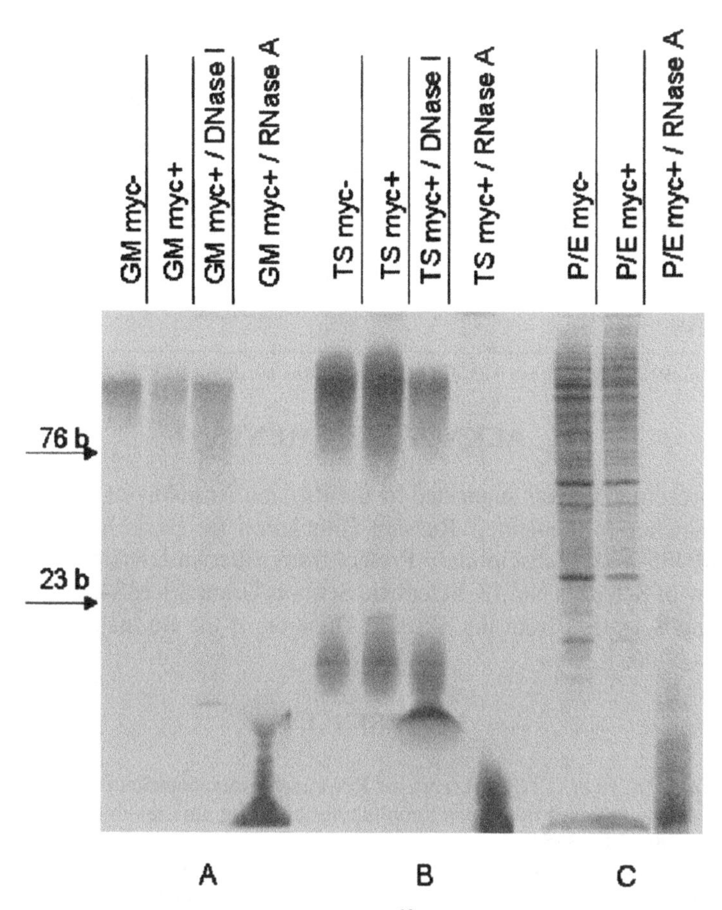

FIGURE 2. Electrophoretic analysis of 5′-[^{32}P]-labeled free and cell-surface-bound extracellular RNA. Nucleic acids isolated 3 h after the PBS/EDTA treatment from growth medium (GM, *panel A*), trypsin eluate (TS, *panel B*), PBS/EDTA eluate (PE, *panel C*).

The influence of mycoplasma infection upon the composition and distribution of extracellular RNA among cell-free and cell-surface-bound fractions suggest the existence of selective mechanisms providing binding of RNA to the cell surface and its excretion out of cells. Therefore, mycoplasma infection should be considered in the analysis of circulating RNA isolated from biological samples.

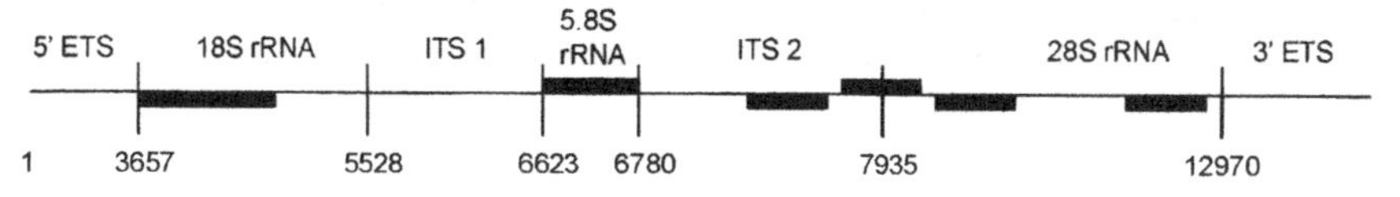

FIGURE 3. Localization of the PCR fragments in human rRNA gene.

TABLE 2. The distribution of rRNA fragments among extracellular fractions of myc− and myc+ HeLa cells

	HeLa myc−			HeLa myc+		
rRNA fragments	Growth medium	P/E	Trypsin	Growth medium	P/E	Trypsin
5′-28S rRNA	+[a]	+	+	+	+	+
3′-28S rRNA	+	+	+	+	+	+
5.8S rRNA	+	+	+	+	+	+
18S rRNA	−	+	+	+	+	+
ITS2	−	−	−	−	+	+
ITS2/5′-28S	−	−	−	−	+	+

[a]Detected rRNA fragments are indicated by "+" and those not detected by "−."

ACKNOWLEDGMENTS

The present work was supported by the Russian Academy of Sciences Program "Science to medicine," Russian Foundation for Basic Research grant 06-04-49485-a, Inter disciplinary Project from Siberian Division of Russian Academy of Sciences No. 13, Scientific Schools Grant SS-1384.2003.4 Grant for Young Scientists from the Siberian Division of the Russian Academy of Sciences 2006.

REFERENCES

1. STROUN, M. *et al.* 1978. Presence of RNA in the nucleoprotein complex spontaneously released by human lymphocytes and frog auricles in culture. Cancer Res. **38:** 3546–3554.
2. LAKTIONOV, P.P. *et al.* 2004. Cell-surface-bound nucleic acids: free and cell-surface-bound nucleic acids in blood of healthy donors and breast cancer patients. Ann. N. Y. Acad. Sci. **1022:** 221–227.
3. MOROZKIN, E.S. *et al.* 2004. Extracellular nucleic acids in cultures of long-term cultivated eukaryotic cells. Ann. N. Y. Acad. Sci. **1022:** 244–249.
4. SHARP, P.A. 2001. RNA interference—2001. Genes Dev. **15:** 485–490.
5. KOLODNY, G.M. *et al.* 1972. Secretion of RNA by normal and transformed cells. Exp. Cell Res. **73:** 65–72.
6. BÖTTCHER, K. *et al.* 2006. Investigation of the origin of extracellular RNA in human cell culture. Ann. N. Y. Acad. Sci. This volume.
7. LAKTIONOV, P.P. *et al.* 2002. Russian patent 2002126328. Date of application: October 2.
8. LAKTIONOV, P.P. *et al.* 2002. Russian patent 20022134341. Date of application: December 19.
9. MOROZKIN, E.S. *et al.* 2003. Fluorometric quantification of RNA and DNA in solutions containing both nucleic acids. Anal. Biochem. **322:** 48–50.
10. HOWE, C.J. & E.S. WARD 1989. RNA Sequencing. Nucleic Acids Sequencing. IRL Press. Oxford England.

Optimized Real-Time Quantitative PCR Measurement of Male Fetal DNA in Maternal Plasma

BERNHARD G. ZIMMERMANN,[a,b] WOLFGANG HOLZGREVE,[a]
NEIL AVENT,[b] AND SINUHE HAHN[a]

[a]*Laboratory for Prenatal Medicine, Department of Research/University Women's Hospital, University of Basel, Switzerland*

[b]*Centre for Research in Biomedicine, University of the West of England, Bristol, UK*

ABSTRACT: DNA of fetal origin is present in the plasma of pregnant women. The quantitative measurement of circulatory fetal DNA (cfDNA) by real-time quantitative PCR (qPCR) has been applied to investigate a possible correlation between increased levels and pregnancy-related disorders. However, as the levels of cfDNA are close to the detection limit (LOD) of the method used, the measurements may not be reliable. This is also problematic for the evaluation of preanalytical steps, such as DNA extraction and cfDNA enrichment by size separation. We optimized a protocol for the qPCR analysis of the multi-copy sequence DYS14 on the Y chromosome. This was compared with an established assay for the single-copy SRY gene. Probit regression analysis showed that the limit of detection (LOD) of the DYS14 assay, (0.4 genome equivalents (GE)) and limit of quantification (LOQ) were 10-fold lower in comparison to SRY (4 GE). The levels of cfDNA obtained from the first trimester of pregnancy could be quantified with high precision by the DYS14 assay (CV below 25%) as opposed to the SRY measurements (26–140%). Additionally, fetal sex was correctly determined in all instances. The low copy numbers of fetal DNA in plasma of women in the first trimester of pregnancy can be measured reliably, targeting the DYS14 that is present in multiple copies per Y chromosome.

KEYWORDS: limit of detection; real-time quantitative PCR; DYS14; SRY

INTRODUCTION

Elevations in the levels of circulatory fetal DNA (cfDNA) have been noted in pregnancy-associated disorders, such as fetal trisomy 21 or pre-eclampsia.[1–3]

Present address for correspondence: Bernhard G Zimmermann, Ph.D., University of California, Los Angeles, Dental Research Institute, 73-017 Center for Health Sciences, 10833 Le Conte Avenue, Los Angeles, CA 90095-1668. Voice: 310-206-1138; fax: 310-825-7609.
e-mail: bgz@ucla.edu

**Ann. N.Y. Acad. Sci. 1075: 347–349 (2006). © 2006 New York Academy of Sciences.
doi: 10.1196/annals.1368.047**

However, the measurement of DNA concentrations is not straightforward as preanalytical factors and quantification procedures by quantitative PCR (qPCR) can have marked effects.[4] The resultant variability makes the comparison of numbers obtained by different researchers difficult. Also, as the measurements of the low levels found in the first two trimesters of pregnancy are very close to the detection limit, it is therefore not surprising that contradictory reports about the elevation of cfDNA levels in pregnancies affected by Down's syndrome exist.[5]

We developed a qPCR assay that allows the accurate evaluation of preanalytical steps, as well as a more precise measurement of cfDNA levels.[5] We further present the utilization of the probit analysis[6] to assess the limits of detection and quantification (LOD, LOQ) of measurements.

METHODS

We designed a new qPCR assay for the amplification of an 84-basepair sequence of the multi copy DYS14 (also known as TSPY gene family) on the short arm of the Y chromosome.[7] According to the BLAST search of the human chromosome database, the assay amplifies 9 copies per Y chromosome with a 100% match.

DNA from plasma of pregnant women was extracted as described.[5] Genomic DNA in low concentrations and cfDNA from plasma of the first trimester of pregnancy were analyzed in triplicate by qPCR as described.[5] Results were calculated in genome equivalents per PCR and per mL of plasma. With the methods used, the DNA from 32 μL of plasma was amplified in one PCR reaction. Thus, 31 GE of cfDNA for SRY and 3.5 GE for DYS14 in 1 mL of plasma are the theoretical LOD at 1 copy per PCR (TABLE 1).

RESULTS

The 95% LOD for both assays was determined by probit analysis[6] from the PCR data of dilute genomic DNA and all plasma samples from pregnancies with female fetuses (TABLE 1). We defined the LOQ as five times the 95% LOD.

Duplicate extractions of 12 pregnancies were quantified with assays targeting the single-copy SRY and the DYS14 assay. The cfDNA levels in the six plasmas with male fetuses were clearly above the LOQ for DYS14. For SRY

TABLE 1. Detection (LOD) and quantification (LOQ) limits for the DYS14 and SRY qPCR of plasma based on the protocols used in this study (values are in GE per mL plasma)

	Theoretical LOD	95% LOD	LOQ
DYS14	3.5	12.5	62.5
SRY	31	125	625

the template copies were below the LOQ and several replicates without amplification occurred. Accordingly, the CV for DYS14 measurements was all below 25% and it was much higher for SRY (26–140%).

Further samples were analyzed with the DYS14 assay, and fetal sex was identified correctly by applying a cutoff at 1 GE per PCR to distinguish between male (25 extractions) and female (33) sex.

DISCUSSION

Our comparison of the qPCR assays shows that measurements of cfDNA in the first trimester of pregnancy by targeting single-copy sequences have a high variability, while the ten-fold lower LOQ of the DYS14 assay allows accurate quantification of all samples. The novel assay will allow an accurate evaluation of preanalytical steps, such as extraction and size separation of fetal and maternal plasma DNA.[8] Furthermore, we showed that it allows a reliable determination of fetal sex.

The DYS14 sequence is a subgroup of the TSPY gene family, which is variable in copy number between individuals.[9] However, we were not able to observe discrepant quantitative results from parallel analyses by SRY and DYS14 with genomic DNA and a larger cohort of pregnancy plasmas (data not shown). Thus, for research purposes, the evaluation of cfDNA quantifications for the assessment of pregnancy-associated disorders is unlikely to be affected by variable copy numbers.

REFERENCES

1. Zhong, X.Y. & W. Holzgreve. 2001. Circulatory fetal and maternal DNA in pregnancies at risk and those affected by preeclampsia. Ann. N. Y. Acad. Sci. **945:** 138–140.
2. Lo, Y.M. *et al.* 1999. Increased fetal DNA concentrations in the plasma of pregnant women carrying fetuses with trisomy 21. Clin. Chem. **45:** 1747–1751.
3. Zhong, X.Y. *et al.* 2000. Fetal DNA in maternal plasma is elevated in pregnancies with aneuploid fetuses. Prenat. Diagn. **20:** 795–798.
4. Johnson, K.L. *et al.* 2004. Interlaboratory comparison of fetal male DNA detection from common maternal plasma samples by real-time PCR. Clin. Chem. **50:** 516–521.
5. Zimmermann, B. *et al.* 2005. Optimised real-time quantitative PCR measurement of male fetal DNA in maternal plasma. Clin. Chem. **51:** 1598–1604.
6. Finney, D.J. 1971. Probit Analysis, 3rd ed. Cambridge University Press. London.
7. Arnemann, J. *et al.* 1987. A human Y chromosomal sequence expressed in testicular tissue. Nucl. Acids Res. **15:** 8713–8724.
8. Li, Y. *et al.* 2004. Size separation of circulatory DNA in maternal plasma permits ready detection of fetal DNA polymorphisms. Clin. Chem. **50:** 1002–1011.
9. Dechend, F. *et al.* 2000. TSPY variants in sixloci on the human Y chromosome. Cytogenet. Cell Genet. **91:** 67–71.

Index of Contributors

Printed and bound by CPI Group (UK) Ltd, Croydon, CR0 4YY

16/04/2025

14658825-0004